CLIMATE CHANGE ADAPTATION

for Health and Social Services

CLIMATE CHANGE ADAPTATION

for Health and Social Services

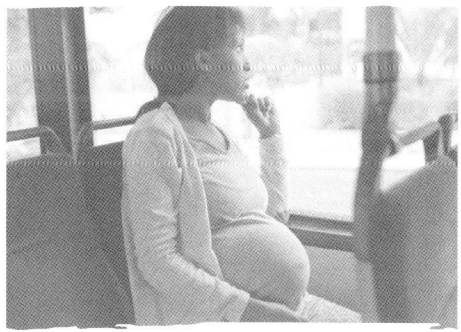

EDITORS: RAE WALKER AND WENDY MASON

CSIRO

PUBLISHING

National Library of Australia Cataloguing-in-Publication entry
    Climate change adaptation for health and social services /
    Rae Walker and Wendy Mason.

    9781486302529 (paperback)
    9781486302536 (epdf)
    9781486302543 (epub)

    Includes bibliographical references and index.

    Environmental health.
    Climatic change – Health aspects.
    Climatic changes – Health aspects – Risk assessment – Australia.

    Walker, Rae, editor.
    Mason, Wendy, editor.

    616.98

Published by
CSIRO Publishing
Locked Bag 10
Clayton South VIC 3169
Australia
Telephone: +61 3 9545 8400
Email: publishing.sales@csiro.au
Website: www.publish.csiro.au

Front cover: images by iStockphoto.com

Set in 10.5/12 Minion and Stone Sans
Edited by Karen Pearce
Cover design by Karen de Heer
Typeset by Thomson Digital
Index by Bruce Gillespie
Printed in China by 1010 Printing International Ltd

CSIRO Publishing publishes and distributes scientific, technical and health science books, magazines and journals from Australia to a worldwide audience and conducts these activities autonomously from the research activities of the Commonwealth Scientific and Industrial Research Organisation (CSIRO). The views expressed in this publication are those of the author(s) and do not necessarily represent those of, and should not be attributed to, the publisher or CSIRO. The copyright owner shall not be liable for technical or other errors or omissions contained herein. The reader/user accepts all risks and responsibility for losses, damages, costs and other consequences resulting directly or indirectly from using this information.

Original print edition:
The paper this book is printed on is in accordance with the rules of the Forest Stewardship Council®. The FSC® promotes environmentally responsible, socially beneficial and economically viable management of the world's forests.

A proportion of the royalties from sales of this book will go towards supporting Enliven.

Contents

4   Communicating about climate change   51

Susie Burke

Part 2: Vulnerable populations and appropriate adaptations   71

5   People with disability and their carers   73

Rae Walker

6   People who are elderly or have chronic conditions   93

Margaret Loughnan and Matthew Carroll

7   Women and children   117

Debra Parkinson, Brad Farrant and Alyssa Duncan

8   Climate change: impact on country and Aboriginal and Torres Strait Islander culture   141

Kerry Arabena and Jonathan 'Yotti' Kingsley

9   Support for adaptation in culturally and linguistically diverse communities   159

Alana Hansen, Scott Hanson-Easey and Peng Bi

10   Rural communities experiencing climate change: a systems approach to adaptation   179

Glenda Verrinder and Lyn Talbot

Part 3: Organisational adaptation   201

11   Community-based health and social services: managing risks from climate change   203

Karl Mallon and Emily Hamilton

12   Engaging communities in climate change adaptation   227

Helena Bishop, Aileen Thoms and Wendy Mason

List of contributors

Kerry Arabena
Kerry Arabena is the Chair for Indigenous Health and Director of the Indigenous Health Equity Unit at the University of Melbourne. She has a background in social work with experience in sexual health, reproductive health, public health, remote area health service administration and community development. Kerry is a descendant of the Meriam people of the Torres Strait, and has a PhD in human ecology. Her research interests include gender issues, social justice, human rights, service provision, harm minimisation, and citizenship rights and responsibilities. Kerry was the founding Co-Chair of the National Congress of Australia's First Peoples and Director of Indigenous Health Research at Monash University. She is an internationally renowned consultant and presenter. Her roles influencing policy and practice include nine ministerial appointments, chairing one international, five national and two state committees, current membership of three boards and Director of seven Aboriginal Corporation Boards of Management.

Guy Barnett
Guy Barnett is an urban ecologist in CSIRO's Land and Water Flagship in Canberra, where he leads an interdisciplinary research team investigating the interactions between the urban environment, climate change and human health. He managed the delivery of the recently completed CSIRO Flagship Collaboration Cluster on 'Urbanism, Climate Adaptation and Health' involving nine research partners, and was an invited member of the Organising Committee for the Australian Academy of Science's 2014 Theo Murphy High Flyers Think Tank 'Climate Change Challenges to Health: Risk and Opportunities'. He is currently leading a project for the New South Wales Environmental Trust on 'Green Infrastructure for Climate Adaptation in Western Sydney'. Guy is a member of several international research networks, including the Resilience Alliance, Stockholm Resilience Centre and the Urban Biosphere Initiative.

Peng Bi
Peng Bi is Professor of Public Health Medicine in the School of Population Health at the University of Adelaide, where he leads a research team in environmental health. Peng is also Node Leader of 'Heat and Health' for the National Climate Change Adaptation Research Facility. His areas of research expertise include public health, environmental epidemiology, ecosystem health and infectious disease epidemiology. He has wide experience in identifying and managing environmental health problems, especially the impact of climate change on population health and risk communication for climate change adaptation among communities. Peng's team has studied the impact of climate variability on transmission of vector- and food-borne diseases, heatwaves and population health, emergency response mechanisms to heatwaves, and vulnerability among groups including outdoor workers, the elderly, and culturally and linguistically diverse communities. His findings have provided important information to policymakers, local authorities and communities in the reduction of the impact of climate change.

Helena Bishop

Helena Bishop has formal qualifications in health science, disability and social work. The majority of her work to this point has been in the area of health promotion. Helena's current role is at Women's Health in the South East, working with women from a range of diverse backgrounds across the Southern Metropolitan Region of Melbourne. Helena combines her passion for prevention with her desire to help people when they are at a point of crisis in their lives. Her work spans across a range of complex health issues affecting women, such as the impacts of climate change on women from a range of culturally and linguistically diverse backgrounds, including refugee women. Helena has worked with communities to implement realistic and relevant adaption and mitigation strategies that have immediate and lasting impacts on their health and wellbeing.

Kathryn Bowen

Kathryn Bowen is a social epidemiologist working at the nexus of global change, health and governance issues. She is an Associate at the Melbourne Sustainable Society Institute, University of Melbourne. Kathryn has worked in global health research and policy since 1999, across public, private and university sectors. She is a Research Fellow within the Earth System Governance Project and a Fellow of the Centre for Sustainability Leadership. Along with five friends, she was a founding board member of Just Change, a climate change and equity organisation. Currently, Kathryn's main program of work is a range of consultancies for the World Health Organization (country and regional offices) and other United Nations agencies on climate change and health, as well as working with City of Melbourne on a climate change adaptation project. She has just completed a four-year multi-country study funded by AusAID, which investigated adaptive capacity in the Asia-Pacific.

Susie Burke

Susie Burke is a psychologist working at the Australian Psychological Society, looking at ways of using psychological knowledge to enhance community wellbeing and promote social justice. A significant part of her work in the 'Psychology in the Public Interest' team is to examine and promote the role that psychology can play in helping us understand the causes, impacts and solutions to climate change. She produces materials for the public and the media on a range of topics including environmental threats, talking with children about the environment, natural disasters, violence, racism and refugees.

Matthew Carroll

Matthew Carroll is a Senior Research Fellow within the School of Rural Health – Churchill at Monash University. He has had extensive involvement in developing and supporting major research platforms looking at the impact of mobile phone use on brain tumours, the establishment of the '45 and Up' study, and managing a major national research network targeting ageing well. His work at Monash has focused on engaging older people in ageing well, whether in terms of resilience and participation following the Black Saturday fires, involvement in lifelong learning, understanding and responding to heat-health risks, avoidance of social isolation, or engaging in physical activity.

Alyssa Duncan

Alyssa Duncan is a Research Assistant at Women's Health in the North. She completed her undergraduate studies in geography at the University of Melbourne and is currently studying law at Monash University. She has previously worked on the impact of short- and long-term environmental disasters on women. This work is the result of original research into men's Black Saturday experiences in Victoria, conducted under the auspices of Women's Health Goulburn North East, with the guidance of Monash University and Monash University's Injury Research Institute, and funding from the National Disaster Resilience Grants Scheme. Alyssa's Gender and Disaster Taskforce work is also supported by Women's Health In the North.

Brad Farrant

Brad Farrant is a Post-doctoral Fellow at the Telethon Kids Institute and is the Australian Research Alliance for Children and Youth's representative on the Climate and Health Alliance's committee of management. Coming from the perspective of developmental psychology, Brad is interested in how broader ecological factors like biodiversity loss, population growth, peak water and climate change will interact to affect children's development now and in the future.

Sharon Friel

Sharon Friel is Director and Professor of Health Equity at the Regulatory Institutions Network at Australian National University (ANU). She is also Director of the Menzies Centre for Health Policy at ANU. In 2010 she was awarded an Australian Research Council Future Fellowship to investigate the interface between health equity, social determinants and climate change, based at the National Centre for Epidemiology and Population Health, ANU. Between 2005 and 2008 she was the Head of the Scientific Secretariat (University College London) of the World Health Organization Commission on Social Determinants of Health. In 2008 she chaired the Rockefeller Foundation Global Research Network on Urban Health Equity, which reported in 2010. She was a member of the Lancet UCL Commission on Climate Change and principal investigator of various competitive grants concerned with climate change and health including the CSIRO Climate Adaptation Flagship Cluster 'Climate, cities and human health – impacts, vulnerability and adaptation'.

John Gardner

John Gardner is a social psychologist in CSIRO's Land and Water Flagship, working in the areas of environmental behaviour, climate change mitigation and climate adaptation. He is involved in a range of projects in these fields, investigating the social, contextual and individual factors that motivate individuals, communities and businesses to change behaviour, to adopt new technology, and to prepare for climate change impacts. John has previously worked as an academic at the University of Queensland Business School, and as a research consultant to public and private sector organisations.

Emily Hamilton

Emily Hamilton is a Policy Officer with the Australian Council of Social Service (ACOSS), the national peak body for community-based social and welfare services in Australia and a voice for people affected by poverty and inequality. She holds a first-class honours degree in social work from the University of New South Wales and has 10 years' professional experience in direct casework practice, human rights advocacy, research and policy development, and project management. Emily joined ACOSS in 2012 to deliver a research project funded by the National Climate Change Adaptation Research Facility (NCCARF), which examined the vulnerability and adaptive capacity of community service organisations (CSOs) to climate-driven extreme weather events. In 2014–15, she is responsible for delivering a project to build the resilience of CSOs to extreme weather events, which is funded by the Commonwealth Attorney General's Department and aims to implement the recommendations from the NCCARF research.

Alana Hansen

Alana Hansen is a Research Associate in the Discipline of Public Health at the University of Adelaide. With a background in environmental health, she was awarded her PhD in 2010. Her research has focused on the effects of weather and air pollution on population health and on the health impacts of extreme heat and climate change in vulnerable populations. Alana's research interests include barriers to adaptation in older persons and the adaptive capacity of migrants and refugees to extreme heat and climate change. Her research involves collaborations with stakeholders in government and non-government organisations and her work has been translated into policy and practice. She is also involved in research to assist in building capacity to curb the public health impact of emerging and re-emerging infectious diseases due to climate change in China.

Scott Hanson-Easey

Scott Hanson-Easey is a social psychologist and Research Associate in the Discipline of Public Health at the University of Adelaide. He was awarded his PhD in psychology from the university in 2011, and has published 10 peer-reviewed journal articles and book chapters. He has examined topics such as public understanding of climate change, the language of racism, environmental risk perception, behavioural adaptation and health communication. His work has drawn from various theoretical and epistemological traditions, including social cognition, social representations theory, discursive social psychology, thematic analysis and interpretive phenomenological analysis. He has collaborated closely with government and non-government stakeholders on pragmatic research problems, which have informed the development of policies and strategies on risk communication. His current research investigates how disaster preparedness and risk messages to culturally and linguistically diverse communities could be better formulated and culturally targeted.

Greg Hunt

Greg Hunt was a science and environmental studies teacher in Victorian secondary schools for several years before becoming an education bureaucrat, writing policy for environmental education. He was Principal at the Zoo Education Service based at Victoria's three zoos at Healesville, Parkville and Werribee and then Education Manager at Melbourne Museum in Carlton. After leaving education he took up working for the environment, first with Waterkeepers Australia which provided support for community advocates in caring for their local waterways. Greg is currently the Executive Officer of the South East Councils Climate Change Alliance. In this role he conducts mitigation and adaptation programs through local government to Melbourne's south-east. In his spare time, he is an active environmental educator speaking to community groups on how to present issues of climate change and promote engagement in environmental care and protection.

Jonathan Kingsley

Jonathan Kingsley is a Research Fellow in the Indigenous Health Equity Unit at the University of Melbourne. For over 10 years, Jonathan has worked in Aboriginal Community Controlled Health Organisations, government bodies, academic institutes and non-government organisations across Australia in the public health and community development field. Recently, Jonathan received a PhD from the School of Health and Social Development at Deakin University. Jonathan sees our ecosystems as central to health and as having the capacity to bridge health inequalities (the basis of his honours, masters, PhD and previous Visiting Scholar position at Cambridge University). Jonathan views himself as not only an academic but an ecological activist through his teaching, publishing and steering committee experiences across several different disciplines, which has led him to being nominated for and winning several environmental awards.

Margaret Loughnan

Margaret Loughnan is a health geographer with expertise in climate health relationships. She specialises in heat stress and spatial determinants of population vulnerably. Margaret's recent work includes identifying heat-health thresholds for Melbourne and 10 Victorian regions: developing maps of population vulnerability for Melbourne for the Victorian Department of Health; identifying heat-health thresholds for all Australian capital cities; and describing the spatial vulnerability of urban populations and related emergency service demands during extreme heat events. She has recently extended this research to examine heat adaptive behaviours and community resilience in Victorian regional towns. Margaret was a visiting scientist at the United States National Centre for Atmospheric Research Research Applications Laboratory in 2010. She has authored one book, two book chapters, 17 peer-reviewed papers, three government reports and 48 national and international conference presentations, all related to climate change, heat stress, population vulnerability and human health.

Karl Mallon

Karl Mallon originally trained in physics in the United Kingdom and then started postgraduate work in particle physics before moving to Australia to undertake PhD research in renewable energy systems. This was followed by a shift into international climate change policy in Europe, North America and Asia, including editing and co-authoring the book *Renewable Energy Policy and Politics*. In 2005 Karl co-founded Climate Risk to work specifically on quantifying the impacts of climate change for business, government and the community, and to establish business cases for adaptation based on sound science, engineering and financial principles. Karl has always maintained that adaptation should not be based on the ability to pay, and has championed the need for government support for community service organisations' adaptation as a means to protect society's most vulnerable and marginalised.

Wendy Mason

Wendy Mason has worked as a practitioner, project manager and as a Chief Executive Officer or senior manager in a diverse range of services in both metropolitan and rural settings including community health, women's health, aged care, disability, mental health, not-for-profit services, hospital networks, general practice and local government. Wendy is well versed in partnership development, stakeholder engagement, strategic and area-based planning and conceptual modelling. Since 2007, as Executive Officer of Enliven (formerly South East Healthy Communities Partnership), Wendy has facilitated dialogue and worked with people in health and social services in the south-east of metropolitan Melbourne. In this role she raises awareness and better understanding of the impacts and influences of changes in climate on organisations and vulnerable populations, taking into consideration challenges such as changes in population growth, the ageing population, disability and rising incidence in chronic disease. The key focus of this work has been to present accurate information and tools that people need, not just to hold services to account but to go further and take actions themselves to facilitate healthier and more resilient communities.

Jacqui Meyers

Jacqui Meyers is a quantitative ecologist with CSIRO's Land and Water Flagship. She specialises in analysing the interactions between people and the environment with a particular focus on resilience, vulnerability and adaptation in urban systems. Her recent research has focused on the climate change impact of more frequent and severe heat events on human health. As part of this work, Jacqui has been modelling the benefits of tree shade for reducing both heat-related health risks and energy usage in residential buildings.

Debra Parkinson

Debra Parkinson is a social researcher, committed to feminism and social justice. As the researcher for both Women's Health Goulburn North East and Women's Health in the North, she has completed research on women leaving violent relationships, partner rape and women's unequal access to resources and the legal system. Most relevant to this publication is her work on environmental justice, particularly her role as co-researcher and author with Claire Zara of 'The Way He Tells It' about women's experiences and workers' observations of the Black Saturday bushfire and its aftermath. This was followed by research with men on their experiences, 'Men on Black Saturday'. Debra was awarded her PhD in Social Sciences from Monash University in 2015 and is currently an Adjunct Research Fellow with Monash Injury Research Institute.

Lyn Talbot

Lyn Talbot had an extensive career in tertiary education in the fields of nursing and public health at La Trobe University, Bendigo. She was a Senior Lecturer in Public Health, Health Education, Health Promotion, Program Planning and Evaluation and Environmental Health. She is the co-author, with Glenda Verrinder, of a widely used health promotion textbook *Promoting Health. The Primary Health Care Approach*. Lyn has undertaken and published research that gives insight into the impacts and outcomes of long-term systematic change in rural communities, and is now the Corporate and Community Planner at the City of Greater Bendigo. Her role includes assisting the small towns and communities in the municipality to develop a local community plan that can assist them to adapt to changing social and environmental circumstances and to achieve their local goals.

Aileen Thoms

Aileen Thoms has been the Health Promotion Manager at Kooweerup Regional Health Service in Victoria since 2007. She developed and manages the community climate change adaptation program at the health service, using a community development intervention framework that has brought community members and community organisations together around a community garden. Aileen is a Registered Nurse who graduated from Queen Margaret University College in Edinburgh, Scotland and migrated to Australia in 1983. Aileen has completed post-graduate education in mental health (psychiatry, Dundee Scotland), gerontology (Monash University) and health sciences/health promotion (Deakin University). She is currently studying for a Master of Health Promotion/Public Health at Deakin University.

Glenda Verrinder

Glenda Verrinder is a Senior Lecturer in the Department of Public and Community Health in the La Trobe Rural Health School. Prior to this role she worked in community-based health agencies for 20 years. Her teaching, research and publications reflect her interest in promoting health, ecological sustainability and healthy rural communities. She has received two Excellence in Teaching Awards from La Trobe University for online learning in the subject 'Rural Health'. She sits on several committees, including the Executive Committee of the Environment and Ecology Health Special Interest Group of the Public Health Association of Australia. Glenda is the author, co-author and editor of several texts and other publications including the chapter 'Climate change and health' in *Health, Illness and Wellbeing: Perspectives and Social Determinants.*

Rae Walker

Rae Walker is an Emeritus Professor in the Department of Public Health at La Trobe University in Melbourne, Australia. She has a background in health, education and social sciences. Rae's research has focused on community-based services with a particular emphasis on service delivery and inter-organisational relationships. Her research has also touched on multi-scale policy development and some hospital service delivery activities. She has authored 50 book chapters, refereed journal papers and several reports, and authored or edited several books. Since 2009 Rae has worked with Enliven (www.enliven. org.au), a non-government organisation, and its network of members to develop an appropriate climate change adaptation intervention framework using the primary health care approach. This was enhanced with systematic literature reviews of evidence on key issues and vulnerable population groups relevant to community-based health and social service organisations.

Introduction: climate change adaptation in the health and social services sector

Rae Walker and Wendy Mason

The origins of this book

In 2007 many leaders in the health and social services sector were concerned about climate change and its impacts on their communities, but were unclear about the role of their organisations and ways of responding that were most appropriate for the sector. The Board of Enliven Victoria, a not-for-profit organisation that includes a Primary Care Partnership in the south-east of Melbourne, resourced several projects with members, creating the tools and assembling the evidence that would support their work. The discussions within Enliven led to several literature reviews by Rae Walker, an Emeritus Professor at La Trobe University, which explored the following key questions:

- What does an appropriate intervention framework for the community based health and social services sector look like?

- What direct and indirect impacts of climate change are most relevant to this sector?

- How does climate change impact on people with disability and their carers, and what can the sector do about it?

It became clear that a practical and relevant perspective on adaptation in the community sector was emerging, and that it needed further development with a broader group of researchers and professionals working in the climate change adaptation field.

This book is a collaboration between Rae Walker and Wendy Mason, then the Executive Officer of Enliven and the driving force in the development of its climate change initiative, and a large number of academic and professional experts from across Australia. The origins of it, however, lie with the organisations participating in Enliven, the leaders of which really wanted to take meaningful action on climate change in their localities.

Context

Climate change impacts on health in number of ways. Some impacts are a result of the changing climate directly altering the weather patterns and environments. For example, higher temperatures and less rain dry the landscape, increasing the bushfire risk. Bushfires have serious health consequences. Higher temperatures increase the number and severity of heatwaves. Heatwaves are a major killer, especially for older people. Other impacts are due to climate policy choices that enhance or hinder the health and wellbeing of particular population groups. For example, rising utility prices have a detrimental impact on low-income populations, and increase health inequalities (Chapman & Boston 2007). Health and social impacts of climate change are already being felt in Australia, and they will be felt more as climate change becomes more severe. In this book we focus on the direct impacts we currently experience, knowing that in the future additional impacts will become more visible and will also need informed and wise adaptive action.

Community-based health and social service organisations are already dealing with the early direct effects of climate change impacts within their organisations and communities. The organisations are experiencing the effects of extreme weather such as storms, fires, floods, heatwaves and droughts on organisational infrastructure and operations (Mallon *et al.* 2013). Identification and mitigation of these risks is an increasingly important task for boards and senior management of organisations. Organisations also are recognising the need for services to respond to the impacts of extreme weather on clients and communities. Facilitation of planning by vulnerable clients and communities to reduce the impacts of heatwaves, fires, floods and storms, for example, is a prudent initiative for organisations to take. In addition, the experience of recent extreme weather, such as the Victorian and New South Wales bushfires, widespread drought and the Queensland floods, has raised awareness of the demands on services in the response and recovery phases of these events. There is much to be done. What does existing evidence indicate should be done? This book assembles the evidence to answer this question for the most vulnerable population groups.

Climate change stresses communities, particularly those with the least capacity to adapt. Climate change amplifies existing health and social inequalities, by making demands for change on people who do not have the resources to respond in the most constructive ways. The population groups we chose to include in this book are those typically identified in the research and policy literature as most vulnerable to the current impacts of climate change.

A significant amount of research on the impacts of climate change on health and social wellbeing, and appropriate response to it, has been done in recent years. However, for health and community service organisations it has been inaccessible. The goal for this book was to assemble this evidence in a way that would inform adaptive action by health and social service organisations. A dedicated group of academics and professionals from across Australia have volunteered their time and expertise to achieving this goal. Each of them is an expert in their field.

Structure

In **Part 1** of this book we explore the issues that help us understand climate change, the necessary responses to it, and how we speak about it in our communities. We explore the evidence of climate change impacts on health and social equity and their implications for a developed country such as Australia. The principles of community-scale adaptation, and their application by community-based health and social service organisations, are discussed in ways that support adaptation by the sector. Although this book is about climate change adaptation, it makes no sense to ignore the upstream causes of it. The purpose of adaptation is to minimise the negative impacts of climate change. We recognise that to be effective in that task organisations also need to reduce their carbon footprint, and that of their communities, to reduce the scale of the problem to which they are trying to adapt. When community-based health and social service organisations choose to act on climate change they necessarily have to communicate with staff, clients and communities. This is not always easy, so the evidence about, and practice of, good climate change communication is discussed in a very accessible way in Chapter 4.

In **Part 2** of this book we focus on a small number of population groups for whom the impacts of climate change are of particular concern. Of course everybody is impacted by climate change but population groups experiencing social disadvantage are impacted most, and typically have least capacity to adapt (e.g. Costello *et al*. 2009). The population groups included in this book are those commonly referred to in the Australian and international climate change literature as being at most risk (e.g. Costello *et al*. 2009; Garnaut 2008).

Community-based health and social services are often funded to provide services to specific population groups. Examples include Aboriginal health, aged care, and disability, migrant and women's services and programs. Others are funded to work with multiple population groups, for example, local government, community health services and general practice. For each of the high-risk population groups we asked authors to assemble the evidence to answer the following questions:

- What are the characteristics of the population group relevant to climate change impacts?
- What is the evidence of current climate impacts on the group?
- What is the evidence describing appropriate interventions for the group?

The assembled evidence can be used to inform agency analyses of the risks climate change has created for their client population, and appropriate interventions. The chapter on Aboriginal populations (Chapter 8) is a little different from the others in that the climate impacts Aboriginal people experience are shared with other population groups but their cultural orientation to caring for country provides special insights into adaptation from which we all can learn.

Some themes run through the population group chapters. Within each population group there are patterns that indicate specific and widespread needs common for members of the group. For example, for culturally and linguistically diverse communities, the biggest climate change adaptation issues are the language of communication and culturally appropriate adaptation. For older people and those with chronic disease, the specific threat is from heatwaves and the key adaptation issue is maintenance of cool places for shelter from the heat. For people with disability, extreme weather events are the greatest threat and the key adaptation issue is inclusive emergency planning, response and recovery. However, population groups are also diverse. For example, some older people will have culturally and linguistically diverse backgrounds. Where this is the case services will need to communicate in the relevant languages and culturally appropriate ways about the risks of heatwaves and sensible responses to them. The clients of any service may have particular needs related to demographics, personal attributes or location, for example, that draw on the evidence assembled in more than one chapter. This simply means that adaptation planning needs to be a thoughtful and informed response to the client population and community.

In **Part 3** we focus on organisations, the risks climate change raises for them and some approaches to managing those risks. For organisations, one part of their response to climate change is identification and management of the risks that arise for the organisation (Mallon *et al*. 2013). The other part is adapting services to meet the climate-change-related needs of the populations they serve. To illustrate this point two organisations collaborated to write Chapter 12, describing their organisations' responses to client needs. In the appendix we have included a quality assessment tool that has been developed and used by a network of organisations in Victoria to organise their thinking about, and adaptation to, the climate change impacts they face.

We trust that the information in this book will be able to be used in a practical and meaningful way by those who work in health and social services, local government and, more broadly, by members of communities and students. We continue to learn and understand through adopting new ideas, connecting with our local communities and sharing our collective research and evidence of climate change impacts and how to best respond in different situations and with different populations. We wish readers the very best in their efforts to understand and respond to the challenges of climate change. On this issue we really are all in it together.

References

Chapman R, Boston J (2007) The social implications of decarbonising the New Zealand economy. *Social Policy Journal of New Zealand* **31**, 104–136.

Costello A, Abbas M, Allen A, Ball S, Bell S, Bellamy R, Friel S, Groce N, Johnson A, Kett M, Lee M, Levy C, Maslin M, McCoy D, McGuire B, Montgomery H, Napier D, Pagel C, Patel J, de Oliveira JAP, Redclift N, Rees H, Rogger D, Scott J, Stephenson J, Twigg J, Wolff J, Patterson C (2009) Managing the health effects of climate change. *Lancet* **373**, 1693–1733. doi:10.1016/S0140-6736(09)60935-1

Garnaut R (2008) *The Garnaut Climate Change Review: Final Report*. Cambridge University Press, Port Melbourne.

Mallon K, Hamilton E, Black M, Beem B, Abs J (2013) 'Adapting the community sector for climate extremes: extreme weather, climate change and the community sector – risks and adaptations'. National Climate Change Adaptation Research Facility, Gold Coast.

PART 1:

ISSUES

1

Health and social impacts of climate change

Kathryn Bowen and Sharon Friel

Key points

- At its heart, climate change is an equity issue. The effects of climate change will be disproportionately felt by those in society who are already vulnerable, such as those with lower socio-economic and health status.
- Many climate change risks to health and society will arise indirectly from a variety of non-health sectors, including agriculture, water, disaster management and planning.
- An understanding of approaches to reduce vulnerability to the health and social risks of climate change must incorporate an understanding of equity issues, as well as focus on multi-sectoral activities.
- Community-based and social service organisations play a vital role in responding to climate change risks, given the need for these responses to be framed at a local level.

Introduction

Climate change poses a major threat to society, human health and health inequities. The health and social impacts of climate change are becoming clearer. There is a general consensus that climate change will negatively affect health and social wellbeing, predominantly due to the exacerbation of existing health conditions such as infectious diseases and malnutrition. Climate change may also increase the risk of non-communicable diseases and mental ill health, directly via increasing frequency and intensity of extreme temperatures and weather events, fires and air pollution, and indirectly via changes to food and water security. These health effects will occur globally.

The distribution of health impacts will be inequitable with the most vulnerable communities, in rich and poor countries, experiencing the worst effects. The degree of a community's vulnerability to the health effects of climate change is determined by a multitude of factors, including spatial segregation and the associated poor physical living conditions, low income, low levels of education, poor governance structures and processes, and already poor health status.

Responses to reduce vulnerability to the effects of climate change must therefore be equity-focused, and are necessarily multi-sectoral given that the majority of the health impacts will arise via other sectors, such as planning, water and agriculture.

The participation of different types of organisations within this response is crucial. It is insufficient to purely rely on government agencies to develop and implement strategies; the roles of 'actors beyond the state' are now receiving much greater attention as being key to successful strategies. These actors include community-based health and social service organisations and community groups, which are already playing a large role in responding to climate change. Climate change mitigation and adaptation activities are often local, hence the involvement of local organisations is vital to adequately and appropriately respond to climate change.

Currently, the health community (both government and non-government, including community-based health and social service organisations) has an opportunity to more strongly influence climate change policy and programs, due to an increasing awareness of health within the climate change community, as well as via the strategic framing of climate change and health. This opportunity is both in relation to mitigation activities, where strategies that provide co-benefits for health and the environment should be prioritised, and adaptation activities, where strategies to reduce vulnerability to health impacts of climate change need to be focused on communities that are most at risk.

In this chapter we outline the drivers of climate change health impacts and discuss how these are spatially, economically and temporally differentiated. We also examine climate change strategies and suggest ways to strengthen the priority given to health and social policy and projects.

Climate change and global health

Climate change and its current and future harmful impacts have again been illustrated in the latest report from the Intergovernmental Panel on Climate Change (IPCC), the Fifth Assessment Report (IPCC 2014). This report provides further evidence for the need to act now and act quickly on climate change in relation to both adaptation and mitigation efforts. The report's main findings are sobering. It is predicted with high confidence that climate change will impact human health predominantly by exacerbating health problems that already exist. Globally, the health impacts from climate change include: injury, disease and death due to more intense and more frequent heatwaves and fires; increased levels of under-nutrition as a result of reduced food production (primarily in poor regions); reduced labour productivity; and increased risks from food-, water- and vector-borne diseases. These impacts will be distributed inequitably across the world, with the poorest communities, countries and regions most adversely affected. Australia, although a wealthy country, will also experience the negative effects of climate change, across rural and urban settings, and different social and economic groups.

Main health and social effects of climate change in Australia

The IPCC's Fifth Assessment Report (Reisinger *et al.* 2014) identifies several projected changes in Australia's climate that have implications for health, including an intensification of drought in south-east Australia, an increase in fire weather, and reduced inflow in south-western Australian river systems. Other key climate change impacts that have risks for health include an increased frequency and intensity of flood damage to infrastructure and settlements (due to extreme precipitation), and increased risks to coastal infrastructure and low-lying ecosystems due to sea-level rise and damaging cyclones. These are indicated as current risks that will increase over time. In addition to these, it has been reported that other concerns, such as air pollution, will be worsened due to our reliance on carbon emissions. Further, there are other effects that are less tangible and harder to quantify, including mental health (Berry *et al.* 2010), solastalgia (Albrecht *et al.* 2007), and the health and social effects of dislocation (McMichael *et al.* 2010). These must also be considered in climate change responses.

Changes in the climate will affect the health of Australian communities in varying ways, depending on their underlying economic, geographic, social and health status. The combination of these issues will play a large part in determining the vulnerability of particular individuals and communities to the threats of climate change and health, and the subsequent design of appropriate responses to alleviate and (ideally) prevent such impacts. Many climate change impacts on health will be expressed at the community level, necessitating community-scale action. Alone this will likely be insufficient to address the causes of climate change and climate-related health risks, so it must be situated within a wider, multi-scale, concerted approach to climate change adaptation and mitigation.

The vulnerability of communities is mediated by their adaptive capacity, which is a measure of how well they can respond to climate change. Research on adaptive capacity focuses on the social vulnerability of communities (and the organisations within them), regions and nation states, and considers a range of factors that may enable them to adjust to changing environmental and social conditions (Adger & Kelly 1999; Berkhout 2012; O'Brien *et al.* 2004). Adaptive capacity constitutes a wide range of variables, including economic, social and environmental indicators. The factors that are most commonly considered include information and skills, economic wealth, technology, infrastructure, equity and institutions (IPCC 2001). Equally important but less studied factors are those that are less tangible, such as community cohesion, social inclusion and governance structures. This same concept can be applied in the Australian context to get an idea of which communities are most vulnerable to the health impacts of climate change in order to design responses that are relevant and appropriate to particular communities.

Climate change as an equity issue

In terms of absolute burden, however, it seems clear that it [climate change] most threatens the poorest and most vulnerable populations in all societies, probably in close inverse proportion to income, wealth, and power. The rich will find their world to be more expensive, inconvenient, uncomfortable, disrupted, and colorless—in general, more unpleasant and unpredictable, perhaps greatly so. The poor will die. (Smith 2008)

The disruption to key life-supporting environmental systems that is caused by climate change has been mostly generated by a small fraction of modern society, and is arguably one of the biggest ethical issues of our time (Gardiner 2004). Modern day climate change looks likely to worsen inequities along several axes – socio-economic, spatial and temporal.

The recent report from the IPCC confirmed that climate change is affecting everyone (IPCC 2014). However, it will have the greatest, and generally earliest, impact on the poorest and most disadvantaged populations in countries across the world, including in Australia, thereby further polarising the haves and the have-nots (Friel *et al.* 2011a).

The impacts of past and continuing climate change will be felt by future people. The effect of climate change on children now and in the future raises another challenging ethical issue. Because of their immature organ systems, neurobiology and dependence on caregivers, children are particularly susceptible to heat stress, gastroenteritis and natural disasters, as well as to family stresses linked to droughts, loss of livelihood and familial dislocation. This may have long-term social and health consequences for many children (Strazdins *et al.* 2011).

Climate change will be experienced in differing ways, depending on where people live (IPCC 2014). Heat stress, extreme rain and flooding events, landslides, air pollution, drought and water scarcity pose risks for people, assets, economies and ecosystems (IPCC 2014). Although the IPCC notes that these are in the (near-term and beyond) future, they are currently being felt, as we have experienced in Australia during the almost decade-long period of long-term drying (2003–12). In both urban and rural settings, risks are exacerbated for communities and individuals that have insufficient essential services and infrastructure. This has been shown in one recent study of several small communities in Australia, where the greatest economic difficulty experienced by disadvantaged groups was the increasing cost of utility bills (Sevoyan *et al.* 2013). This pressure will only increase due to the predicted rise in warmer weather. Another concerning finding in this study was that lower-income households did not have strong social networks that could help moderate the lack of economic resources (Sevoyan *et al.* 2013).

Climate change pathways to health inequities

As discussed earlier in the chapter, changes in climate conditions and variability affect human wellbeing, safety, health and survival in many direct and indirect ways. Some of the vulnerability among different population groups is due to geography and the associated level of external climatic exposure, as seen in the regional variation in predicted rates and types of climatic change (IPCC 2014). However, much of the currently emerging variation in the health impacts of climate change is due to existing health and social inequities and the differing capacities to adapt to changing climatic conditions among different social groups. Climate change related health risk, by and large, sits on top of pre-existing infectious and non-communicable disease burdens, which are already proportionately higher in socially disadvantaged communities across Australia and other developed countries (McMichael & Lindgren 2011). The brunt of adverse health impacts of climate change, to begin with, will mostly be borne by low-income and geographically vulnerable populations. In general, however, the greatest health risks are experienced by those contributing least to the underlying environmental damage (i.e. the least economically advanced countries and lower social status groups within rich and poor countries alike) (Friel *et al.* 2008). We now describe some of the ways in which climate change will contribute to health inequities.

Risks to health equity from extreme weather events and sea-level rise

Sea-level rise poses both direct and indirect risks to health equity. Coastal inundation, more extensive episodes of flooding, increasingly severe storm surges (especially at times of high tide) and damage to coastal infrastructure (roads, housing and sanitation systems) would all pose direct risks to health. There is, too, a range of indirect risks including the mental health consequences of property loss, break-up of communities, displacement and emigration. Low-lying cities and towns near coasts will most probably face increased risks from more frequent and more intense hurricanes, cyclones and storm surges, causing flooding, direct injury and damage to infrastructure such as roads, housing, water and sanitation systems. Poorer households are usually at a higher direct health risk due to weaker structures, less safe locations and building sites, and the weaker resilience of infrastructure in poorer cities and towns to withstand damage (Costello *et al.* 2009). Poorer households also often lack the economic resources to evacuate in the face of climate-related disasters, or to rebuild damaged structures.

The experiences of severe storms and floods that occur in the Philippines most years as well as the 2010 earthquake in Haiti, have demonstrated the vulnerability of poor urban areas to major disasters. The flooding of New Orleans in 2005, and its effects on elderly rest home patients and poor people who could not evacuate because of lack of transport, gave a striking example of what might happen among socially disadvantaged communities, even in rich countries (Sharkey 2007). In the wake of Hurricane Katrina, children from lower-income groups in the United States were at increased risk of developing severe mental health symptoms (McLaughlin *et al.* 2009). Nearly half of the children surveyed six months after the 2003 bushfires in Canberra, Australia had elevated symptoms of post-traumatic stress disorder (McDermott *et al.* 2005).

Research into the 2003 bushfires in Victoria, Australia, which destroyed homes, agricultural assets and public infrastructure as well as adversely affected the health and livelihoods of many local people, found that people's vulnerability arose from the circumstances of their everyday lives (Whittaker *et al.* 2012).

Extreme temperatures and their impact on health inequities

Increased temperatures caused by climate change, and amplified in cities by the heat island effect, are causing increased health risks (Huq *et al.* 2007). Heatwave mortality and morbidity increases have been reported across the United States (IPCC 2007; Luber & McGeehin 2008) and in developing countries (Hajat *et al.* 2005; Kjellstrom 2009; Kovats & Akhtar 2008). The notorious August 2003 heatwave in Western Europe caused an estimated 40 000–50 000 deaths, especially in older people (APA 2010; Kovats & Hajat 2008; Robine *et al.* 2008). In a study of the effects of weather on hospital admissions of people with pre-existing cardiac disease within Melbourne, Australia, the authors found that the impact crossed the social divide, with heart attacks increasing during hot weather in people from both advantaged and disadvantaged neighbourhoods (Loughnan *et al.* 2010).

Heat-related health risk is socially graded. Lower socio-economic and minority ethnic groups are more likely to live and work in warmer neighbourhoods and in buildings that are poorly ventilated and absorb heat, increasing the risk of heat stress and associated morbidity and mortality. Densely populated neighbourhoods with few trees have maximum temperatures during the day 1–3°C higher than places with parks or open landscape areas (US EPA 2007). Poor neighbourhoods with weak infrastructure, buildings and unplanned developments with few green spaces are likely to be more exposed to high temperatures compared to more affluent neighbourhoods (Kovats & Akhtar 2008). The climate conditions in workplaces are a major concern because climate change makes many workplaces hotter during the hottest part of the year (Kjellstrom 2009). Temperature extremes affect physiological functioning, mood, behaviour (accident-proneness) and workplace productivity. The already poorer health outcomes experienced among lower occupational grades will be exacerbated by temperature extremes, especially in outdoors workers and those working in poorly ventilated hot factory conditions (Kjellstrom 2009).

Drought

Droughts are predicted to become more frequent and severe in many regions of the world under climate change. They cause hunger, starvation, displacement and misery; farming jobs are lost and suicide rates can increase, especially in farmers (Berry *et al.* 2011; Judd *et al.* 2006; McMichael *et al.* 2008). Agricultural systems are intrinsically linked with environmental conditions, which are already under threat in much of southern Australia because of rising heat and protracted drying. Increasing drought periods in Australia may challenge the viability of agriculture in some regions, and hence those communities that depend on primary production. A worst-case scenario may herald the collapse of some communities. Human health impacts arising from such a transition, including impacts on mental health, would be profound (Berry *et al.* 2010; Hanna *et al.* 2011). Recent analysis of Australian data showed that during a seven-year period of major and widespread drought, one pattern of relative dryness (extreme cumulative number of months in drought culminating in a recent period of dryness lasting a year or more) was associated with a 6.2% increase in distress for rural dwellers but not urban dwellers (O'Brien *et al.* 2014). Rural vulnerability to mental health problems is greatly increased by socio-economic disadvantage. Related factors, such as reduced access to health services as communities decline and a 'stoical' culture that inhibits help seeking, may compound this (Berry *et al.* 2011).

Impacts on infectious diseases and the risk to health equity

Climate change will significantly influence, and mostly increase, the range of infectious diseases (food-borne, water-borne and vector-borne) (Campbell-Lendrum & Corvalan 2007; Patz *et al.* 2008). Poor living conditions are the breeding ground for climate-sensitive infectious diseases such as diarrhoea, malaria and dengue (Campbell-Lendrum & Corvalan 2007). When basic infrastructure is inadequate, existing conditions of poor sanitation and drainage and impure drinking water are further stressed in extreme weather events and flooding. This leads to more opportunities for the transmission of infectious diseases, which puts households at higher than usual risk.

Pressures on food security

There is growing recognition of the additional stress on food insecurity presented by climate change. The drought-prone and long-term drying conditions in Australia and in other subtropical regions around the world, higher temperatures, rising sea levels, increasing frequency of flooding, and acidification of oceans contribute to impaired yield, quality and affordability of food in many countries (UNDP 2007). Further, climate change induced disturbances to traditional living and eating patterns among rural and remote Indigenous populations may also affect food security through reduced options for physical mobility and increased reliance on imported energy-dense processed foods, thus potentially amplifying obesity, cardiovascular disease and diabetes (Furgal & Seguin 2006).

Climate change is affecting the availability and price of food in Australia (Bradbear & Friel 2013; Friel 2010). Prolonged drought and the frequency of climate change induced extreme weather events have affected the supply of fruits and vegetables and therefore the consumer price. Modelling estimates suggest that between 2005 and 2007 (approximately the middle of the 10-year drought in Australia) there was a 33% increase in the price of vegetables and a 43% increase in fruit prices because of the drought (Quiggin 2010). Rising food prices most affect the poor. In Australia, the cost of consuming a healthy diet based on national health guidelines already uses ~40% of the disposable income of welfare-dependent families compared to 20% of an average income household (Kettings *et al.*

2009). Climate change related additional price increases will add potentially unmanageable financial pressures on some households, leading to food insecurity and also physical and mental distress.

Importance of multi-sectoral partnerships in strengthening health and wellbeing

The health impacts of climate change will occur predominantly via other sectors, such as water and agriculture. They will arise from effects such as floods and droughts, with flow-on impacts such as sanitation issues, destruction of infrastructure and compromised food security. For this reason, partnerships and cooperation between sectors is essential to enhance and support health and wellbeing in the face of these challenges (Bowen *et al.* 2014). To date, the level of partnerships across different scales and sectors in Australia has been limited; however, efforts are underway to improve this. Cooperation will bring benefits that extend beyond the issue of climate change: if partnerships are embedded across sectors then this will also be valuable for general emergency management, health systems strengthening and urban planning goals. Some sectors are advanced in their climate change response planning, such as the planning for sea-level rise and the planning for reduced water availability in southern Australia (IPCC 2014). Others are less so.

Local communities are essential partners when forming responses to climate change, particularly adaptation (IPCC 2014). This is because climate change activities will be primarily driven from this scale; it is at the level of the local community where the full brunt of climate change and its effects will be faced, risk information will be managed and financing sought (IPCC 2014).

Local governments in Australia are showing varying levels of sophistication in the ways that they are leading decisions around mitigation and adaptation to climate change. Many are developing and implementing proactive climate change plans, with an impressive array of policies and programs. The City of Melbourne, for example, has a well-developed suite of policies that have undergone community consultation and public critique, and some of which now form the council's policies. Some examples include doubling tree canopy cover for the city's urban forest, upgrading drainage infrastructure, funding more energy efficient buildings, implementing planning processes to minimise climate risk, and installing various water-sensitive urban design initiatives. Another goal for the council is that Melbourne is carbon neutral by 2020 (with the council itself internally meeting this goal in 2012).

Climate change adaptation at the local government scale has been supported and provided with more flexibility due to legislative reforms that allow local councils to develop adaptation activities within their local jurisdictions (DCCEE 2010). Many councils are now actively developing climate change activities. In Victoria, the state government recently funded 35 projects representing 32 local councils to develop activities relating to climate change at a local government level. Examples of these include a project led by Darebin Council which aims to identify and promote innovative mechanisms to assist vulnerable households to adapt to climate change risks and address rising energy costs; a project led by Greater Geelong Council which aims to build the capacity of coastal communities, state and local government and service authorities to assess risk and prepare adaptation responses; and a project led by the City of Melbourne to produce an economic framework to value the benefits of green infrastructure in order to develop the business case to adapt. Many of these projects include a wide range of cross-council partnerships, which will strengthen their application and transferability to many Victorian regions. However, there are few cross-state collaboration or partnerships, which is an area worth developing in order to synthesise and synergise learnings.

It is also important to note here that the role of national governments is still important for climate change activities at a local level. National governments can coordinate adaptation efforts of local and sub-national governments by, for example, protecting vulnerable groups, providing financial support, and assisting with information, policy and legal frameworks (IPCC 2014).

Given the multitude of climate change activities occurring at the local scale, it is vital that these projects include health and social variables in their planning and implementation. The examples provided are incorporating a broad approach in the way they conceptualise climate change – as an issue that is not just environmental or economic – and it is important that this is supported and the information is fed back into the local community to realise its full potential. It is a timely opportunity to ensure that the current and future effects of climate change thoroughly incorporate all social and health dimensions in order to increase our resilience to climate change.

Case study: communities coming together to strengthen health and social outcomes in conditions of adversity

On 9 February 2014 extreme weather conditions in the Gipplsand region of Victoria, Australia triggered a fire near the Strzelecki Highway that headed to the Hazelwood Power Station and its open cut mine in Morwell. This fire raged out of control for four weeks, causing many residents to flee due to the thick blanket of smoke and ash. Levels of small particle pollution were 10 to 15 times higher than the recommended daily minimum.

Essential services were temporarily suspended, with schools, mail services and the courthouse shutting down, and residents were advised to either stay indoors or find accommodation elsewhere. Residents were not informed of the health risks or preventive measures to take until more than two weeks after the fire began. Effects of the fire are anecdotal at this point and include hospital admissions due to carbon monoxide exposure (predominantly firefighters). The fire was brought under control on 25 March.

The Victorian Government launched an inquiry into the fire, examining its origin, the steps that were taken to respond to the emergency, the role of the mine's operator, GDF Suez, and the adequacy of the health information that was disseminated to Morwell's 12 000 residents (State Government of Victoria 2014). The board of the inquiry reported to the government in August 2014. The board found that the impact of the Hazelwood mine fire on the Latrobe Valley community was significant, particularly in relation to adverse health effects presently and into the future. The board estimated that the total cost of the mine fire borne by the Victorian Government, the local community and the mine operator exceeded A\$100 million. Importantly, a the board also announced a health study to look at the long-term health effects of the fire over a 10-year period.

The role of the community has been on display in this particular example of an extreme weather event, the likes of which are predicted to increase in their frequency and intensity. In particular, a community group – Voices of the Valley – emerged as a result of the fire. The group has been active in critiquing the government inquiry, as well as contributing submissions. They have also been part of a GetUp (an advocacy non-government organisation) campaign to focus attention on the community's plight to receive answers to the events of February and March. The emergence of this volunteer-led and community-based group is an example of the strength of social capital (Putnam 1995) and that it can have broad and positive consequences on the resilience of a community, beyond just the issue at hand (in this case, the fire in the power station). The quality of the social capital within a community has been suggested as a measure of its potential for adopting a cooperative approach to addressing local problems (Fukuyama 2001; Pilkington 2002). However, it often takes points of crises, like the fire in the power station, for communities to come together strongly. The key here is to sustain the community's momentum to build a healthy and equitable community for the long term. This is particularly important for a rural area like Morwell, which displays a high level of health and social inequity – it is ranked as the fourth most disadvantaged area in Victoria as measured by the index of disadvantage, which includes variables such as income, education and employment (ABS 2011).

When adaptation may jeopardise our wellbeing – the concept of maladaptation

Adaptation efforts have been implemented across various sectors, with mixed results. The most alarming examples can be explained as forms of 'maladaptation', which has been defined as 'action taken ostensibly to avoid or reduce vulnerability to climate change that impacts adversely on, or increases the vulnerability of other systems, sectors or social groups' (Barnett & O'Neill 2010). Two examples of this concept that have been used to illustrate its relevance include the Wonthaggi water desalination plant and the Sugarloaf Pipeline, both in Victoria, Australia. Both of these infrastructure projects will have (or are already having) high social and environmental costs (for an extensive discussion of these see Barnett & O'Neill 2010). Adaptation plans and policies must therefore be adequately assessed to ensure that they do not in fact increase the vulnerability of systems, sectors or social groups. Barnett and O'Neill (2010) suggest five criteria with which to assess potential pathways to maladaptation: increasing greenhouse gas emissions; burdening the most vulnerable; high opportunity costs; reducing adaptation incentives; and limiting choices available to future generations.

Discussion

The chapter has outlined the main health and social effects of climate change relevant to Australian communities, and used the lens of equity to examine how these impacts will be felt differently, depending on underlying levels of vulnerability. Hence, a first step towards adaptation to future climate change is reducing vulnerability and exposure to present climate variability (IPCC 2014). Some suggestions are provided in the following discussion.

Policy and practice to address climate-related health inequities

We have demonstrated how climate change will work through existing social inequities to exacerbate health and health inequities. Equity-oriented climate change adaptation means attention to the economic and social conditions in which communities live – this is not just a climate change policy issue, it requires inter-sectoral action, involving health, planning and social sectors. For example, policies and programs in urban planning and design, rural development, and workplace health and safety can help communities and vulnerable populations to adapt to existing climate change, mitigate further climate change and, if done well, improve health equity (Friel *et al.* 2011b; Friel *et al.* 2008). Similarly, equity-focused health organisations and social services will not only improve health but they will help the communities most affected by climate change. They will also build community resilience to predicted health risks from heatwaves, bushfires, infectious diseases, diminished air quality and the mental health impacts of climate change (Blashki *et al.* 2011).

There are a few areas in which the health sector and community health organisations can focus their attention to improve climate-related health inequity (Friel 2013):

Evidence-informed practice
Evidence-informed practice requires good data on the extent of the problem and up-to-date evidence on the causes of health inequities and on what works to reduce them. It also requires an understanding of the evidence such that the causes of health inequities are acted on. In a study of the capacity of metropolitan Community Health Services in Victoria to respond to climate change, Olaris (2008) observed that the provision of clear information about climate change and related evidence based practice would increase the sector's ability to undertake effective action.

People-centred practice
All members of society – including those most disadvantaged and marginalised – are entitled to participate in the identification of priorities and targets that guide deliberations underlying public health practice. That focus is stimulated by, and feeds into, local conditions of inclusion and fair representation. This could be supported by the development of a statutory local climate change health equity action plan that is regularly monitored and reviewed, and ensures monitoring and reporting against a set of specific health equity focused outcomes.

Prevention-focused practice
If public health practitioners are to address climate change and its impact on health equity there needs to be a refocusing of activities towards the removal of barriers to access and use of quality primary health care, and on the conditions in which people grow, live, work and age. This means prioritising services that prevent or ameliorate the health damage caused by living and growing up in disadvantaged circumstances rather than on behaviour change. It requires the development and improvement of good quality, integrated local services co-produced with the public to achieve needs-driven outcomes.

Conclusion

Climate change throws into sharp relief many of the issues to do with inequities in living standards, political power, resource use, levels of exposure to environmental stresses, and health and life expectancy (Friel *et al.* 2008). Addressing the common drivers of climate change and population health will not only improve health, but advances will be made in social equity such that communities and nations will be better able to resist current climate change and avert further damage to the environment.

References

Adger NW, Kelly PM (1999) Social vulnerability to climate change and the architecture of entitlements. *Mitigation and Adaptation Strategies for Global Change* **4**, 253–266. doi:10.1023/A:1009601904210

Albrecht G, Sartore G, Connor L, Higginbotham N, Freeman S, Kelly B, Stain H, Tonna A, Pollard G (2007) Solastalgia: the distress caused by environmental change. *Australasian Psychiatry* **15**, S95–S98. doi:10.1080/10398560701701288

APA (American Psychological Association) (2010) 'Psychology and global climate change: addressing a multifacted phenomenon and set of challenges'. American Psychological Association. Washington, D.C.

ABS (Australian Bureau of Statistics) (2011) *Census of Population and Housing: Socio-Economic Indexes for Areas (SEIFA)*. Cat. no. 2033.0.55.001. Australian Bureau of Statistics, Canberra.

Barnett J, O'Neill S (2010) Maladaptation. *Global Environmental Change* **20**, 211–213. doi:10.1016/j.gloenvcha.2009.11.004

Berkhout F (2012) Adaptation to climate change by organizations. *WIREs Climate Change* **3**, 91–106. doi:10.1002/wcc.154

Berry H, Bowen K, Kjellstrom T (2010) Climate change and mental health: a causal pathways framework. *International Journal of Public Health* **55**(2), 123–132. doi:10.1007/s00038-009-0112-0

Berry HL, Hogan A, Owen J, Rickwood D, Fragar L (2011) Climate change and farmers' mental health: risks and responses. *Asia-Pacific Journal of Public Health* **23**(2 Suppl), 119S–132S. doi:10.1177/1010539510392556

Blashki G, Armstrong G, Berry HL, Weaver HJ, Hanna EG, Peng B. *et al.* (2011) Preparing health services for climate change in Australia. *Asia-Pacific Journal of Public Health* **23**(2 Suppl), 133S–143S. doi:10.1177/1010539510395121

Bowen KJ, Ebi K, Friel S (2014) Climate change adaptation and mitigation: next steps for cross-sectoral action to protect global health. *Mitigation and Adaptation Strategies for Global Change* **19**, 1033–1040. doi:10.1007/s11027-013-9458-y

Bradbear C, Friel S (2013) Integrating climate change and health into food policy: an analysis of how climate change can affect food prices and population health. *Food Policy* **43**, 56–66. doi:10.1016/j.foodpol.2013.08.007

Campbell-Lendrum D, Corvalan C (2007) Climate change and developing-country cities: implications for environmental health and equity. *Journal of Urban Health* **84**(S1), 109–117. doi:10.1007/s11524-007-9170-x

Costello A, Abbas M, Allen A, Ball S, Bell S, Bellamy R, Friel S, Groce N, Johnson A, Kett M, Lee M, Levy C, Maslin M, McCoy D, McGuire B, Montgomery H, Napier D, Pagel C, Patel J, de Oliveira JAP, Redclift N, Rees H, Rogger D, Scott J, Stephenson J, Twigg J, Wolff J, Patterson C (2009) Managing the health effects of climate change. *Lancet* **373**, 1693–1733. doi:10.1016/S0140-6736(09)60935-1

DCCEE (Department of Climate Change and Energy Efficiency) (2010) 'Climate change adaptation actions for local government'. Department of Climate Change and Energy Efficiency, Canberra.

Friel S (2010) Climate change, food insecurity and chronic diseases: sustainable and healthy policy opportunities for Australia. *NSW Public Health Bulletin* **21**(5–6), 129–133. doi:10.1071/NB10019

Friel S (2013) Improving equity. In *Oxford Handbook of Public Health Practice*. (Eds C Guest, W Ricciardi, I Kawachi and I Lang) 3rd edn. pp. 406–416. Oxford University Press, Oxford.

Friel S, Marmot M, McMichael AJ, Kjellstrom T, Vågerö D (2008) Global health equity and climate stabilisation - need for a common agenda. *Lancet* **372**(9650), 1677–1683. doi:10.1016/S0140-6736(08)61692-X

Friel S, Butler CD, McMichael AJ (2011a) Climate change and health: risks and inequities. In *Global Health Ethics*. (Eds S Benatar and G Brock) pp. 198–209. Cambridge University Press, Cambridge.

Friel S, Hancock T, Kjellstrom T, McGranahan G, Monge P, Roy J (2011b) Urban health inequities and the added pressure of climate change: an action-oriented research agenda. *Journal of Urban Health* **88**(5), 886–895. doi:10.1007/s11524-011-9607-0

Fukuyama F (2001) Social capital, civil society and development. *Third World Quarterly* **22**(1), 7–20. doi:10.1080/713701144

Furgal C, Seguin J (2006) Climate change, health and vulnerability in Canadian northern Aboriginal communities. *Environmental Health Perspectives* **114**, 1964–1970

Gardiner S (2004) Ethics and global climate change. *Ethics* **114**, 555–600. doi:10.1086/382247

Hajat S, Armstrong BG, Gouveia N, Wilkinson P (2005) Mortality displacement of heat-related Deaths: a comparison of Delhi, Sao Paulo, and London. *Epidemiology* **16**(5), 613–620. doi:10.1097/1001.ede.0000164559.41092.2a

Hanna EG, Bell E, King D, Woodruff R (2011) Climate change and Australian agriculture: a review of the threats facing rural communities and the health policy landscape. *Asia-Pacific Journal of Public Health* **23**(2 Suppl), 105S–118S. doi:10.1177/1010539510391459

Huq S, Kovats S, Reid H, Satterthwaite D (2007) Reducing risk to cities from disasters and climate change. *Environment and Urbanization* **19**, 3–15. doi:10.1177/0956247807078058

IPCC (Intergovernmental Panel on Climate Change) (2001) *Climate Change 2001: Impacts, Adaptation, and Vulnerability. Contribution of Working Group II to the Third Assessment Report of the Intergovernmental Panel on Climate Change*. (Eds J McCarthy, OF Canziani, N Leary, D Dokken and K White). Cambridge University Press, Cambridge, UK.

IPCC (Intergovernmental Panel on Climate Change) (2007) *Climate Change 2007: Impacts, Adaptation and Vulnerability. Contribution of Working Group II to the Fourth Assessment Report of the Intergovernmental Panel on Climate Change*. (Eds ML Parry, OF Canziani, JP Palutikof, PJ van der Linden and CE Hanson) Cambridge University Press, Cambridge, UK.

IPCC (Intergovernmental Panel on Climate Change) (2014) IPCC 2014: Summary for Policy-makers. In *Climate Change 2014: Impacts, Adaptation, and Vulnerability. Part A: Global and Sectoral Aspects. Contribution of Working Group II to the Fifth Assessment Report of the Intergovernmental Panel on Climate Change*. (Eds CB Field, VR Barros, DJ Dokken, KJ Mach, MD Mastrandrea, TE Bilir, M Chatterjee, KL Ebi, YO Estrada, RC Genova, B Girma, ES Kissel, AN Levy, S MacCracken, PR Mastrandrea and LL White) pp. 1–32. Cambridge University Press, Cambridge, UK.

Judd F, Jackson H, Fraser C, Murray G, Robins G, Komiti A (2006) Understanding suicide in Australian farmers. *Social Psychiatry and Psychiatric Epidemiology* **41**, 1–10. doi:10.1007/s00127-005-0007-1

Kettings C, Sinclair A, Voevodin M (2009) A healthy diet consistent with Australian health recommendations is too expensive for welfare-dependent families. *Australian and New Zealand Journal of Public Health* **33**, 566–572. doi:10.1111/j.1753-6405.2009.00454.x

Kjellstrom T (2009) Climate change, direct heat exposure, health and well-being in low and middle income countries. *Global Health Action* **2**, 1–3.

Kovats S, Akhtar R (2008) Climate, climate change and human health in Asian cities. *Environment and Urbanization* **20**(1), 165–175. doi:10.1177/0956247808089154

Kovats R, Hajat S (2008) Heat stress and public health: a critical review. *Annual Review of Public Health* **29**, 41–55. doi:10.1146/annurev.publhealth.29.020907.090843

Loughnan M, Nicholls N, Tapper N (2010) The effects of summer temperature, age and socio-economic circumstance on acute myocardial infarction admissions in Melbourne, Australia. *International Journal of Health Geographics* **9**, 41. doi:10.1186/1476-072X-9-41

Luber G, McGeehin M (2008) Climate change and extreme heat events. *American Journal of Preventive Medicine* **35**(5), 429–435. doi:10.1016/j.amepre.2008.08.021

McDermott B, Lee E, Judd M, Gibbon P (2005) Posttraumatic stress disorder and general psychopathology in children and adolescents following a wildfire disaster. *Canadian Journal of Psychiatry* **50**, 137–143.

McLaughlin K, Fairbank J, Gruber M, Jones R, Lakoma M, Pfefferbaum B, Sampson NA, Kessler RC (2009) Serious emotional disturbance among youths exposed to Hurricane Katrina 2 years postdisaster. *Journal of the American Academy of Child and Adolescent Psychiatry* **48**, 1069–1078. doi:10.1097/CHI.0b013e3181b76697

McMichael AJ, Lindgren E (2011) Climate change: present and future risks to health, and necessary responses. *Journal of Internal Medicine* **270**(5), 401–413. doi:10.1111/j.1365-2796.2011.02415.x

McMichael A, Butler C, Weaver H (2008) 'Climate change and AIDS: a joint working paper'. UNEP & UNAIDS, Nairobi.

McMichael AJ, McMichael C, Berry HL, Bowen KJ (2010) Climate change, displacement and health: risks and responses. In *Climate Change and Displacement: Multidisciplinary Perspectives.* (Ed. J McAdam) pp. 191–220. Hart Publishing, Oxford.

O'Brien K, Eriksen S, Schjolden A, Nygaard L (2004) 'What's in a word? Conflicting perceptions of vulnerability in climate change research'. CICERO Working Papers. CICERO, Oslo.

O'Brien L, Berry H, Coleman C, Hanigan I (2014) Drought as a mental health exposure. *Environmental Research* **131**, 181–187. doi:10.1016/j.envres.2014.03.014

Olaris K (2008) Community health services and climate change: exploring the sector's capacity to respond. *Environmental Health* **8**(2), 28–41.

Patz J, Olson S, Uejio C, Gibbs H (2008) Disease emergence from global climate and land use change. *The Medical Clinics of North America* **92**, 1473–1491. doi:10.1016/j.mcna.2008.07.007

Pilkington P (2002) Social capital and health: measuring and understanding social capital at a local level could help to tackle health inequalities more effectively. *Journal of Public Health* **22**(1), 7–20.

Putnam RD (1995) Bowling alone: America's declining social capital. *Journal of Democracy* **6**, 65–78. doi:10.1353/jod.1995.0002

Quiggin J (2010) 'Drought, climate change and food prices in Australia'. Australian Conservation Foundation, Melbourne.

Reisinger A, Kitching RL, Chiew F, Hughes L, Newton PCD, Schuster SS, Tait A, Whetton P (2014) Australasia. In *Climate Change 2014: Impacts, Adaptation, and Vulnerability. Part B: Regional Aspects. Contribution of Working Group II to the Fifth Assessment Report of the Intergovernmental Panel on Climate Change.* (Eds VR Barros, CB Field, DJ Dokken, MD Mastrandrea, KJ Mach, TE Bilir, M Chatterjee, KL Ebi, YO Estrada, RC Genova, B Girma, ES Kissel, AN Levy, S MacCracken, PR Mastrandrea and LL White) Cambridge University Press, Cambridge, UK.

Robine J, Cheung S, Roy S, Oyen HV, Griffiths C, Michel J, Herrmann FR (2008) Death toll exceeded 70,000 in Europe during the summer of 2003. *Comptes Rendus Biologies* **331**, 171–178. doi:10.1016/j.crvi.2007.12.001

Sevoyan A, Hugo G, Feist H, Tan G, McDougall K, Tan Y, Spoehr J (2013) 'Impact of climate change on disadvantaged groups: issues and interventions'. National Climate Change Adaptation Research Facility, Gold Coast.

Sharkey P (2007) Survival and death in New Orleans: an empirical look at the human impact of Katrina. *Journal of Black Studies* **37**, 482–501. doi:10.1177/0021934706296188

Smith K (2008) Mitigating, adapting, and suffering: how much of each? *Annual Review of Public Health* **29**, 11–25. doi:10.1146/annurev.publhealth.29.020907.090759

State Government of Victoria (2014) 'Deputy Premier announces terms of reference for the Hazelwood mine fire inquiry'. <http://www.premier.vic.gov.au/media-centre/media-releases/9453-deputy-premier-announces-terms-of-reference-for-hazelwood-mine-fire-inquiry.html>

Strazdins L, Friel S, McMichael A, Woldenberg-Butler S, Hanna E (2011) Climate change and children's health: likely futures, new inequities? *International Journal of Public Health* **2**(4), 493–500.

UNDP (United Nations Development Programme) (2007) 'Human development report: fighting climate change: human solidarity in a divided world'. United Nations Development Programme, New York.

US EPA (United States Environmental Protection Agency) (2007) 'Reducing urban heat islands: compendium of strategies – trees and vegetation'. United States Environmental Protection Agency, Washington D.C.

Whittaker J, Handmer J, Mercer D (2012) Vulnerability to bushfires in rural Australia: a case study from East Gippsland, Victoria. *Journal of Rural Studies* **28**(2), 161–173. doi:10.1016/j.jrurstud.2011.11.002

2

Adaptation: living with a changing environment

Guy Barnett, John Gardner and Jacqui Meyers

Key points

- Irrespective of future climate change mitigation efforts, a likely minimum of 2°C of global warming is now 'locked in', demanding a greater focus on climate adaptation.
- Uncertainty is no excuse for inaction. When adaptation decision-making is focused on near-term decisions with short lifetimes, uncertainty about the future matters less.
- Adaptation may be easier for organisations with longer planning horizons, links to climate adaptation knowledge and existing assessments of their climate vulnerability.
- Community-based health and social service organisations exist to provide for the care and wellbeing of their clients. Reducing climate change vulnerability is core business.

Introduction

Australia has one of the most variable climates in the world, with naturally occurring cycles of wet and dry periods that often result in extreme events such as heatwaves, bushfires, floods and drought. Climate extremes are considered a natural part of the Australian landscape and are ingrained in the culture of its people: the stoic bush family facing drought, or the heroic volunteer firefighter tackling a blaze (Head *et al.* 2014). Thus, the concept of adapting to a variable and changing climate is not new in Australia, as its people have done so since human occupation began some 50 000 years ago. Modern examples of climate adaptation include the introduction of engineering standards for housing design following Cyclone Tracy (Kiem *et al.* 2011) and investing in water security infrastructure such as household rainwater tanks and desalination facilities in response to prolonged drought (Preston & Stafford Smith 2009). While not all past adaptation activities have been successful, there are lessons to be learnt that can inform future adaptation to contemporary, anthropogenic climate change (Dovers 2009).

Despite this familiarity with, and capacity for, adaptation in Australia, the challenge posed by anthropogenic climate change is enormous and beyond previous human experience. Without large and immediate reductions in global greenhouse emissions, it is unlikely that the world will avoid 'dangerous climate change', defined as an increase in global average temperature of more than 2°C above pre-industrial levels (Peters *et al.* 2012a; Stafford Smith *et al.* 2011). Current thinking in Australia is that we should be planning to adapt to 3–4°C of global warming, with 4°C now a distinct possibility by the end of this century (Palutikof *et al.* 2013; Stafford Smith *et al.* 2011). The key message is that irrespective of future mitigation efforts, a likely minimum of 2°C of global warming is 'locked in', which demands increasing emphasis on climate adaptation. This does not mean that mitigation efforts are unimportant, as global emissions reduction will help to minimise the long-term impacts of climate change. Instead, it recognises that climate change is impacting Australia now, and will continue to do so, with a need for adaptation to reduce social and environmental damage resulting from these impacts.

Several comprehensive reviews of climate adaptation in Australia not only detail the current state of knowledge, but also identify the frontiers of climate adaptation science and practice (see Ash & Stafford Smith 2013; Palutikof *et al.* 2013; Preston & Stafford Smith 2009). What we can take from these reviews is that effective adaptation implies substantial change. Short-term, small-scale and reactive responses to climate impacts are more properly termed 'coping', while 'adaptation' is generally used to reflect larger, longer-term and often more proactive responses. In practice, however, it can be rather difficult to effectively differentiate between these two climate impact responses (i.e. where coping ends and adaptation begins).

The most dramatic adaptation responses are considered 'transformational'. With the growing prospect of 4°C of global warming this century, there are now increasing calls to move from 'incremental' to 'transformational' adaptation (Dovers 2009; Head *et al.* 2014; Park *et al.* 2012). The key difference is the extent of change that is required, with incremental adaptation primarily focused on maintaining current systems and processes, and transformation referring to the creation of fundamentally new systems or processes (Park *et al.* 2012). Using the Australian wine industry as an example, incremental adaptation might include the use of water-efficient irrigation technologies and planting vines with drought-tolerant rootstock, whereas transformational adaptation may involve relocation of vineyards to cooler regions.

Adaptation operates at nested and interacting scales (Preston & Stafford Smith 2009). For example, the capacity of an individual household to respond to a climate impact depends on its own resources, but it is also influenced by the capacity of the local community, which in turn is influenced by the adaptive capacity of the region and so on. Consequently, there are situations where local-scale organisations seeking to undertake climate adaptation actions have been constrained by a lack of either motivation or support from institutions at higher scales. At the same time, the options that are available for adaptation action at higher scales can be impacted by the level of readiness for action at lower scales, with governments or industry sometimes unable to implement policies until the broader population at large or organisations in the sector are ready for them. To complicate matters further, the same adaptation response might take different forms at different scales (Adger & Vincent 2005). For example, adapting to reduced rainfall on an individual farm might involve switching to more drought-tolerant crops, but at a regional level, may involve changes in water resource management regimes such as the allocations of irrigation rights.

The role that community-based health and social service organisations can play in managing the impacts of climate change on vulnerable individuals and communities is relatively poorly understood in Australia (Mallon *et al.* 2013), and addressing this knowledge gap was clearly a key motivation for this book. In this chapter, we draw on a range of research undertaken in the CSIRO Climate Adaptation National Research Flagship in recent years, providing an introduction to the concept of climate adaptation and its application in Australia. We then outline the nature of the climate adaptation challenge that confronts the nation, including the role of climate science and the development of projections of future climate to inform impact and vulnerability assessments. The main findings of an adaptation benchmarking survey of Australian organisations undertaken by Gardner *et al.* (2010) are discussed, including the drivers of and barriers to adaptation planning. The chapter concludes with a discussion of how to measure adaptation success, drawing on some lessons learnt from two case studies.

The adaptation challenge

A major emphasis within climate adaptation research has been the development of climate change projections in order to understand likely impacts. These projections can help different institutions and enterprises address the fundamental question of *'what are we adapting to?'* (Preston & Stafford Smith 2009). What follows is an introduction to the science of climate projections and their use in assessing climate change impacts, including a discussion of scientific uncertainty and how this uncertainty can be addressed in decision-making. We then outline the direct and indirect impacts of climate change in Australia and discuss the concept of vulnerability, which is widely used to frame the challenge of adaptation. Finally, we discuss the issues of equity and fairness (or the lack thereof) in the distribution of climate change impacts and the capacity for adaptation.

Climate projections and uncertainty

Projections of future climate are a fundamental input into climate change impact assessments that are used to inform adaptation planning. These projections are typically derived from a range of global climate models (GCMs), which are based on a wide range of assumptions about the earth system and climate feedbacks, as well as scenarios of future greenhouse emissions. It must be acknowledged that although we continue to see advances in the modelling of future climates, the accuracy of these projections will always be limited by fundamental, irreducible uncertainties (Dessai *et al.* 2009). It is difficult, for example, to accurately predict changes in human activity which will determine the level of future greenhouse emissions. While some uncertainties can be quantified, others cannot. This has led to calls for more funding to improve the accuracy of projections (Dessai *et al.* 2009).

These uncertainties and assumptions raise the question of whether improved projections are in fact necessary for making better adaptation decisions (Dessai *et al.* 2009; Howden *et al.* 2013; Hulme *et al.* 2009). Some argue that, while climate science has been an important part of defining the case for global action on climate change, it has proven far less insightful for developing effective adaptation strategies (Howden *et al.* 2013). In support of this argument, Dessai *et al.* (2009) point out that climate is only one of many change drivers that will influence the success or otherwise of future adaptation efforts. These authors advise that our ability to predict many other important change drivers, such as economic markets, cultural preferences, globalisation processes and so on, are often more limited than our ability to predict future climate. Yet despite this uncertainty, decision-making still occurs.

Coming back to the science of climate projections, there are many GCMs, each with their own strengths and weaknesses, which makes it difficult to know which to use in impact and vulnerability assessments. For example, OzClim, the online tool developed by CSIRO to provide climate projections for Australia (Ricketts & Page 2007), contains data from over 20 GCMs that were used in the Fourth Assessment Report of the Intergovernmental Panel on Climate Change (IPCC). Many researchers combine the results from multiple GCMs to produce a single projection with a mean value and an associated range of uncertainty, for example, a warming of 1.5°C with a range of 1–2°C (Hennessy et al. 2012). This approach is now considered undesirable, with concerns that the internal consistency that is preserved within individual GCMs is lost when projections from different GCMs are combined.

To address this issue, Whetton et al. (2012) developed the Representative Climate Futures (RCF) approach, which can be used to describe plausible future climates in a region. These RCFs are constructed from the output of individual GCMs, which are then classified into a common set of climate futures with estimates of their likelihood, for example, 'most likely', 'high risk' and 'least change'. For instance, Hennessy et al. (2012) have used this approach in eastern Victoria, where a future climate in 2030 of *warmer, but with little change in rainfall'* was supported by 20 of the 24 climate models. This climate future is therefore considered to be the 'most likely'. Also of interest in impact and vulnerability assessments are climate futures that, while less likely, may have the highest potential for impact. As such, the construction of RCFs and the selection of specific GCMs to be used in climate change impact and vulnerability assessments should be tailored to ensure they are 'fit for purpose'.

It is also important to note that the level of uncertainty in climate projections varies greatly across sectors and across decision contexts, depending on the particular climate driver and time period being considered. For example, over the next 50 years, rises in atmospheric carbon dioxide concentrations, rises in average and maximum temperatures and increased sea levels are virtually certain, compared with greater uncertainty regarding changes in rainfall and the number of cyclones (Reisinger et al. 2014). Also, the magnitudes of expected changes are more certain for time periods that occur in the near future (e.g. 2030) than for those in the distant future (e.g. 2100). As noted by Reisinger et al. (2014), emerging best practice in Australia tackles uncertainty through a focus on the decision lifetime and form of adaptation response being considered (Stafford Smith et al. 2011); that is, focusing particularly on the decisions that need to be made in the short term, the 'lifetime' of those decisions, and the risk posed by climate change over the course of that decision lifetime. In support of this approach, Reisinger et al. (2014) provide the example of a farmer who needs to make a decision about which crop to plant next year. While climate is an important consideration in the decision, the decision lifetime is only a year or two, providing an opportunity to quickly adjust or recover. When adaptation decision-making is framed around near-term decisions with short lifetimes, uncertainty about the future matters less (Reisinger et al. 2014; Stafford Smith et al. 2011).

There will also be decisions with a long 'lifetime', such as those associated with the location and design of long-lived infrastructure (e.g. buildings and bridges) or future urban development. In these cases, greater consideration must be given to uncertainty over the course of the decision lifetime. Climate change is likely to be one of many drivers of uncertainty that must be addressed in these situations. Nonetheless, there are a range of well-known techniques for decision-making under complex uncertainty, such as 'the precautionary principle', 'adaptive management', 'no regrets strategies' or 'risk hedging'

(Reisinger *et al.* 2014). Even long-term infrastructure decisions are often the cumulative effect of many shorter-term decisions, linked to political and financial cycles (Stafford Smith *et al.* 2011). Consequently, the 'nesting' of shorter-term decisions within a longer-term framework that accounts for climate change and other uncertainties is important for adaptation planning.

Direct and indirect impacts

Australia is a large country with a diverse and variable climate, encompassing the hot and humid tropical north through to the deserts of the arid interior and cool temperate and alpine zones in the south. Similarly, the impacts of climate change vary widely across the nation. A critical prerequisite for climate adaptation is understanding the nature and distribution of current and future climate change impacts, as it helps to address the fundamental question of *'what are we adapting to?'* (Preston & Stafford Smith 2009). Over the past decade, there has been a rapidly growing number of climate change impact studies undertaken in Australia, many of which have been included in the Fourth Assessment Report (Hennessy *et al.* 2007) and Fifth Assessment Report (Reisinger *et al.* 2014) of the IPCC. In the following section, we outline some of the main findings of this body of work.

Climate change has already been observed in Australia. Since 1950, mean temperature has warmed by 0.4–0.7°C, sea levels have risen by ~70 mm, regional rainfall patterns have changed, and droughts and heatwaves have become more frequent and more intense (Hennessy *et al.* 2007). The most recent evidence is provided in the *State of the Climate 2014* report, which is based on historical climate observations and monitoring (Bureau of Meteorology & CSIRO 2014). In this report, 2013 was identified as Australia's warmest year on record. In fact, 7 of the 10 warmest years on record (since records began in 1910) have occurred since 1998. Extreme fire weather and the length of the fire season have increased across large parts of Australia. Trends in rainfall have been more difficult to detect due to high variability, although large increases have been observed in north-west Australia since 1970, while winter rainfall in the south-west of Australia has been in steady decline. The rates of sea-level rise vary across Australia, with the south and east experiencing rates that are similar to the global average (1.7 mm per year). Higher rates have been observed on Australia's northern coastline.

Since the release of the Fourth Assessment Report, there is increasingly strong evidence that observed carbon dioxide emissions in the atmosphere are tracking high growth emission scenarios (Le Quere *et al.* 2009; Peters *et al.* 2012b; Raupach *et al.* 2007). Over the past 50 years, growing greenhouse gas concentrations have contributed to warming of Australian air and sea-surface temperatures, and this trend is set to continue over the 21st century. The Fifth Assessment Report (Reisinger *et al.* 2014) identified eight key impacts of climate change in Australia this century. In many cases, these impacts are not particularly new, but rather represent a further exacerbation of the already observed impacts of climate change. The eight key impacts are:

- significant change in the composition and structure of coral reefs
- loss of mountain ecosystems and some associated native species
- increased frequency and intensity of flood damage to communities
- increased constraints on water resources within southern Australia
- increased illness, death and infrastructure damage during heat waves
- increased damage, economic loss, and risk to human life from bushfires

- increased risk to coastal infrastructure from continuing sea level rise
- significant reduction in agricultural production due to continued drying.

The consequences of many of these impacts of climate change on human health and wellbeing have been well documented in Australia (Bambrick *et al.* 2011; Berry *et al.* 2010; Hughes *et al.* 2011; McMichael *et al.* 2006). A useful classification of the health effects of climate change is whether they are primary, secondary or tertiary (Butler & Harley 2010).

Primary health effects are essentially the direct physical consequences of climate change, such as heatwaves, bushfires, floods and so on. Harm to human health is usually rapid and obvious, typically resulting in injuries or death. For example, the 'Black Saturday' bushfires in Victoria in February 2009 caused 173 deaths (Cameron *et al.* 2009) and a further 374 deaths were attributed to the associated heatwave (Department of Human Services 2009). In late January and early February 2011, a heatwave in the Greater Sydney region of New South Wales resulted in 132 excess deaths, as well as an eight-fold increase in hospital emergency department visits for direct heat effects and a three-fold increase in visits for dehydration (Schaffer *et al.* 2012). Just a few weeks earlier, 24 people lost their lives in southern Queensland due to major flooding of the Brisbane River and its tributaries (van den Honert & McAneney 2011). These examples highlight the human vulnerability to extreme climate events, which are expected to increase in frequency and severity with climate change.

Secondary effects are more indirect, resulting from changes in biophysical and ecological processes and systems. For example, climate impacts may lead to changes in the vectors of infectious disease, or changes in the parasite and host animal ecology of zoonotic diseases (Butler & Harley 2010). There is growing evidence in Australia that mosquitoes that carry vector-borne diseases such as dengue fever could extend their range and activity with climate change (Hu *et al.* 2012). While dengue fever is currently confined to northern Queensland, there is concern that installation of domestic water storage tanks in response to prolonged drying could extend the geographic range of dengue vectors (Beebe *et al.* 2009; Kearney *et al.* 2009). Outbreaks of food- and water-borne diseases are also expected to become more frequent and intense in the presence of climate change. Secondary effects can also result from the impacts of climate change on food and water availability (McMichael 2013). For example, reductions in rainfall can reduce food yields and increase food prices.

The tertiary health effects of climate change are the most diffuse. They are the product of interconnections between the environment and a wide range of cultural, political, economic and social factors. Examples of tertiary health effects include the impact of drought on mental health in rural communities (Berry *et al.* 2010) and the consequences of rising tension and conflict around the world due to declines in natural resources (Butler & Harley 2010).

Vulnerability assessments

The concept of vulnerability is used widely across a range of scientific disciplines (Turner *et al.* 2003), but the focus here is on its application in relation to climate change. Vulnerability assessments are typically a first step in establishing which populations and which assets are most vulnerable to the effects of climate change. Vulnerability assessments are often seen as an end in themselves, yet in many cases are poorly suited to addressing questions of what adaptation action is needed (Ash & Stafford Smith 2013). More advanced approaches are now being developed that enable consideration of the spatial distribution of vulnerability (Preston *et al.* 2011), as well as the multi-scaled nature of vulnerability and options available for adaptation (Barnett *et al.* 2013). Combining vulnerability assessments with adaptation assessments can identify weaknesses in a system and intervention options, and improve evidence and understanding of the linkages between climate and health (Ebi *et al.* 2006). Combined assessments can also provide a baseline against which changes in disease risk and protective measures can be monitored and which opportunities for adaptive capacity can be built.

The most common application of the vulnerability concept in climate science is where it is seen as a product of the interactions between exposure to a climate hazard and the sensitivity of the populations affected, taking into account adaptive capacity for ameliorating potential impacts (Smit & Wandel 2006). Broadly, the term 'adaptive capacity' reflects the ability of a human or natural system to deal with a stressor. Such responses involve the mobilisation of resources within that system. Consequently, the availability of resources and the ability to deploy them are considered critical components of adaptive capacity (Adger *et al.* 2004).

The adaptive capacity of a system reflects its ability to respond to climate change impacts as well as to any other impacts or stressors that may arise. Thus, climate change adaptation must be considered within the context of other (non-climate) impacts. Such impacts (e.g. global economic changes, population growth and conflict) that draw on the resources of a system may reduce the capacity to cope with climate impacts. For example, before the global financial crisis (GFC), many urban developers would market sustainability features to differentiate their products from their competitors, but immediately after the GFC, price point became the overriding factor and *'nice-to-have green measures'* were omitted (Shearer *et al.* 2013). As such, having adaptive capacity does not guarantee that an adaptive response will be deployed into action (Adger & Vincent 2005). Adaptive capacity simply reflects the potential for undertaking adaptation.

The concept of adaptive capacity can also be applied to vulnerable population groups and individuals. In this context, the findings of a recent study by Unsworth *et al.* (2013) on the psychological drivers of climate adaptation are particularly salient. These authors found that adaptive behaviours from individuals are not necessarily dependent on understanding climate change as a threat to oneself and one's way of life, but rather stem from a belief that the behaviour helps them to achieve important personal goals. An example in practice might be a person who is deaf or requires the use of a wheelchair. Their capacity to survive a flood or a bushfire would require thinking about the options available in their environment (adaptive capacity), as well as technologies that might enhance their individual capacities to hear or physically move around. In this way, adaptation becomes less about climate change and more about improving quality of life.

Issues of equity and fairness

It is widely acknowledged that the impacts of climate change are not (and will not be) evenly distributed across society (Adger *et al.* 2006). For instance, consider the vulnerability of low-income households to extreme heat in Australia. We know that climate change is likely to bring more frequent and severe heatwave events. What is less commonly understood is that heat-related health risks are more prevalent in low-income households, that low-income households are often geographically concentrated in the hottest parts of our cities and that they have fewer resources to invest in adaptation (Barnett *et al.* 2013).

Complicating this further is the issue of housing tenure. The adaptation of low-income private rental properties may be exacerbated by the 'split incentive', where landlords are unlikely to invest in housing retrofits such as the installation of ceiling insulation to reduce the climate vulnerability of their tenants, because they may not realise any direct benefit or return on that investment.

Geography is also important. The majority of the Australian population lives along the coast. An estimated 711 000 homes are located within 3 km of the coast and less than 6 m above sea level, rendering many households vulnerable to sea-level rise (Hennessy *et al.* 2007). Rural, regional and remote communities are also particularly vulnerable, due to their reliance on and level of connectedness with the natural environment. This problem also extends to industries that are heavily dependent on natural resources, particularly agriculture, mining and tourism.

Climate change is a pressing moral and ethical issue for reasons of equity and fairness. In this context, it is important to distinguish between groups that cause climate change and those that are exposed to its impacts (Sovacool 2013). In Australia, a classic example are the communities of Aboriginal and Torres Strait Islander people for which this paradox has been well described (Head *et al.* 2014). At the same time, there remains significant potential for contributions from Indigenous knowledge systems, which are typically highly adaptive by their very nature, for improving both understanding of and responses to climate change.

Pathways to adaptation

Adaptation to climate change does not 'just happen', although there is little discussion in the literature of the mechanisms by which it occurs. Much of the existing literature offers either a conceptualisation of the barriers to adaptation without providing empirical evidence, or case studies of adaptation activities that are difficult to generalise due to differences in approach. The result is a lack of conceptual clarity regarding the process of adaptation, which likely reflects both the diversity of actions that can be considered to be adaptation (as outlined earlier) and the relative immaturity of research and practice on climate change adaptation.

To help fill this knowledge gap Gardner *et al.* (2010) undertook a survey to ascertain the current level of adaptation planning in Australian public and private sector organisations. A total of 740 organisations were sampled across 24 different organisation types that spanned industry, local government, infrastructure and non-government organisations. The survey was undertaken in two phases between 2008 and 2010, with the results revealing that while most organisations were aware of climate change and considered both mitigation and adaptation important, both the nature and extent of adaptation activity was highly variable.

The lessons learnt through this survey are presented as a linear process model describing the common pathways to adaptation (Fig. 2.1). There are four key stages through which an organisation will usually pass before commencing adaptation planning. These stages are:

- clear understanding of climate change as an important issue
- awareness of the organisation's own climate change vulnerability
- a sense of responsibility for developing a solution or response
- willingness to engage in adaptation planning and to mobilise resources.

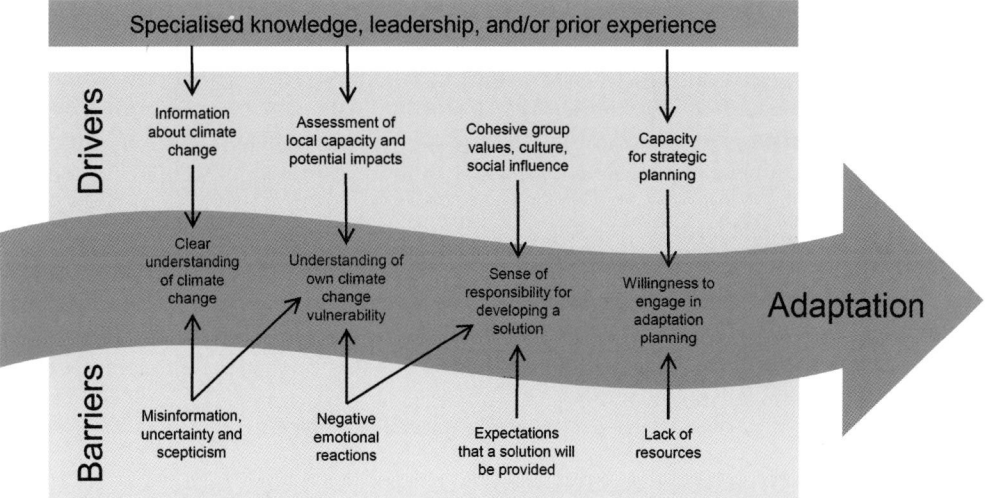

Figure 2.1.   A pathway to adaptation with associated drivers and barriers (adapted from Gardner *et al.* 2009).

There has been some discussion of the factors that determine how and why adaptation will (or will not) occur. Adaptation is unlikely to occur efficiently through the action of market forces alone, suggesting public policy intervention will be needed (Stern 2008). Historically, however, governments have been slow to apply public policy to 'creeping' environmental problems (Moser 2005). We argue that an organisation's decision to undertake adaptation planning represents the end of a pathway or chain of pre-conditions, as shown in Figure 2.1.

For an organisation to engage in adaptation planning requires a willingness to do so, which in turn requires a sense of responsibility for developing a solution, which first requires that there is a recognition of the problem (in this case the organisation's own vulnerability to climate change). Recognition of the problem itself will typically require an accurate understanding of climate change issues. Different organisations will find themselves located at some point along this pathway – the organisation's position will determine the nature and extent of engagement required in order for the organisation to progress towards an adaptation plan. A discussion of the role of effective engagement and the drivers of, and barriers to, adaptation now follows.

Towards effective engagement

There are some general points about beginning the process of engaging organisations or communities in adaptation planning. First, the groups of people that will be engaged are unlikely to be homogenous. Some participants will be extremely sceptical and others will be already convinced. Some will be familiar with strategic planning, while others may not be. Engagement processes need to be well designed to acknowledge and address the differences between participants. This addresses a key feature of effective engagement: the capacity to incorporate different opinions, backgrounds, views and agendas among participants. These different beliefs and perspectives about climate change and long-term planning form part of a wide range of diversity that engagement practitioners must both capture and incorporate.

Acceptance of anthropogenic climate change is not necessarily a pre-requisite for making progress on climate adaptation. An organisation may have good strategic planning practices in place and only need to be convinced that some future changes related to climate 'might occur' in order to include these potential climate risks in their existing planning processes. There may also be situations where organisations already possess the specialised knowledge, leadership and/or prior experience to readily incorporate climate adaptation into their general planning processes. Organisations like this may have an 'easy sell' in terms of engagement to promote adaptation planning. On the other hand, groups without this specialised knowledge may require more in-depth engagement in order to communicate information about climate change and to establish their vulnerability, sense of responsibility and willingness to act.

Drivers that promote adaptation

Adaptation planning requires a capacity for strategic planning, which is not present in all organisations. Those that have previous experience in strategic planning, and those with a longer planning horizon, will more easily incorporate adaptation planning. Organisations without such experience would need to either import or develop this strategic planning capacity.

A sense of responsibility for finding a solution is more likely in situations where the values and culture of the organisation are conducive to taking on such responsibility. In some cases, this sense of responsibility has been promoted by a perception of social influence or social support. For example, a local council is more likely to engage in adaptation planning if they perceive their constituents expect them to take action. While engagement processes would be unlikely to change pre-existing values or to generate social influence, they may help to highlight existing values or increase awareness of the prevailing expectation for developing a response.

Having said this, a core value of community-based health and social service organisations is the health and wellbeing of their clients. Service providers are familiar with the process of assessing client risks and meeting their needs, with the impacts from climate change just another dimension to consider. As such, there exists an inherent capacity for adaptation to the direct impacts of climate change in community-based health and social service organisations (e.g. asking elderly clients if they have somewhere to go during a heatwave). Adaptation to the indirect impacts of climate change may be more of a challenge. This is where it becomes important for an organisation or community to understand the degree of their vulnerability to both the direct and indirect impacts of climate change, through a formal vulnerability assessment.

Vulnerability assessments are usually conducted with the assistance of outside expertise. Not only is such an assessment necessary to form the basis of any effective adaptation plan, it is also necessary to provide an accurate measure of vulnerability. This is important, since an organisation's perceptions of their own vulnerability might be inaccurate. For example, in a project undertaken with several local Sydney councils, it was found there was a poor correlation between the council's perceived vulnerability and their actual vulnerability as measured through a detailed climate change vulnerability assessment (Preston *et al.* 2007). There are several subsequent chapters in this book that explore the vulnerability of specific population groups to climate change and readers should refer to these for further examples.

Barriers that hinder adaptation

Progress towards adaptation can be hampered by a range of barriers that may be encountered along an adaptation pathway. These barriers can often be addressed through a well-designed and effective engagement process, as many barriers will typically stem from misinformation, uncertainty or scepticism about climate change. If these issues around the communication and understanding of climate change are not addressed, they are likely to retard progress towards adaptation. It is also worth noting that as time progresses, the number of people who remain sceptical about climate change may decrease, as the attribution of extreme climatic events to climate change is increasingly accepted and the body of empirical research evidence builds. Nonetheless, it is important to recognise that negative emotional reactions to climate change of fear and despair are normal, but interfere with development of an understanding of climate vulnerability and a sense of responsibility for taking action (Moser 2008). These emotional responses can be acknowledged and addressed through well-designed engagement processes.

The expectation that an external agency (typically the state or federal government) will take responsibility for finding a solution to an organisation's vulnerability is a widely held view in some contexts. Community groups or organisations that are used to waiting for direction from the federal government are unlikely to suddenly take on responsibility for their own strategic planning and protection. Many groups expect and prefer that an outside agency will provide the required drive and leadership for climate adaptation planning on their behalf. Yet as argued earlier, community-based health and social service organisations exist to provide quality care and to meet the needs of their clients. Thinking about climate adaptation as an extension of this requirement, it becomes core business that should not be dependent on others. Community-based health and social service organisations have the mandate and the responsibility to provide leadership in adapting their local communities to climate impacts.

Finally, once a group is willing and capable of adaptation planning, a lack of resources may present a final important barrier to any actual planning. This is not necessarily something that is easily overcome by engagement processes. An organisation may simply have insufficient funding or time to adequately assess vulnerability and plan for climate adaptation. In opposition to this argument, many community groups and organisations are now required to routinely undertake risk assessments, and consideration of the impacts of climate change can often be done as part of these existing processes with little need for additional resources.

Measuring success

There are several key characteristics of organisations that predict climate adaptation activity: long planning horizons, links to outside organisations with climate adaptation knowledge, and having undertaken a vulnerability assessment. It cannot be assumed, however, that just because there is action on climate adaptation, that this action has been effective or successful. A definition of success is required to determine if the activity has been sufficient to address climate change impacts or whether adjustments are required (Ash & Stafford Smith 2013).

Uncertainty should also be a consideration when measuring adaptation. Where there is greater certainty regarding potential climate impacts, more emphasis can be placed on measuring specific adaptation actions. Where there is less certainty of potential climate impacts, the focus could more usefully be placed on measures of adaptive capacity and risk management that might confer resilience to a range of possible shocks and surprises.

Monitoring and evaluation

While there are many initiatives promoting adaptation action, there remains little consensus on the objective of adaptation and the criteria and indicators used to define success or failure (Doria *et al.* 2009). This lack of consensus may simply reflect the diverse and often contested values in society, which result in different perspectives on the goals of adaptation (Adger *et al.* 2005; O'Brien & Wolf 2010). In the context of community-based adaptation, engagement processes are critically important in this regard. Engagement processes that are clear and transparent, involving a range of stakeholders, can be used to identify and prioritise local adaptation actions according to agreed criteria (Ebi & Semenza 2008). Stakeholders should agree on the metric to be used to measure the benefits of an adaptation action, but at a minimum the benefits should exceed the cost. Even if there are people who do not all agree with the final list and prioritisation of adaptation actions, if the process has been transparent they should at least understand the criteria and metrics used to inform the decision-making.

Successful adaptation can be defined as any adjustment that reduces the risks associated with climate change, or reduces the vulnerability to climate change, to some predetermined level, without compromising economic, social and environmental sustainability (Doria *et al.* 2009). With uncertainty around climate projections and consequent risk, successful adaptation is most likely when decision-makers choose strategies, policies and activities that are robust to a wide range of plausible climate futures (Dessai *et al.* 2009), and that account for other change drivers such as urbanisation and population growth. Not only is the robustness of decision-making important, but so too is the recognition of the interconnectedness of different sectors, activities and policies (e.g. health, housing, cities, energy, climate change). For example, as noted by Howden-Chapman and Chapman (2012), it may be that adaptation action in the energy sector is difficult to justify on the basis of carbon savings alone, but if the health and social outcomes of the action are also considered, the benefits outweigh the costs.

In some cases, however, it must be acknowledged that adaptation is not always desirable or possible. There may be limits to adaptation that are largely determined by the biophysical system. For example, if sea level rises far enough, low-lying islands might become uninhabitable, and no amount of preparation can prevent such an outcome. In other situations adaptation may not be desirable, a situation often termed 'maladaptation'. This occurs when actions to reduce the effect of climate change on one system inadvertently increase the impacts or exposure within another system (see Barnett & O'Neill 2010). For example, building energy-intensive desalination plants in Australia to ensure future supplies of fresh water has a maladaptive side-effect of increasing Australia's reliance on centralised high-emission electricity sources (see Adger & Barnett 2009). An example that is more relevant to community health is the widespread installation of government subsidised rainwater tanks intended to drought-proof towns and cities, which may inadvertently also help extend the range of the mosquitos that carry dengue fever (Beebe *et al.* 2009; Jansen & Beebe 2010).

Learning from case studies

An advantage of describing case studies is that they demonstrate examples of organisations and communities engaging in climate change vulnerability and adaptation assessments. The following examples are drawn from experiences within CSIRO's Climate Adaptation National Research Flagship in working with organisations and communities, and demonstrate the importance of partnerships for building shared understanding of climate change and the role of vulnerability and adaptation assessments. Reflections are also provided on the drivers of and barriers to adaptation, which are commonly encountered at the nexus of science, policy and practice.

Case study: adapting coastal communities to sea-level rise

This case study is focused on the Eurobodalla Shire Council in New South Wales and its experience with adaptation to sea-level rise (Fazey *et al.* 2015; Gorddard *et al.* 2012). The council commenced the process of adaptation planning by commissioning a vulnerability assessment to ascertain the risk posed by sea-level rise and coastal inundation to low-lying urban infrastructure along the council's 130 km of coastline. The assessment was based on the latest climate science and projections of future climate change. This provided the council with an understanding of its climate vulnerability, which together with advice that it may be legally liable if it did not take action to safeguard the community, prompted development of an interim sea-level rise policy (Fazey *et al.* 2015). The goal of the policy was to protect important infrastructure such as the harbour, while promoting the managed retreat in other vulnerable parts of the shire with less people, hence resulting in fewer potential economic impacts.

The responses from stakeholders to the adoption of the interim policy were highly polarised (Gorddard *et al.* 2012). Those people residing in properties set back from the beach were largely silent, but supportive of the policy's intent to protect the shire's beaches and coastal ecosystems. A small number of beachfront property owners and developers objected strongly to the policy on the grounds that it violated their rights and unfairly imposed costs on them (Fazey *et al.* 2015). This led to tension between community groups and the council. The issue was further complicated by a change in state government, which led to planning changes that enabled property owners to undertake temporary capital works to protect their properties, which undermined aspects of the council's interim sea-level rise policy. Furthermore, it was found that the coastal planning system lacked the flexibility required by council to implement its interim policy in a fair and effective manner (Abel *et al.* 2011) and that the roles and responsibilities of different levels of government were unclear.

Several key insights can be gleaned from this case study. First, the impacts of climate change were not distributed evenly in the shire, with owners of beachfront properties most affected. Second, there was a diversity of values, interests and beliefs in the community on the issue of sea-level rise and coastal inundation, which resulted in tension and conflict over the various pathways to adaptation (Fazey *et al.* 2015). Finally, the existing planning system and governance structures were found to be insufficient for effective and fair adaptation, with vexing issues around property rights, liability and compensation remaining

(Abel *et al.* 2011). Overall, this case study highlights that even when an organisation such as the Eurobodalla Shire Council develops a clear understanding of climate change impacts and the adaptation responses required, the process of converting this knowledge into action can be challenging.

Case study: adapting low-income households to extreme heat

This second case study highlights lessons from a collaborative research project between the CSIRO Climate Adaptation National Research Flagship and two social housing authorities responsible for ~35% of Australia's social housing portfolio (Barnett *et al.* 2013). A catalyst for the project was a growing concern for the vulnerability of low-income households to extreme temperatures following the death toll associated with the February 2009 heatwave in southern Australia. This concern motivated the social housing authorities to engage CSIRO to identify the potential impacts of climate change on their housing assets and residents, and to build the research evidence required to engage their staff in a process of adaptation planning.

As is common with projects of this nature, a significant barrier that needed to be overcome was building sufficient trust and understanding with the housing authorities in order to secure access to sensitive data on housing assets. Data protocols were collaboratively developed and agreed to preserve confidentiality and privacy. At the same time, there were also concerns about communication of the project and managing reputational risks, but this was dealt with through processes to ensure the anonymity of the housing authority partners. With these early barriers overcome, an analysis of climate vulnerability and opportunities for adaptation was undertaken for 103 809 social housing assets, distributed across most major climate zones in Australia (Barnett *et al.* 2013). The assessment results formed a key input into a workshop process that was designed and run by the social housing authorities. The workshops aimed to facilitate information exchange between different organisational silos, and to help identify various existing organisational processes that might be suited to either intervention or the inclusion of adaptation measures. The purpose was to normalise climate change as a legitimate business concern and engage staff in developing the pathways to adaptation. As an early measure of success, the social housing authorities are using this information in strategic planning and explicitly including climate change considerations in their decision-making.

There are several lessons that can be learnt from this case study. First, adaptation planning often takes considerable time and resources. Undertaking an assessment of climate change vulnerability generally requires specialist skills and expertise, with the outcome contingent on the quality of the data made available to the assessment. This issue is confronting for organisations with little research experience, which may need to build relationships, understanding and trust with research providers. Second, there may be some organisations that are concerned about the ramifications of revealing their climate vulnerability, either from the perspective of issues management or that an expectation will be created that they must now act on this knowledge. These risks can be managed and there will often be considerable opportunity to 'mainstream' adaptation into existing organisational processes. Finally, adaptation planning must be highly participatory, engaging diverse interests across the organisation. This not only generates both awareness and understanding of the 'climate problem', but joint ownership of the 'solutions'.

Conclusion

Adaptation is a broad term, which can be applied to a wide array of activities beyond climate change. Adaptation can variously be targeted at reducing the sensitivity of the system to climate impacts (e.g. by using drought-resistant crops), reducing exposure to impacts (e.g. by moving to an area with higher projected rainfall), increasing capacity to cope with impacts (e.g. by improving emergency response systems to deal with bushfires), and by making changes that improve the efficiency of subsequent adaptation actions (e.g. by building climate impact projections into development planning). Having said this, adapting to climate change is not easy. Responses to extreme events or crises are often short-term or reactionary and can lead to maladaptation. For example, maladaptation to climate change can include actions that increase greenhouse emissions in the atmosphere, place a growing burden on the vulnerable, reduce incentives to adapt, and set paths that limit the choices available to future generations.

Community groups or organisations that are more likely to embrace climate adaptation are those with longer planning horizons, links to other organisations that possess climate adaptation knowledge, and those that have undertaken a vulnerability assessment. Distilling the lessons from the two case studies presented earlier, we see that the impacts of climate change are not distributed evenly in society – whether we are talking about sea-level rise and beachfront property owners or the health impacts of extreme heat and low-income households. This inequity can result in tension and conflict concerning adaptation pathways, which are best addressed by an inclusive and transparent engagement process. In some cases, the outcomes can be readily 'mainstreamed' into existing organisational processes, as was the case with the social housing authorities. Yet in others, such as the case of the Eurobodalla Shire Council, this approach was much less effective due to deficiencies in the existing planning systems and governance structures. Although adaptation can be complex and difficult, it is much better than the alternative. Not taking action to help prepare for the impacts of climate change means that we are accepting larger, more unpredictable costs in the future, instead of smaller, more controllable and predictable costs now.

Overcoming scepticism around the need to implement climate adaptation strategies and a more general resistance to changing regulatory regimes can present barriers to the uptake of climate adaptation practices. Where these barriers exist, there are 'precautionary' or 'no-regrets' approaches that can still deliver results. A no-regrets approach places emphasis on actions that should be taken *regardless* of climate change because they improve sustainability or other important outcomes, while at the same time putting organisations in a better position to adapt to the effects of climate change. Overall, there is value in all efforts. Undertaking a vulnerability assessment and considering adaptation options is a useful first step, rather than ignoring the problem all together. Partial efforts are better than nothing at all, because they place you in a better position when considering further adaptation actions at a later point in time. Finally, attempting adaptation planning and getting it wrong is better than no attempt at all, as it builds understanding of adaptation and an opportunity to learn from past experience.

References

Abel N, Gorddard R, Harman B, Leitch A, Langridge J, Ryan A, Heyenga S (2011) Sea level rise, coastal development and planned retreat: analytical framework, governance principles and an Australian case study. *Environmental Science & Policy* **14**, 279–288. doi:10.1016/j. envsci.2010.12.002

Adger WN, Barnett J (2009) Four reasons for concern about adaptation to climate change. *Environment & Planning A* **41**, 2800–2805. doi:10.1068/a42244

Adger WN, Vincent K (2005) Uncertainty in adaptive capacity. *Comptes Rendus Geoscience* **337**, 399–410. doi:10.1016/j.crte.2004.11.004

Adger WN, Brooks N, Bentham G, Agnew M, Eriksen S (2004) 'New indicators of vulnerability and adaptive capacity'. Tyndall Centre for Climate Change Research, Norwich.

Adger WN, Arnell NW, Tompkins EL (2005) Successful adaptation to climate change across scales. *Global Environmental Change* **15**, 77–86. doi:10.1016/j.gloenvcha.2004.12.005

Adger WN, Paavola J, Huq S, Mace MJ (Eds) (2006) *Fairness in Adaptation to Climate Change*. MIT Press, Cambridge, USA.

Ash AJ, Stafford Smith M (2013) Adaptation research: community science or discipline? In *Climate Adaptation Futures*. (Eds J Palutikof, SL Boulter, AJ Ash, M Stafford-Smith, M Parry, M Waschka and D Guitart) pp. 47–55. John Wiley and Sons, Oxford.

Bambrick HJ, Capon AG, Barnett GB, Beaty RM, Burton AJ (2011) Climate change and health in the urban environment: adaptation opportunities in Australian cities. *Asia-Pacific Journal of Public Health* **23**, 67S–79S. doi:10.1177/1010539510391774

Barnett J, O'Neill S (2010) Maladaptation. *Global Environmental Change* **20**, 211–213. doi:10.1016/j.gloenvcha.2009.11.004

Barnett G, Beaty RM, Chen D, McFallan S, Meyers J, Nguyen M, Ren Z, Spinks A, Wang X (2013) 'Pathways to climate adapted and healthy low income housing'. National Climate Change Adaptation Research Facility, Gold Coast.

Beebe NW, Cooper RD, Mottram P, Sweeney AW (2009) Australia's dengue risk driven by human adaptation to climate change. *PLoS Neglected Tropical Diseases* **3**, e429. doi:10.1371/journal.pntd.0000429

Berry HL, Bowen K, Kjellstrom T (2010) Climate change and mental health: a causal pathways framework. *International Journal of Public Health* **55**, 123–132. doi:10.1007/s00038-009-0112-0

Bureau of Meteorology and CSIRO (2014) 'State of the Climate 2014'. Commonwealth of Australia, Canberra.

Butler CD, Harley D (2010) Primary, secondary and tertiary effects of eco-climatic change: the medical response. *Postgraduate Medical Journal* **86**, 230–234. doi:10.1136/pgmj.2009.082727

Cameron PA, Mitra B, Fitzgerald M, Scheinkestel CD, Stripp A, Batey C, Niggemeyer L, Truesdale M, Holman P, Mehra R (2009) Black Saturday: the immediate impact of the February 2009 bushfires in Victoria, Australia. *The Medical Journal of Australia* **191**, 11–16.

Department of Human Services (2009) 'January 2009 heatwave in Victoria: an assessment of health impacts'. Victorian Government, Melbourne.

Dessai S, Hulme M, Lempert R, Pielke R (2009) Do we need better predictions to adapt to a changing climate? *Eos, Transactions, American Geophysical Union* **90**, 111–112. doi:10.1029/2009EO130003

Doria MF, Boyd E, Tompkins EL, Adger WN (2009) Using expert elicitation to define successful adaptation to climate change. *Environmental Science & Policy* **12**, 810–819. doi:10.1016/j. envsci.2009.04.001

Dovers S (2009) Normalizing adaptation. *Global Environmental Change* **19**, 4–6. doi:10.1016/j. gloenvcha.2008.06.006

Ebi KL, Semenza JC (2008) Community-based adaptation to the health impacts of climate change. *American Journal of Preventive Medicine* **35**, 501–507. doi:10.1016/j.amepre.2008. 08.018

Ebi KL, Kovats RS, Menne B (2006) An approach for assessing human health vulnerability and public health interventions to adapt to climate change. *Environmental Health Perspectives* **35**, 1930–1934.

Fazey I, Wise RM, Lyon C, Campeanu C, Williams J (2015) Past and future adaptation pathways. *Climate and Development* (online), doi:10.1080/17565529.2014.989192

Gardner J, Dowd A-M, Mason C, Ashworth P (2009) 'A framework for stakeholder engagement on climate adaptation'. CSIRO Climate Adaptation National Research Flagship, Brisbane.

Gardner J, Parsons R, Paxton G (2010) 'Adaptation benchmarking survey: initial report'. CSIRO Climate Adaptation National Research Flagship, Brisbane.

Gorddard R, Wise RM, Alexander K, Langston A, Leitch A, Dunlop M, Ryan A, Langridge J (2012) 'Striking the balance: coastal development and ecosystem values'. Australian Department of Climate Change and Energy Efficiency and the CSIRO Climate Adaptation National Research Flagship, Canberra.

Head L, Adams M, McGregor HV, Toole S (2014) Climate change and Australia. *Wiley Interdisciplinary Reviews: Climate Change* **5**, 175–197.

Hennessy K, Fitzharris B, Bates BC, Harvey N, Howden M, Hughes L, Salinger J, Warrick R (2007) Australia and New Zealand. In *Climate Change 2007: Impacts, Adaptation and Vulnerability: Contribution of Working Group II to the Fourth Assessment Report of the Intergovernmental Panel on Climate Change*. (Eds ML Parry, OF Canziani, JP Palutikof, PJ van der Linden and CE Hanson) pp. 507–540. Cambridge University Press, Cambridge, UK.

Hennessy K, Clarke J, Whetton P, Kent D (2012) 'An introduction to internally consistent climate projections'. The Centre for Australian Weather and Climate Research, CSIRO and the Bureau of Meteorology, Canberra.

Howden M, Nelson RA, Crimp S (2013) Food security under a changing climate: frontiers of science or adaptation frontiers? In *Climate Adaptation Futures*. (Eds J Palutikof, SL Boulter, AJ Ash, M Stafford-Smith, M Parry, M Waschka and D Guitart) pp. 56–68. John Wiley and Sons, Oxford.

Howden-Chapman P, Chapman R (2012) Health co-benefits from housing-related policies. *Current Opinion in Environmental Sustainability* **4**, 414–419. doi:10.1016/j.cosust.2012.08.010

Hu W, Clements A, Williams G, Tong S, Mengersen K (2012) Spatial patterns and socioecological drivers of dengue fever transmission in Queensland, Australia. *Environmental Health Perspectives* **120**, 260–266. doi:10.1289/ehp.1003270

Hughes L, McMichael T (2011) 'The critical decade: climate change and health'. Climate Commission Secretariat (Department of Climate Change and Energy Efficiency), Canberra.

Hulme M, Pielke R, Dessai S (2009) Keeping prediction in perspective. *Nature Reports Climate Change* **3**, 126–127. doi:10.1038/climate.2009.110

Jansen CC, Beebe NW (2010) The dengue vector *Aedes aegypti*: what comes next. *Microbes and Infection* **12**, 272–279. doi:10.1016/j.micinf.2009.12.011

Kearney M, Porter WP, Williams C, Ritchie S, Hoffmann AA (2009) Integrating biophysical models and evolutionary theory to predict climatic impacts on species' ranges: the dengue mosquito *Aedes aegypti* in Australia. *Functional Ecology* **23**, 528–538. doi:10.1111/j.1365-2435.2008.01538.x

Kiem AS, Verdon-Kidd DC, Boulter S, Palutikof J (2011) 'Learning from experience: historical case studies and climate change adaptation'. National Climate Change Adaptation Research Facility, Gold Coast.

Le Quere C, Raupach MR, Canadell JG, Marland G, Bopp L, Ciais P, Conway TJ, Doney SC, Feely RA, Foster P, Friedlingstein P, Gurney KR, Houghton RA, House JI, Huntingford C, Levy PE, Lomas MR, Majkut J, Metzl N, Ometto J, Peters GP, Prentice IC, Randerson JT, Running SW, Sarmiento JL, Schuster U, Sitch S, Takahashi T, Viovy N, van der Werf GR,

Woodward FI (2009) Trends in the sources and sinks of carbon dioxide. *Nature Geoscience* **2**, 831–836. doi:10.1038/ngeo689

Mallon K, Hamilton E, Black M, Beem B (2013) 'Adapting the community sector for climate extremes'. National Climate Change Adaptation Research Facility, Gold Coast.

McMichael AJ (2013) Globalization, climate change, and human health. *The New England Journal of Medicine* **368**, 1335–1343. doi:10.1056/NEJMra1109341

McMichael AJ, Woodruff RE, Hales S (2006) Climate change and human health: present and future risks. *Lancet* **367**, 859–869. doi:10.1016/S0140-6736(06)68079-3

Moser SC (2005) Impact assessments and policy responses to sea-level rise in three US states: an exploration of human-dimension uncertainties. *Global Environmental Change* **15**, 353–369. doi:10.1016/j.gloenvcha.2005.08.002

Moser SC (2008) More bad news: the risk of neglecting emotional responses to climate change information. In *Creating a Climate for Change: Communicating Climate Change and Facilitating Social Change*. (Eds SC Moser and L Dilling) pp. 64–80. Cambridge University Press, Cambridge, USA.

O'Brien KL, Wolf J (2010) A values-based approach to vulnerability and adaptation to climate change. *Wiley Interdisciplinary Reviews: Climate Change* **1**, 232–242.

Palutikof J, Parry M, Smith MS, Ash AJ, Boulter SL, Waschka M (2013) The past, present and future of adaptation: setting the context and naming the challenges. In *Climate Adaptation Futures*. (Eds J Palutikof, SL Boulter, AJ Ash, M Stafford-Smith, M Parry, M Waschka and D Guitart) pp. 1–30. John Wiley and Sons, Oxford.

Park SE, Marshall NA, Jakku E, Dowd AM, Howden SM, Mendham E, Fleming A (2012) Informing adaptation responses to climate change through theories of transformation. *Global Environmental Change* **22**, 115–126. doi:10.1016/j.gloenvcha.2011.10.003

Peters GP, Andrew RM, Boden T, Canadell JG, Ciais P, Le Quéré C, Marland G, Raupach MR, Wilson C (2012a) The challenge to keep global warming below 2°C. *Nature Climate Change* **3**, 4–6. doi:10.1038/nclimate1783

Peters GP, Marland G, Le Quere C, Boden T, Canadell JG, Raupach MR (2012b) Rapid growth in CO_2 emissions after the 2008–2009 global financial crisis. *Nature Climate Change* **2**, 2–4. doi:10.1038/nclimate1332

Preston BL, Stafford Smith M (2009) 'Framing vulnerability and adaptive capacity assessment: discussion paper'. CSIRO Climate Adaptation National Research Flagship, Canberra.

Preston B, Smith T, Brooker C, Gorddard R, Measham T, Withycombe G, McInnes K, Abbs D, Beveridge B, Morrison C (2007) 'Mapping climate change vulnerability in the Sydney Coastal Council Group'. Prepared for the Sydney Coastal Councils Group and the Australian Greenhouse Office. CSIRO, Melbourne.

Preston BL, Yuen EJ, Westaway RM (2011) Putting vulnerability to climate change on the map: a review of approaches, benefits, and risks. *Sustainability Science* **6**, 177–202. doi:10.1007/s11625-011-0129-1

Raupach MR, Marland G, Ciais P, Le Quéré C, Canadell JG, Klepper G, Field CB (2007) Global and regional drivers of accelerating CO_2 emissions. *Proceedings of the National Academy of Sciences of the United States of America* **104**, 10288–10293. doi:10.1073/pnas.0700609104

Reisinger A, Kitching RL, Chiew F, Hughes L, Newton PCD, Schuster SS, Tait A, Whetton P (2014) Australasia. In *Climate Change 2014: Impacts, Adaptation, and Vulnerability. Part B: Regional Aspects. Contribution of Working Group II to the Fifth Assessment Report of the Intergovernmental Panel of Climate Change*. (Eds VR Barros, CB Field, DJ Dokken, MD Mastrandrea, KJ Mach, TE Bilir, M Chatterjee, KL Ebi, YO Estrada, RC Genova, B

Girma, ES Kissel, AN Levy, S MacCracken, PR Mastrandrea and LL White) pp. 1371–1438. Cambridge University Press, Cambridge, UK.

Ricketts JH, Page CM (2007) A web based version of OzClim for exploring climate change impacts and risks in the Australian region. In *MODSIM 2007 International Congress on Modelling and Simulation*. (Eds L Oxley and D Kulasiri) pp. 560–566. Modelling and Simulation Society of Australia and New Zealand, Christchurch.

Schaffer A, Muscatello D, Broome R, Corbett S, Smith W (2012) Emergency department visits, ambulance calls, and mortality associated with an exceptional heat wave in Sydney, Australia, 2011: a time-series analysis. *Environmental Health* **11**, 3. doi:10.1186/1476-069X-11-3

Shearer H, Taygfeld P, Coiacetto E, Dodson J, Banhalmi-Zakar Z (2013) 'The capacities of private developers in urban climate change adaptation'. National Climate Change Adaptation Research Facility, Gold Coast.

Smit B, Wandel J (2006) Adaptation, adaptive capacity and vulnerability. *Global Environmental Change* **16**, 282–292. doi:10.1016/j.gloenvcha.2006.03.008

Sovacool BK (2013) Adaptation: the complexity of climate justice. *Nature Climate Change* **3**, 959–960. doi:10.1038/nclimate2037

Stafford Smith M, Horrocks L, Harvey A, Hamilton C (2011) Rethinking adaptation for a 4°C world. *Philosophical Transactions. Series A, Mathematical, Physical, and Engineering Sciences* **369**, 196–216. doi:10.1098/rsta.2010.0277

Stern N (2008) The economics of climate change. *The American Economic Review* **98**, 1–37. doi:10.1257/aer.98.2.1

Turner BL, Kasperson RE, Matson PA, McCarthy JJ, Corell RW, Christensen L, Eckley N, Kasperson JX, Luers A, Martello ML, Polsky C, Pulsipher A, Schiller A (2003) A framework for vulnerability analysis in sustainability science. *Proceedings of the National Academy of Sciences of the United States of America* **100**, 8074–8079. doi:10.1073/pnas.1231335100

Unsworth K, Russell S, Lewandowsky S, Lawrence C, Fielding K, Heath J, Evans A, Hurlstone M, McNeill I (2013) 'What about me? Factors affecting individual adaptive coping capacity across different populations'. National Climate Change Adaptation Research Facility, Gold Coast.

van den Honert RC, McAneney J (2011) The 2011 Brisbane floods: causes, impacts and implications. *Water* **3**, 1149–1173. doi:10.3390/w3041149

Whetton P, Hennessy K, Clarke J, McInnes K, Kent D (2012) Use of representative climate futures in impact and adaptation assessment. *Climatic Change* **115**, 433–442. doi:10.1007/s10584-012-0471-z

3

Mitigation of climate change

Greg Hunt

Key points

- Our often unthinking energy use has led us to a changed climate, with very few of the changes for the better. We must use less carbon-based energy to mitigate these changes, though this will take time.
- There are already many climate impacts to which we have to adapt.
- There are pathways to follow in mitigation and adaptation, and the changes we need are possible.
- There are examples where we have changed the ways we do things – we need to look and learn from these.

Introduction

Climate change is major change. The consequences of the build-up of greenhouse gases are working their way through every aspect of our lives, our communities, our economy and the environment. If we are to minimise the unwanted consequences, we need to know how this increase in greenhouse gases has come about, what effects these increases in greenhouse gases are having on our natural and social systems, and how we can reduce their production. This chapter looks at responses to these questions.

In the cold and damp of a Melbourne winter, the gainfully employed can sit down to a dessert of fresh Mexican mangoes and plump Oregon cherries after a main course that included Peruvian asparagus. They can then retire to their home theatre and watch on the metres-wide plasma screen a streamed movie downloaded in seconds over a broadband internet connection. The kitchen is full of energy-guzzling appliances, the bathroom has hair dryers and heated towel racks, the laundry has a programmable washing machine and a clothes dryer. There are computers, printers, game consoles and stereos in the family room. The garage is stuffed with boy's toys – a carbon-fibre bike, a jet ski and a four-wheel drive because hey, the kids must get to school and a chap has to get to work safely.

As for getting to work, look at the vehicle occupancy rate on a city freeway in the morning peak. In the workplace shirtsleeves are the workwear of choice in the high-rise office block, so much more comfortable around the meeting table or in front of a screen, though outside is chill, damp and gloomy.

This is not just an Australian issue, for approaches to work and play are similar across the developed world. A winter in southern Australia hardly compares with the conditions of a North American winter, where T-shirts in big shopping malls are only possible due to the massive energy consumption that keeps the internal temperature in the low 20s (Celsius) while outside it is also in the low 20s (Fahrenheit). Cities across northern Europe are little different. In developed economies in the tropics, the external environment might be different but the response is the same: use energy to modify the internal environment. Air-conditioned shopping malls are common in the oil states and in the cities of south-east Asia. In an extremely energy-intensive recreation venue, a ski dome in Dubai, skiers in the Middle East can ski a black run while outside the temperature might be 40°C.

All of this is only possible through the availability of abundant and relatively cheap energy – energy to create environments, to transport goods and people around the world, to own and use manufactured goods and to allow behaviour patterns that result in high personal greenhouse gas emissions.

Talk to grandparents about living and working and you might start to feel sorry for them, for there were none of these labour-saving and oh-so-life-enriching appliances. We simply expect that energy will be available when we flick the switch and we flick the switch often – and too often heedlessly.

It was the harnessing of steam, the development of internal combustion engines and the exploitation of vast deposits of fossil fuels that allowed the population increases and the consumption patterns that mark the early 21st century. To live in these times and to have the income to take full advantage of what the times offer is to be one of life's winners.

Not all are so fortunate, however. Levels of inequality between countries, and between communities within countries, consign many people to living as they always have: with low consumption, poor health and high vulnerability to the calamities that nature offers. A substantial change in the way we generate energy and the way we use energy is needed.

Carbon dioxide and fossil fuels

Life on earth is only made possible through the energy that earth receives from the sun. It is the earth's distance from the sun that provides the temperature range within which life has evolved. It is the light energy from the sun that fuels the photosynthetic reactions that occur in plants to start life's food chains, whether for plants which we eat directly or for the animals that eat the plants and that in turn we eat. Photosynthesis enables trees to grow to provide wood for heating and cooking.

Most of the energy used around the world is currently generated from fossil fuels. For millions of years, photosynthesis has transformed the sun's light energy into chemical energy in the form of the carbon–hydrogen bond. This chemical energy is transformed into heat energy when plant starches, sugars, the lignin of wood (all with carbon–hydrogen bonds) are burnt. Burning is simply combining the fuel with oxygen to release the energy in the form of heat. Photosynthesis in the Carboniferous period (around 350–300 million years ago) supported vast forests that were transformed into the coal that we use now. Equally, microscopic plants and animals accumulated and were transformed to become the vast oil and gas reservoirs that are exploited today. Fossil fuels are literally that – the remnants of photosynthesis all those millions of years ago.

Fossil fuels, whether in the solid form of coal, liquid form of oil, or as natural gas, are extremely energy dense and readily transportable. It is these characteristics that have led to the worldwide use of fossil fuels as the basis for development and the standards of living enjoyed by fortunate fractions of the world's population. It is also the reason why weaning the world off fossil fuels is proving so difficult.

Yet wean we must, for there is another extremely unfortunate characteristic of the use of fossil fuels to contend with. When carbon is combined with oxygen in the release of the energy in the carbon–hydrogen bond, carbon dioxide is produced. This colourless, odourless gas is released into the atmosphere where it forms a gaseous blanket, trapping heat. The use of fossil fuels has, in the past 100 years, resulted in steadily rising average temperatures, which we call global warming. The rising global temperature shifts the physical climate system, a process we call climate change. The changing climate shows up as changing patterns of rainfall, drought, extreme storms and heatwaves, for example.

Not all of the earth is warming at the same rate. For example, some parts of the earth cool because a warmer atmosphere can hold more water vapour, which is moved by global circulation patterns to result in deeper snowfalls. The manifestations of global warming and climate change differ between places.

In Australia, these changing circulation patterns are the most likely cause of the long-term reductions in rainfall and streamflow that we observe. The evidence suggests that the storms that used to bring rain over south-eastern Australia are now being pushed further south and so miss the Australian continent (CSIRO 2012). While rainfall is decreasing, long-term trends show average temperatures are increasing, but with significant regional variations. For example, the mean air temperature in Australia has increased 0.9°C since 1950 (Bureau of Meteorology & CSIRO 2014).

Language

The way we talk about climate change is important if people are to understand what is occurring. Humans typically think short-term, in days, months, years, or decades, not in the long timescales over which major change is significant (Gleik 2012). Humans experience weather in short timescales – cold days, a dry month or a wet year – while climate, the aggregate of weather, is a feature of millennia. Weather contributes to climate, but is it not the same as climate.

Our understanding of the ways that different groups in the community use language can also cloud our thinking. To most of us, 'uncertainty' means not knowing. To scientists, however, uncertainty is how well something is known. And, in science, there's often not absolute certainty. So when a researcher suggests that the impacts of climate change are uncertain, it is not intended that there is doubt about climate change occurring, rather that there should be caution as to whether the temperature will increase by 0.5°C or 0.6°C, for example, and with what degree of confidence there can be that it will be one rather than the other.

When climate scientists talk about a warming climate they are referring to measurements (of air or ocean temperature) that fluctuate from year to year but have a trend line that shows the changing temperature over the period of years in the graph. The long-term rising trend line for average temperature suggests that there is movement from a cooler climate to a warmer climate, whereas the peaks and troughs in the graph from year to year tell us that weather is naturally variable. There are cooler years interspersed with warmer years, yet the general trend is clearly a warming. It is the trend we refer to when we speak of global warming.

Impacts of climate change

Australia has experienced recent extremes of weather. Reports from the Climate Commission, now the Climate Council, summarise the major weather extremes of the last two summers. In the summer of 2012/13 these extremes included:

- the hottest summer on record, the hottest day and month ever recorded in Australia to that time
- the highest rainfall totals in some locations to that time, major flooding in south-east Queensland and northern New South Wales, and five rivers exceeding their previous flood heights
- the lowest monthly rainfall records for central and eastern Australia
- bushfires in every state and territory, with particularly severe ones in New South Wales and Tasmania
- tornadoes that hit Bundaberg and other coastal towns in Queensland
- tropical cyclones, which brought extreme rainfall to Queensland and northern Western Australia (Steffen 2013, p. 4).

As energy increases in the atmosphere, weather cycles are more extreme. The heating of the oceans means that they expand and sea level rises, storm surges add to coastal flooding and erosion, there are higher winds, the incidence of extreme fire weather increases and heatwaves occur more often. These changes to weather are almost all in one direction; hence, we can say that the climate is changing.

The impacts of climate change are experienced as the disruptions that are caused by fire, floods and heatwaves. The 2009 heatwave is seared into the consciousness of Melbournians. During this heatwave, which lasted from 26 January to 1 February 2009, the maximum temperatures on 28, 29 and 30 January were 44°C, 45°C and 46°C, respectively.

This three-day period had significant impacts on health and demand for services. Examples of these impacts on health services were described by the Department of Human Services (2009) and are given in Box 3.1.

Floods

While health services are severely affected by climate change, there are also major consequences for planners. Land use planning and the operations of planning schemes are the statutory responsibility of Victoria's local governments. The identification of hazards resulting from storm tides and storm surges have led to local governments preventing developments on land subject to coastal flooding. A recent application for a permit for residential development of eight dwellings in Lakes Entrance, a coastal regional town in eastern Victoria that is subject to flooding, was granted by the council but rejected upon an appeal to the Victorian Civil and Administrative Tribunal (VCAT). The following excerpt is taken from a sustainability and climate change update issued by the law firm Maddocks, which has a climate change practice:

According to VCAT's analysis, if climate change risks – including sea level rise, storm surges and other associated coastal hazards – are likely to affect a proposed development in the future, approval of such development should be avoided until 'responses are put in place to address and minimise these risks'.

Box 3.1: Impacts of Victoria's January 2009 heatwave on health and health services

Ambulance Victoria metropolitan emergency case load
- 25% increase in total emergency cases and 46% increase over the three hottest days
- 34-fold increase in cases with direct heat-related conditions (61% in those 75 years or older)
- 2.8-fold increase in cardiac arrest cases.

Emergency department presentations
- 12% overall increase in presentations, with a greater proportion of acutely ill patients and 37% increase in those 75 years or older
- 8-fold increase in direct heat-related presentations (46% in those aged 75 years and over)
- Almost 3-fold increase in patients dead on arrival (69% being 75 years or older).

Total all-cause mortality
- There were 374 excess deaths over what would be expected: a 62% increase in total all-cause mortality.
- The greatest number of deaths occurred in those 75 years or older, representing a 64% increase.

(Department of Human Services 2009)

Figure 3.1.   Warneet Boat Hire in an Easter tide, 21 April 2011 (Photograph: David Westlake).

VCAT's decision raises significant questions regarding infill development in coastal municipalities where established settlements already exist. First, the decision implies that coastal development of low lying settlements should not be approved until the relevant climate change assessments have been undertaken and suitable responses to climate change have been developed. (Maddocks 2010, p. 3)

Figure 3.1 shows coastal flooding from a high Easter tide on a calm day. A storm occurring when the tide is this high, with surges whipped up by a south-westerly wind, would drive this water up from the boat hire to the township. Storms with high rainfall bring overland flooding and the coincidence of these events, which are projected to occur more often, will call for more safety callouts by emergency services personnel. The responsible planning authorities, in most cases local governments, are already taking into account the need to implement precautionary planning decisions, as the VCAT example shows.

Fire

We used to look forward to summer, now we are fearful of it. We hold our collective breath, fearful of the fire season and what loss of life and livelihood might occur, let alone what area of forest and grassland, and all that depend upon them, will be burnt. On 7 February 2009, one week after the three-day heatwave described earlier in this chapter, Melbourne endured 'Black Saturday', when bushfires to the city's north and north-east fringes resulted in more than 170 deaths and considerable loss of community infrastructure. Following this event, a Royal Commission was set up to investigate and recommend bushfire responses.

Fires are increasingly concerning for many sectors in the Australian community. Insurance companies are one sector of the business community that is most concerned about climate change impacts, and also one of the best informed. Bushfires that damage property are a major source of insurance claims. In 2010 Munich Re, the insurer's insurer, stated:

A scientific look into the future reveals bleak prospects. A scenario with global warming of 2.9°C by mid-century shows that the danger of catastrophic fire days is to be expected at 85% of the observation stations in south-east Australia, as opposed to the current 46%. In addition, model results suggest that the fire season will start earlier and end slightly later, and will also be more intense. This reduces the window for pre-season controlled back-burning and more resources will be required to maintain fire fighting standards. Shorter intervals between fires can have a major impact on ecosystems, threaten biodiversity and stretch emergency services and communities to their limits. (Munich Re 2010, p. 20)

In a later publication (Munich Re 2013, p. 30) the insurer summarises the worldwide incidence of natural catastrophes over the period 1980–2012. Catastrophes include geophysical events (earthquakes, tsunamis, volcanic eruptions), meteorological events (storms), hydrological events (flood, mass movement of water) and climatological events (extreme temperature, drought, forest fire). While the number of geophysical events is relatively constant over that period, the number of both meteorological and hydrological events has doubled, while the number of climatological events has almost trebled. The greatest increases have been in storms and in extreme temperature. Consider the increased population and increased infrastructure exposed to risk through such events and it is little wonder that the insurance industry pays close attention to climate change impacts and conducts detailed financial analyses of its consequences.

Emergency management

Many of the effects of climate change are felt in these natural disasters that are occurring with increasing frequency. In Australia and Oceania over the period 1980–2010, geological events have not increased in frequency, but events linked with climate change (meteorological, hydrological and climatological) have increased. It is this increase that has insurance companies, such as Munich Re, concerned. The trends are very clear. Catastrophes that result in insurance payouts are on the increase. This is obviously why a reinsurance company is interested in climate change. With continuing contributions of greenhouse gases being made, there is projected to be a further increase in both frequency and intensity of these events.

This increase in natural disasters has also prompted action on the part of government. Australia's National Strategy for Disaster Resilience, developed through an initiative of the Council of Australian Governments, calls for all levels of government, businesses, individuals, the not-for-profit sector and the community to be involved in emergency management (COAG 2011). The strategy provides guidance and direction on the role that each party has to play in managing climate change risks, and their approaches to increasing disaster resilience.

While there is a serious need for mitigation measures to reduce the causes of climate change, a reduction in extreme weather events, and therefore a reduction in the need for adaptation measures to climate change consequences, is still a long way off.

Mitigation and adaptation

Climate change impacts are, science tells us, the result of burning fossil fuels and releasing carbon dioxide into the atmosphere. We will add to the problems that we now face if we continue to do as we are doing, that is, if we continue business as usual. The response logically follows: if carbon is the problem, then less carbon (or where possible, no carbon) is the solution. We must reduce our use of fossil fuels and so release less carbon dioxide into the atmosphere. This is climate change mitigation. Though it will take some time for carbon dioxide levels to start to decrease, the sooner we start our mitigation efforts, that is, slow the carbon dioxide release, the sooner dangerous climate change impacts will decrease.

While there is a serious need for mitigation measures to reduce the causes of climate change, a reduction in extreme weather events is a still long way off. Changing how and where we live because of the dangerous climate change impacts that are occurring is adaptation. We need to both mitigate and adapt, and one must accompany the other. It would be foolish to add to the problem while we try to deal with it by carrying on with business as usual. We must adapt and mitigate, for doing nothing to slow and stop the release of further carbon dioxide would be akin to mopping the floor while leaving the tap running.

Less energy and different energy

If we are to make significant reductions in the amount of carbon-based energy use, that is, if our mitigation efforts are to be successful, the task is very much ahead of us.

There is still a major reliance on fossil fuels within the energy mix in Australia. The Bureau of Resources and Energy Economics (BREE 2014) reports that carbon-based energy in the form of oil, coal and gas accounts for over 94% of our energy use while non-carbon-based renewable energy such as hydroelectricity, solar energy and wind energy provide less than 6% of our energy needs. This will need to change into the future if there is to be a significant reduction in the volume of greenhouse gases that are emitted. Clearly, the proportion of energy derived from renewable, non-carbon-based sources needs to change. This change will not happen quickly as there is major investment in current energy infrastructure that supports business as usual. As this necessary change slowly unfolds, we can make faster change by using less energy altogether.

Mitigation can involve using less energy derived from fossil fuels to slow the release of carbon dioxide into the atmosphere. Houses can be designed to take advantage of orientation away from the north and west where walls absorb heat in summer. Shading with verandas and trees and the use of awnings and exterior blinds means less energy is used to keep cool in summer. Insulation can make a very big difference to how often a heater or air-conditioner is turned on. Double glazing on windows uses the insulating properties of the sandwich of air between the panes to conserve energy and help maintain stable indoor temperatures.

Choosing appliances with high energy-efficiency ratings, often shown as a star rating on the appliance label, is plain common sense. Using low-energy lights, installing a more efficient heater or using a more effective cooling system will also reduce domestic energy consumption. While we can use technology to reduce energy use, there are simple behaviours that can achieve the same end. Our forebears wore jumpers indoors on winter days or opened the blinds at every opportunity to use natural light. Mitigation efforts do not always rely on expensive technology.

Commercial and industrial use of energy is changing also. When energy was relatively cheap, manufacturing processes did not need to be particularly energy efficient. However, the rising price of transport fuels, gas and electricity is now incentive for business to become more energy efficient – an imperative that is only heightened when the need to reduce carbon emissions is taken into account.

Energy auditing companies

A combination of drivers, from rising energy prices and some form of carbon pricing to a desire to claim a marketing advantage through recognition as a good corporate citizen, has driven several companies to reduce their use of carbon-based energy. For example, RepuTex (www.reputex.com) is an energy and emissions market analyst that provides services to companies wishing to improve their corporate social responsibility performance. Companies such as GenesisNow (www.genesisnow.com.au) visit premises and carry out carbon auditing, while Arup (www.arup.com) prepares carbon management plans. In the residential sector, a Yellow Pages online search on 'energy audits residential domestic' yields 1385 listings of companies in Victoria and 2084 listings in New South Wales that could audit home energy use and work with the householder to develop an action plan to reduce energy consumption.

Where there are limits to the efficiencies that can be achieved, greater use of non-carbon based energy must be the response, for renewable energy from the sun or wind does not involve the release of carbon dioxide.

Different energy

Reducing energy consumption and switching to renewable energy sources are not always easy. We have established patterns of activity and behaviour that involve energy consumption that we have built up over our entire lives. We have a 'comfort zone' where our work and play patterns are simply 'what we do'. Changing what we do takes us from that comfort zone and we need a compelling reason and a strong motivation to do that.

Switching to renewable energy might be one of the easier options, as we can remain in our comfort zone by changing energy suppliers. If we want to generate energy from rooftop solar panels there will be a considerable capital outlay to buy and install the panels on the north-facing roof of the house. Do we have the capital? Do we have a north-facing roof on our house? Is it our house or does it belong to a landlord? If generating solar energy from rooftop panels is not an option it is possible to purchase electricity from an energy company that sources renewable energy on behalf of its customers.

Using renewable energy in the home is a response of choice for an increasing number of Australians. Solar hot water systems using the sun's heat energy have been in use for many years. Increasingly, Australia's rooftops are also being adorned with arrays of photovoltaic cells (PVs) to generate electricity using the sun's light energy. According to a *Sydney Morning Herald* (Arup 2013) article in June 2013, more than one million solar panel systems have been installed on the rooftops of homes and businesses across Australia.

That the uptake of solar energy has not been faster is a result of many barriers. Solar panel systems are expensive and there are fewer government rebate programs than were previously available. The tariffs that electricity supply companies would use to buy unused solar-generated electricity have been reduced, and not all homes are suitable sites for solar panel systems. Further, a move away from the centralised generators that have supplied our electricity for almost 100 years to local electricity generation constitutes significant change. For a large, coal-fired energy generator that is providing our electricity, there are powerful incentives for 'them' to continue their own version of business as usual. The fuel supply is secure, the generating equipment is in place and operating, the distribution network is sending electricity across the country. There would be a massive disruption to move to decentralised electricity generation that is not based on fossil fuels. The generation company's share price is likely to be affected and shareholders love stability above most things. We will need to bring about a significant change in how our economy operates to switch from fossil fuels for energy generation – and significant change only occurs when we make it happen. In short, we need a new social norm regarding energy use.

Social change

There have been major changes in some aspects of human behaviour. In 1970, the number of people killed on Victoria's roads peaked at 1061 (State Government of Victoria 2014) whereas in 2013 it had fallen to 242 (Transport Accident Commission 2014). Think about

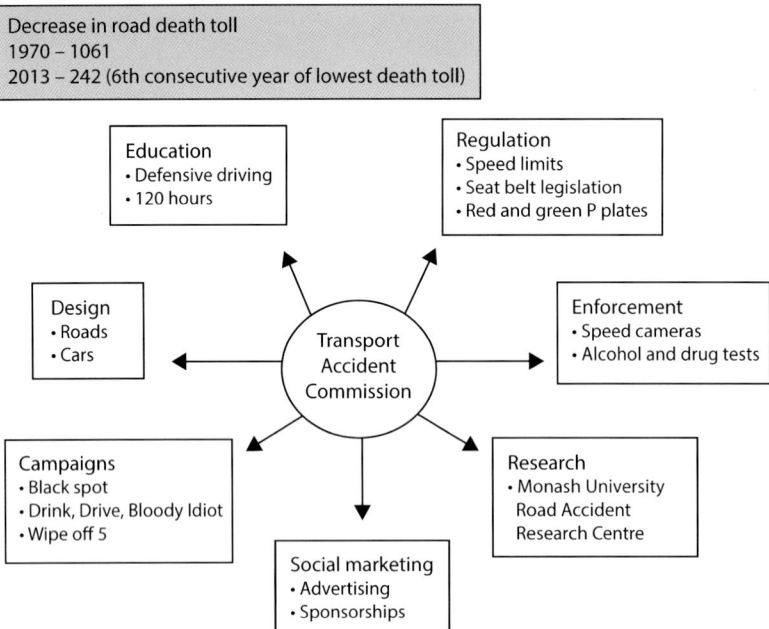

Figure 3.2. Establishing a new social norm (SECCCA 2014).

how the Victorian population has increased over this time period and how many extra cars there are on the roads.

There are now thousands of Victorians alive due to an aggressive and very comprehensive campaign to change the social norms regarding driving (Fig. 3.2). This has involved car and road design, regulation and enforcement, and information and education. Young people must now complete 120 hours of supervised driving before they take the licence test. They are restricted initially in the number of their peers that they can have in the car and they almost always have a designated driver who abstains from drinking alcohol. Safer cars are the result of safety testing, rigorous standards, and airbags and braking systems that have markedly improved survival rates in accidents. Alcohol sponsorships are bought out and replaced with social marketing campaigns to promote responsible driving. If there has been an avenue available to bring about a fundamental change in the way drivers behave on the road, it has been taken.

Yes, this major change in the road toll has been brought about over quite a long time. This long-term commitment across the range of avenues available is made possible through secure funding. A component of the compulsory third-party vehicle insurance that car owners pay is directed to the Transport Accident Commission to allow this to occur. Knowing that there is an assured and adequate funding stream means long-term contracts to deliver long-term benefits can be struck.

Social marketing has been applied in public health, for example, to reduce tobacco consumption (Quit campaign), to prevent skin cancer (Slip, Slop, Slap) and to reduce domestic water use (155 L). Successful programs use the range of social marketing tools available. The programs target specific audiences and use mass communication. They mix incentives with mass advertising, and regulation with sanctions. If we wish to bring about a new social norm in energy use, there are models from which we can learn.

Examples of mitigation programs

Mitigation begins with being aware of energy use. On a greenfield residential estate on Melbourne's urban fringe, Selandra Community Place (City of Casey 2012) is operated by the City of Casey as a centre for the provision of community services to new residents. New parents attend coffee mornings to learn of council support services, residents take evening yoga classes or attend sustainable living workshops. The building also serves as a display home, one of a number in a street of display homes which visitors to the precinct wander through, particularly on weekends, to gather ideas for the home that they would like to buy or build.

The house is also a showcase for sustainable living, for it is an eight-star zero emissions house. It has a 5 kW solar array on the roof, and is fully insulated, double glazed and draught sealed. It has passive solar design throughout and water saving measures in abundance. All of the sustainability features are interpreted, with video screens showing explanatory animations, and displays comparing different levels of insulation and showing the thermal properties of double glazing. Staff at the house on weekends strike up conversations with all visitors who wander in. These visitors did not visit because it was a sustainable house; they visited because it was a part of the display village. They are unaware of its sustainability, which becomes immediately apparent at the start of the conversation. In very many cases, visitors look nonplussed at the mere mention of sustainability and they can contribute little to the conversation. They do not know what questions to ask and it is clear that the display home with its attraction to the general public serves as an extremely useful introduction to sustainability.

The 'Save it for the Game' project, conducted by the South East Councils Climate Change Alliance (SECCCA),[1] set out to reduce energy use in sporting clubs across Melbourne's south-east (SECCCA 2014). Energy auditors examined energy use by bowls clubs, at athletics tracks and in the club rooms and sportsgrounds used by football, netball and tennis clubs. They were surprised by what they found.

While double-door refrigerators were not unusual in club rooms, energy-efficient refrigerators were unusual. Many of the double-door refrigerators were discards, scrounged by a committee member who knew someone who was throwing one out 'but it'll come in handy down at the club rooms'. These refrigerators were generally old and were not serviced. The door seals were often perished, the thermostat was not functioning correctly (such that the refrigerator was almost permanently running) and typically the refrigerator was empty. It was perpetually ready by being in continuous operation, but only stocked up on game day. In giving priority to convenience for food preparation or keeping drinks cold for occasional pie nights, high energy use in a poorly-functioning refrigerator was the consequence.

One bar refrigerator was found contained within a purpose built cubicle to fit under a bench top. Being enclosed meant that the heat exchanger couldn't work efficiently and the refrigerator did not cool down enough for the thermostat to switch the motor off. Another refrigerator had a heated panel on a double-glazed door to prevent frosting so that the contents, if there were any, were readily visible. These refrigerators were using around A$1200 of electricity per year when perhaps A$300 could have been spent on providing refrigerated goods on the occasions during the year when they were needed. For a sports club, spending money on wasted electricity rather than on sport is a consequence of not being energy-aware. But they are sports people, not energy people and they too need to know what questions to ask.

A partner in several of SECCCA's energy efficiency projects specialises in draught-proofing houses. Chimneys, gaps under doors, incorrectly fitted exhaust fans, warped timber window frames – all can serve as places where either warmed or cool air can be lost from a building. If energy has been used to heat or cool that air, then that energy has been wasted. Air Barrier Draught Proofing (www.airbarrierdraftproofing.com.au) tests a building for its 'leakiness' such that its tightness can be improved and therefore its energy efficiency increased. In one building tested, they were able to write specifications for a smaller air-conditioning unit as the volume of air to be treated would be lower. Not only was there a saving in energy, there was a saving in the purchase price for a smaller unit.

Perhaps the best place to start with energy efficiency is in the home – where we live and how we live. Another of SECCCA's projects was to work with councils' community care clients to assist with energy efficiency in the home. Mitigation in the form of energy efficiency can involve technological responses and behavioural responses. Unlike houses being built on greenfield estates, where energy efficiency involving technological responses can be included at the design phase, energy efficiency in existing houses invariably involves retrofitting technology, such as installing insulation in the walls and ceiling, double-glazing windows and draught-proofing around doors and window frames. These activities cost money but are very effective because once retrofitting is done, life can continue much as before. However, if you are not particularly well-off, as is the case with many council community care clients, then there is not the money for retrofitting.

Much cheaper, but requiring an ongoing awareness, is energy efficiency through behavioural responses. If we can practise energy efficiency to the extent that it becomes simply a new habit, considerable savings in energy, and in dollars, can accrue. It is our addiction to convenient energy that means that we might think of an electric blanket rather than a hot-water bottle, that we turn on the heater rather than put on a jumper, or that we turn on the light rather than open the curtains. None of these responses are difficult; they simply require comprehensive energy awareness. A driver for energy awareness is the lower energy bills that result from reduced energy use. It simply makes sense to use less energy.

The mitigation imperative

Whether a deliberate climate change response, or simply a means of saving money, the consequences of reducing the amount of energy that is used and switching to non-carbon-based forms of energy is a contribution to mitigating the long-term impacts of climate change. Reducing emissions of greenhouse gases, as energy efficiency does, is a necessary response to climate change. Without a universal and comprehensive recourse to every possible mitigation opportunity that we can find, the impacts of climate change will accumulate. The rise in average temperatures that we are seeing will continue, fire weather will begin earlier in the season and continue later, average rainfall will decline even as the incidence of intense rainfall with flash flooding rises. We have no choice but to mitigate!

Endnote

1. South East Councils Climate Change Alliance Inc (SECCCA) is a grouping of eight councils in Melbourne's south and south-east formed to collaborate in regional

responses, both mitigation and adaptation, to climate change. Details of SECCCA's work can be found on their website (www.seccca.org.au).

References

Arup T (2013) Solar hits the million mark despite cuts to incentives. *Sydney Morning Herald*. 5 April 2013 <http://www.smh.com.au/national/solar-hits-the-million-mark-despite-cuts-to-incentives-20130404-2h9st.html>

BREE (Bureau of Resources and Energy Economics) (2014) 'Australian energy update, July 2014'. BREE, Canberra.

Bureau of Meteorology and CSIRO (2014) 'State of the Climate 2014'. Commonwealth of Australia, Canberra.

City of Casey (2012) *Selandra Community Place*. City of Casey, Victoria, <http://www.selandra communityplace.com.au>

COAG (Council of Australian Governments) (2011) 'National strategy for disaster resilience: building the resilience of our nation to disasters'. Council of Australian Governments, Canberra.

CSIRO (2012) 'Climate and water availability in south-eastern Australia: a synthesis of findings from Phase 2 of the South Eastern Australia Climate Initiative (SEACI)'. CSIRO, Canberra.

Department of Human Services (2009) 'January 2009 heatwave in Victoria: an assessment of health impacts'. Victorian Government, Melbourne.

Gleik P (2012) Climate change, disbelief, and the collision between human and geologic time. *Forbes*, 16 January 2012 <http://www.forbes.com/sites/petergleick/2012/01/16/climate-change-disbelief-and-the-collision-between-human-and-geologic-time>

Maddocks (2010) VCAT's decision on the impact of climate change in L Taip v East Gippsland Shire Council: A warning for coastal councils. *Sustainability & Climate Change Update* August 2010, Maddocks, Melbourne.

Munich Re (2010) 'Topics geo: natural catastrophes 2009 – analyses, assessments, positions (Asia version)'. Munich Re, Munich, Germany.

Munich Re (2013) '2012 Natural catastrophe year in review'. Munich Re, USA. <http://www.munichre.com/us/property-casualty/events/webinar/2013-01-natcatreview/index.html>

SECCA (South East Councils Climate Change Alliance) (2014) *Save it for the Game*. South East Councils Climate Change Alliance, Narre Warren, Victoria, <http://www.seccca.org.au/projects/save-game>

State Government of Victoria (2014) *Cameras Save Lives*. Victorian Goverment, Melbourne, <http://www.camerassavelives.vic.gov.au/home/road+trauma/>

Steffen W (2013) 'The angry summer 2012/2013'. Climate Commission Secretariat (Department of Climate Change and Energy Efficiency), Canberra.

Transport Accident Commission (2014) *Annual Road Toll*. Victorian Government, Melbourne, <http://www.tac.vic.gov.au/road-safety/statistics/road-toll-annual>

4

Communicating about climate change

Susie Burke

Key points

- Talking about climate change is critical at all levels of society to raise awareness and engage people in solutions.
- Acknowledging your own distress about the issues, gathering support for yourself, and being informed about problems and solutions are important starting places for conversations.
- Framing the messages in your conversation to highlight the particular concerns that you think will resonate with your audience, like health, children/grandchildren's future, or protection from harm like bushfires, is helpful.
- The use of emotion, stories and vivid images helps create meaningful messages.
- Establishing norms that let the public know that people just like them are also heeding the climate warnings, and responding by reducing their own carbon footprints and adapting to the changed environment, is a very important method of change.

Introduction

Talking about climate change is critically important. We need to talk about the threats, the impacts, and what we need to do about it. While vital, these conversations are not always easy to have. Fortunately, there are many lessons from social science about how to overcome barriers and communicate more effectively, and these skills are described in this chapter. This chapter is a little different from the others in this book in that it discusses not only the evidence about good climate change communication, but also models some of these communication skills through the use of a more personal style to discuss this topic.

The world is already in times of dangerous climate change. Many of the risks scientists have warned us about in the past are now happening (e.g. increased duration and frequency of heatwaves, increased risk of bushfires, cyclones, floods, severe storms, shifting rainfall patterns, increased droughts, sea-level rise, coastal inundation, loss of species) and worse is predicted. So, not only does human-caused global warming exist, but it is also growing more and more dangerous, and at a pace that has now made it a planetary emergency.

Climate change has been described as the most serious global health threat of the 21st century (Costello *et al.* 2009). It dramatically disrupts some of life's essential requirements for health – water, air and food. The literature on proven and plausible health impacts of global climate change now covers virtually every organ system in the human body (Sheffield & Landrigan 2010). Wherever you live, climate change threatens your health (Chapter 1 explores this in detail).

The scientific consensus on climate change is clear. A total of 11 944 climate papers published between 1991 and 2011 were analysed and 97.1% endorsed the consensus position that humans are causing global warming (Cook *et al.* 2013). This study is significant in several ways. First, the problem is clearly real, and most climate scientists acknowledge this fact. Second, it locates the cause of the problem squarely with human actions, which means that it is primarily human behaviour change that is required to bring about urgent change in every area of concern where the national and global environment is threatened. Humans need to stop producing greenhouse gases (mitigation discussed in Chapter 3) and must adapt to the changed climate now (adaptation discussed in Chapter 2). Above all, it means that people need to be talking about it. Climate change needs to be talked about at all levels of society, in all levels of an organisation, and in political, professional, work and community roles, as well as in people's personal lives, because:

- it threatens people now, and in the future, and they have a right to know
- people everywhere all need to be urgently engaged in solutions to reduce the human contribution to warming by reducing emissions
- an urgent task is to prepare for the unavoidable threats, like extreme weather events, in order to protect lives and minimise damage.

This chapter is about these conversations.

A variety of climate-related conversations

There is an enormous range of climate-related conversations needed to develop better community understanding of climate change and how people can adapt and mitigate. Within organisations there need to be conversations with executive directors, boards and managers about putting climate change on the organisation's agenda, and developing climate change policies to guide future activities. Within the office it is important to raise awareness of climate change among colleagues, improve sustainability within the work-place, or work with staff to build their own capacity to talk about climate change. With clients, or community groups, conversations will help build their awareness of climate-related problems, for example, conversations about reducing greenhouse gases by choosing green energy and public transport, about local drought, heatwave and fire risks from the changing climate, and about how to adapt homes to cope with these changes. There are also many essential climate change conversations to be had in our personal lives within our families, friendship groups and communities.

This chapter discusses some of the challenges inherent in these climate conversations. The bulk of the chapter, however, is about helpful communication strategies for improving people's understanding and engagement with climate change. This is, after all, the ultimate point of these climate change conversations. Climate change conversations are how we can talk with people so that they recognise that the climate problem is real, urgent, relevant to them, *and* talk in a way that empowers them to respond effectively and do something about the problem that is useful.

Climate change conversations can feel daunting – but don't have to be

Climate change conversations can feel daunting for several reasons. People might fear other's reactions (e.g. 'Maybe they are sceptics and will think I'm wrong'; 'I hate getting into an argument'), worry that they are not knowledgeable enough about the topic (e.g. 'The issues are very complex, I'm not sure that I can properly explain them'), wish to avoid thinking about it themselves (e.g. 'I can't bear thinking about it, it's so upsetting, I feel overwhelmed'); or worry about what others might think about them (e.g. 'People might get sick of me always talking about bad news'; 'What if they think I'm just a crazy greenie?'; 'I worry that people might think I'm morbid/uninformed/naive/boring/paranoid or radical').

While all of these fears are legitimate, they do not need to be a conversation killer. There are many ways of successfully managing these concerns so that the important conversations can begin. What follows are key techniques for overcoming barriers to conversations.

Preparing for the conversations
Gathering support for yourself

Talking about climate change with others means first recognising that it is an enormous problem needing urgent action. These realisations can be distressing, so it is important to gather support for this work.

A group called Psychology for a Safe Climate have written about the psychological challenges of facing the climate change problem in their publication *Let's Speak About Climate Change* (Psychology for a Safe Climate 2013). Climate change stirs up tension between two different feelings within ourselves: between our wish for the way we want life to be, and our recognition that we are already in times of dangerous climate change, with worse forecast. Facing up to the reality of climate change, and bringing up the subject with friends, family and colleagues is difficult psychological work.

Doherty & Clayton (2011) reviewed studies that have explored the range of emotions associated with climate change, like subclinical depressive emotions, guilt, and despair associated with climate change and other global environmental issues (e.g. Buzzell & Chalquist 2009; Norgaard 2009; Nicholsen 2002). Fritze *et al.* (2008, p. 9) discussed how 'at the deepest level, the debate about the consequences of climate change gives rise to profound questions about the long-term sustainability of human life and the Earth's environment'. For some, climate change is a manifestation of a global ecological crisis causing uncertainty and emotional distress (Stokols *et al.* 2009). In this vein, Kidner (2007) described the loss of security engendered by uncertainty about the health and continuity of the earth's natural systems and how the impact of these emotions tends to be underappreciated because of a lack of recognition of subjective feelings of environmental loss in traditional scientific and economic frameworks. Glenn Albrecht has coined a new term 'solastalgia' to describe the grief that is experienced when we recognise that our familiar environment is changing and becoming lost (Albrecht *et al.* 2007).

In a large-scale survey of Australians' perceptions and understandings of climate change in Australia, Reser *et al.* (2012) reported that 20% of people show appreciable distress about climate change. A similar survey in America reported that a large percentage of people surveyed about climate change report feeling disgusted, hopeful, helpless or sad about the issue, and a quarter report feeling depressed or guilty (Maibach *et al.* 2008).

There is good reason to expect that facing up to the reality of climate change is distressing. In these circumstances, support from other people can be very helpful. It can take the form of conversations with close friends and trusted colleagues about feelings and anxieties about climate change. It can also be helpful for people to acknowledge their own tendencies to ignore, deny or avoid thinking about environmental problems. Talking about this will help others to identify and acknowledge similar reactions in themselves.

Often we don't know how strongly we feel about something until we find ourselves speaking about it to someone else. It can be amazingly therapeutic to give voice to feelings, rather than leave them swimming around inside our hearts or heads. (Psychology for a Safe Climate 2013, p. 12)

Forming or joining a group with other concerned climate advocates, and associating with a wider community of people who are taking on climate change, is another way of managing the psychological distress of facing climate change. As well as being a place for sharing concerns and stressors, a group also helps to foster a sense of hope, optimism and efficacy. Associating with like-minded people can create support, ideas, reinforcement and praise. Group approval and identification is a very good source of reward.

And of course, with supportive peers, climate conversations can be rehearsed. Peers could use a role-play conversation between a person promoting climate action and a sceptic, to practice handling these different perspectives.

Vignette

Claire knew that she needed to get climate change on the agenda at her workplace. Working in a rural healthcare clinic, she saw many families who were struggling with the drought. Her knowledge of the climate literature told her that the break of the drought wasn't going to be end of the environmental threats in her area. She knew that her organisation had to start talking about climate change, and face up to the coming threats, at the same time as helping their clients think about the issues and do something about it themselves. But she kept putting it off. 'What if they think I'm crazy?' she would ask herself. 'Why would they listen to me? What if they are sceptics?' Claire decided to find a buddy at her workplace who she thought would share similar views. They talked together about their distress at the environmental changes being driven by rising temperatures, and the slow pace of change being led by the government. They agreed that raising their work colleagues' awareness about the impacts of climate change on people's health was essential, and planned to start a 'green team' of like-minded colleagues who could meet together to talk further about these common concerns. The green team started meeting once a fortnight over lunch to put sustainability on the agenda.

Listening to others' concerns

Of course, communication is not one way, and climate conversations will involve dialogue with the other person. If the person agrees with all that you are saying, the exchange will be relatively easy! But how does one have difficult conversations when the other person is wary, sceptical or disagrees?

One of the most useful skills widely taught by social scientists and communicators is the skill of reflective listening. Many health professionals have well-honed listening skills for dealing with clients. The same skills can be used with great effect in having difficult climate conversations with colleagues, clients, friends and neighbours.

People listen better when they feel listened to. When confronted by someone who disagrees strongly with your views, it can be helpful to first stop and listen carefully to what they are saying. By listening to others it is possible to draw out the values that they hold close. Then, it is possible to find common underlying values that are shared, and which are also supported by climate supportive responses consistent with the others' values. A person who values his or her health is dependent on clean air, water and soil. A person who cares deeply about her children or grandchildren needs a stable climate for these children to live safely in.

Becoming informed

Becoming informed about the climate change problem, and the solutions, is another important preparation for climate change conversations. This whole book is about pulling together information to support community-based health and social service agencies to act in appropriate ways. But it cannot say everything. Other books, news media, film and the internet are good places for getting information about the problems. Friends can often be a good source of information too, and can suggest websites, books or magazines that they have found useful.

Reliable information comes from reputable sources, like peer-reviewed publications, indexed journals and research conducted by experts in the field with many prior publications and citations by other researchers. The Intergovernmental Panel on Climate Change (IPCC) reports, involving thousands of expert reviewers and hundreds of contributing authors from over 100 countries, are good examples of thorough, evidenced-based science about climate change, impacts and solutions. The American Association for the Advancement of Science (AAAS) is the world's largest non-government general science member ship organisation. Their 'What We Know' project summarises the science on climate change to ensure that the reality, risks and responses of climate change are well understood.

Part of being informed is familiarity with common arguments and useful counter-arguments that help with climate conversations. Some websites give examples of common arguments refuting climate change and counter arguments and suggestions for how to deal with them, for example, Skeptical Science (www.skepticalscience.com) and the AAAS website for the 'What We Know' project (whatweknow.aaas.org). It is not necessary to argue every point. Admitting that you do not know does not invalidate your concern about these issues. Instead it shows audiences that acting responsibly does not require a high degree of expertise. Personal and subjective knowledge, the lived experience and being able to talk about them in a personal way is as important as knowing about things. Facts do not change people's minds; stories do. (Useful tools for using stories to communicate are discussed later in this chapter.)

There is no shortage of good data that supports the idea that climate change is a problem. But there is still a long way to go to make sure that we are using it to maximum effect. This chapter is about how to do this.

Establishing the scientific consensus

Recent research by cognitive scientists on climate change communication shows the importance of the role of scientific consensus in acceptance of the science on climate change. According to Lewandowsky *et al.* (2013), when people are told that a scientific consensus exists (in the order of a whopping 97% of climate scientists), then most of the people accept and believe the science on climate change. Telling people about the consensus boosts people's acceptance of the scientific facts relating to climate change by a significant amount, particularly so for people who would normally tune out of evidence alone.

Focusing on the solutions

Talking about the climate problem with people, and sharing powerful facts or stories to illustrate the size of the challenge, is only one part of the story. Research about behavioural change shows clearly that people also need to be aware of clear steps they can take to do something about it. Climate conversations need to end with a principled solution or example that illustrates how the problem can be addressed in a way that inspires hope. This also helps to build a sense of self-efficacy, a concept that psychologists understand is very important in getting people to act on environmental threats. People are encouraged and motivated by knowing that they are making a positive difference. Most people do want to do the right thing, and be 'good, important and useful'. The more they believe that their actions can and do make a genuine difference, the easier it is for them to do these things (Futerra 2010).

Framing the message

Even when people understand climate change issues, and have the support of colleagues to speak about them, there is still the issue of *how* to talk about climate change. Which bit of the complex climate change story should be the focus of the conversation? What do you want the listener to be thinking about as you talk so that they see the value of the topic and take a personal interest in it?

A concept that is often used to help communicate messages is *framing*.

Framing refers to the way in which information is packaged and presented (or 'framed') to influence how the audience perceives the meaning of the information. In an appropriately framed conversation, the words and phrases chosen encourage the audience to focus on particular aspects of a complex topic that have greater relevance to them. Frames can help communicate why an issue might be a problem, who or what might be responsible and, in some cases, what should be done. The intention is not to deceive or manipulate people, but to make the information about climate change more accessible to them. Climate change conversations need to resonate with the participants in the conversation.

The message 'climate change is regarded as the biggest health threat of the 21st century', for example, uses a health frame to talk about the risks of climate change. The listener is invited to think about the negative impact of climate change on people's health. Alternatively, the message 'climate change has already contributed to the extinction of numerous species' uses an ecological frame to invite the listener to think about the severe impacts that rising global temperatures will have on biodiversity.

An excellent guide for understanding more about framing in the context of climate change conversations (plus other extremely useful tips) can be found in *The Psychology of Climate Change Communication – a Guide for Scientists, Journalists, Educators,*

Political Aides, and the Interested Public (CRED 2009). The guide is available on the Center for Research on Environmental Decisions (CRED) website (www.cred.columbia. edu/guide).

The audience's mental models – know your audience

As well as the communicator being able to choose how to frame information to evoke particular associations in the listener, the listener also uses frames to understand and make sense of information. Cognitive scientists have strong evidence that people use mental models in long-term memory to store previous experiences and understandings of the world (Lakoff 2006). These mental models provide people with a quick and easy way to process information. They help shape actions and behaviour, influence what we pay attention to in complicated situations, and define how we approach and solve problems. They also serve as the framework into which people fit new information. In sum, we use existing mental models to make sense of incoming messages. (Of course, sometimes a mental model also serves as a filter, resulting in selective knowledge uptake. People often only seek out, or absorb, the information that matches their mental model, confirming what they already believe about an issue. This poses a potential stumbling block for climate change communicators that will be addressed later in a section on debunking myths.)

Knowing as much as you can about an audience's mental models helps in finding a frame for climate communication that resonates with the audience. One way of doing this is to consider the audience's membership in specific subcultures (e.g. ethnicity, class, age, common beliefs, gender, religion) because this will provide clues about the mental models they may use to receive and interpret information. Is the audience comprised of the organisation's board members, who are concerned about the financial success of the organisation? Is the audience low-income clients who find it difficult to make ends meet and are stressed about rises in living expenses? Are they parents worrying about their children's

Vignette

Joseph needed to talk with his board about getting climate change onto the organisation's agenda. He thought a lot about the audience. Joseph knew that several of the board members were very risk averse, so he decided to frame his communication using a risk management frame. He described the enormous risks that he saw his organisation was taking by not acknowledging the threats that climate change poses to their operation and by failing to take steps to adapt to these threats before they happened. He raised their awareness of how vulnerable their premises would be in the event of a cyclone or tropical storm. If their building was damaged or destroyed how would they continue to operate? What would happen to all of their documentation? How would they continue to provide a service? Would they even survive as an organisation? An outcome of this conversation was that the board began to take serious steps towards adapting to these threats. They added climate change adaptation to their priorities and made changes on a variety of fronts, like extending their insurance against extreme weather events, storing copies of electronic information off site, purchasing a power generator to charge mobile phones in the event of power outages, and seeking funding to be able to upgrade their windows and roofs to make their building more cyclone resistant.

health, or elderly people who suffer during heatwaves? And think about the local impacts of climate change. Where does the audience live? In an urban environment where heatwaves are relentless, or the urban fringe in a place which is vulnerable to bushfires? Maybe the local environment is a coastal region that is regularly threatened by cyclones and severe storms, or a coal mining community concerned about air quality?

The speaker has great power to present information in a particular way (the frame) which will trigger a specific mental model in the mind of the listener, and influence how they will hear the message, how they will interpret the message, and also how they will respond to the message. Messages can be framed to highlight the particular concerns that you think will resonate with your audience, like health, children/grandchildren's future, or protection from harm from events like bushfires. Frames can also be used to emphasise particular values, like equity, or caring for others, or personal success, which is discussed in the next section.

Local frame: making climate change here and now

Many people do not regard climate change as an immediate threat to themselves, but as a threat to future generations and to people in far-away places. However, people are more inclined to take responsibility for the problems that they see as happening in their local environment and in the immediate or near future. Humans tend to discount risks that concern future events. Climate change risks are not only projections for future changes, but they also refer to gradual processes – which may not be perceived as risky because they happen slowly and almost imperceptibly. Unfortunately, this means that the majority of people see the worst problems as happening far away from them, and distant in time, which is also where they feel they have the least responsibility to do anything about them (Uzzell 2000).

In reality, though, climate change is a threat to each of us, now and into the future. No part of the planet is immune from the disruptive effects of rising temperatures and increased variability in weather. Australia, already the driest habitable continent on the earth, is particularly vulnerable to the effects of climate change because of its low rainfall and susceptibility to natural disasters like prolonged drought, bushfires, cyclones and floods.

To counteract this human tendency to pay less regard to things that we see as happening far away and distant in time (social scientists call this temporal and spatial distancing), climate change needs to be communicated in a way that helps people to see that the climate problem is here, now and for sure. By reducing the psychological distance between individuals and environmental impacts, people can start to see both the risks and the remedies as *local* issues. Also, the information being communicated must be relevant to the listener, as well as salient and important. Climate change conversations can then become about local threats to the local community, like drought, increased risk of catastrophic bushfires, failing crops, heatwaves or the rising prices of electricity, water and food. (For a detailed list of risks relevant to community sector organisations, see the Australian Council of Social Service report written by Mallon *et al.* 2013.) Conversations about climate change adaptation can be about learning how to recognise signs of heat stress, putting shade cloth up in outside areas, installing water dispensers in community organisations, building local community gardens, making bushfire survival plans for ourselves and our neighbours, and mapping safe routes for cycling around town instead of depending on cars.

Vignette

Sarah had been running a group for elderly people in the senior citizens' centre in her town for over a year. There were pressures in her region from drought and a depressed economy. Rising food and electricity prices were an ongoing stress for the pensioners; heatwaves were also becoming an issue. She had some ideas about how to help these older people adapt to the long, hot summers and she wanted to make the link with climate change. She asked the group to discuss changes they'd noticed in the summers. Some said they didn't remember such long, hot, dry spells when they were younger. Sarah explained that increasing heatwaves and less rainfall is one of the expected climate impacts. Sarah also asked the group members to talk about what they thought it would be like when their grandchildren grew up. They agreed that if things continued in this way it would be even worse. They then talked about ways in which they could manage the heat so that summer is not such a struggle. They shared practical suggestions like hanging wet towels in front of fans. One woman talked about how she wrings her t-shirt out in a basin of water before wearing it and could be in the running for a wet t-shirt competition! Another woman brought in some insulating film that she was putting on her windows to cut down some of the heat that was coming in. They planned to plant shady trees on the northern side of their houses to give them fruit as well as shade their houses. And if the trees took too long to grow, at least the next owner would benefit. At one meeting they started talking about helping some of their older friends who lived alone. They thought they could get in the practice of ringing each other on hot days to check how each was going. One group member, an enthusiastic walker, decided to make her early morning walk take her via an elderly neighbour's house to help her lower her blinds on hot days. Everyone in the group was able to identify at least one neighbour who they thought they could help out on hot days. One woman whose son was building his first home encouraged him to make sure that he put in double-glazed windows and the best insulation possible in the roof. She started to realise that summers were only going to get more difficult.

Extreme weather frame

Another obvious frame for talking with the public about climate change is an extreme weather frame. Despite the many uncertainties about the causality of individual weather events, scientists increasingly acknowledge that there is a global pattern of extreme weather events consistent with their understanding of climate change (Peterson *et al.* 2013).

For the majority of people in Australia, climate change is most likely to be experienced as changes in the intensity and/or frequency of cyclones, storms, floods, bushfires, heat-waves or drought. To illustrate this point, a large-scale survey of Australians' understanding of climate change found that 45% of respondents reported a direct personal experience with weather events thought to be associated with climate change (Reser *et al.* 2012). Extreme weather events offer a vitally important opportunity to build awareness and conviction around climate change. They are personal, emotional, can be seen, felt and lived through. They may help climate change feel more real and salient.

The UK-based Climate Outreach and Information Network (COIN), founded by George Marshall, recently released a report about how to talk about extreme weather as a way of bringing climate change to the front of the public mind (Marshall 2014). They recognise the potential for extreme weather events to open up the issue of climate change and begin a conversation in the community.

The COIN report argues for an approach that focuses on peer communication at a local level and recommends that any climate campaign linked to extreme weather events is thoroughly trialled and tested before being launched. The key recommendation is that communicators using an extreme weather frame should validate the powerful narratives of resilience, community pride and mutual caring that often emerge during natural disasters. They can build on these to create a broad-based discussion of long-term preparedness and adaptation.

Vignette

Adam was working as a social worker in a community that was recently flooded for the third year in a row. Adam started a regular column in the local newspaper talking about flood recovery. He would talk about how the clean-up was going and let people know about any grants or local initiatives that were helping people get back on track. Each column would have a personal story about neighbours helping each other out, or changes that a local resident was making to his storage of valuable documents so as to be better prepared next time. Over the weeks, Adam started talking about extreme weather as something that we can expect again and again, and drew links with the greater climate variability being experienced all over the world as global temperatures increase. Always, he finished with a good news story of local resilience – he built a picture of a township that was getting better at anticipating further extreme weather, helping each other out with getting better prepared, and building great networks of volunteers to help clean up, and cook and care for people impacted by the next disaster.

(N.B. Marshall also acknowledges that there are many psychological barriers to be aware of in framing climate change in terms of extreme weather events. For example, the victims of extreme weather events have strong personal and social reasons for not wishing to accept that such events may become even stronger and more frequent; for this and other reasons they may have strong reasons to suppress discussion of climate change during recovery. For a complete discussion, refer to Marshall (2014).)

Health frame

Drawing awareness to the connections between climate change and human health is also an effective method for communicating with the public about climate change.

Health depends on clean air (free from smoke and pollution), clean water, clean soil to grow fresh fruit and vegetables, and shelter from the elements. Humans are living in a shared life-support system. By articulating the serious individual health consequences of an unstable climate, extreme weather events, air pollution, lack of access to clean water, etc., communicators can help frame climate change as a concrete, personal concern for everyone (CRED 2009). (So a health frame can also coexist with a local and personal frame.)

The Climate and Health Alliance (CAHA) uses the health frame in their communications about climate change. They use two main health messages – one linking the problems created by climate change to health, and one linking the solutions to climate change to improvements in health. Key CAHA messages include:

- Climate change is affecting people in every nation, on every continent, and it's having an overwhelmingly negative impact on health.
- Renewable energy reduces air pollution and so can prevent diseases like asthma, pneumonia, heart attacks and stroke.

Excellent examples of these messages can be found on the CAHA website (www.caha.org.au), as well as their global counterpart, GCAHA (e.g. see www.climateandhealthalliance.org/resources/ipcc-resources/banner-infographic).

Another useful frame is a psychological health and wellbeing frame. Again, linking the impacts of climate change as well as *solutions* to climate change to psychological health and wellbeing is very useful. Messages could include:

- Riding or walking rather than sitting in cars improves your cardiovascular health and vitality, can reduce rates of dementia and depression, reduce social isolation and increase community connectedness.
- Joining Landcare groups/protecting natural environments/spending time in nature improves our quality of life and provides hope and inspiration.

The Transition Town movement uses these positive health and wellbeing messages very successfully (Hopkins 2008). Transition towns are communities that are actively finding solutions to the twin problems of climate change and peak oil. Transition town proponents argue that the more we anticipate and adapt to having less energy and resources, then the less catastrophic the impacts of peak oil and climate change on individuals and communities. They focus, therefore, on building local solutions to significant environmental and societal problems. This is done by finding ways of meeting people's basic needs (for food, shelter, transport, health care, water, etc.) that require less energy, produce fewer greenhouse gases and use only a fair share of the world's resources. One of the key strategies in transition towns around the world is to build a vision of an abundant future in which people live low-energy, low-impact lifestyles in resilient communities. This vision includes messages like:

- Help communities to transform so that walking, riding and active transport are the normal way to get around, and reduce diseases of affluence (e.g. too much high-energy food, too little activity and a range of related health problems such as heart disease, high blood pressure, diabetes and depression).
- Improve mental health by developing a sense of purpose in contributing to the future, and in considering the wellbeing of other generations and people in the other parts of the world, rather than just meeting their own needs.

Another group which focuses on communicating via a health frame is the Robert Wood Johnson Foundation (RWJF). Their 2010 report on how to communicate the social determinants of health (RWJF 2010) has many useful tips for linking the environment and health. They looked for a common language that would expand people's views about what it means to be healthy to include not just where health ends, but also where it begins. They found that priming audiences about the connection between health and climate with messages they already believe makes the concept more credible. People already value their health, so talking about the importance of your health and your children's health, and linking this to messages about climate change is very effective, for example, 'health starts—long before illness—in the food we eat, in the active ways we get around, and in your neighbourhood, and in the parks and gardens that we live near, in the air we breathe' (RWJF 2010, p. 7).

Fear frame

Much research has been done on the use of a fear frame to talk about environmental problems and solutions. Some researchers argue that fear and worry can motivate people to remove themselves from threatening situations or to change the environment in ways that reduce feelings of being at risk. There is some evidence that worry may directly result in an increased disposition to act, as well as elevating judgments about climate change risk. Other researchers focus on ways in which fear can actually be an obstacle to change (Oskamp 2000). People can become easily overwhelmed at the magnitude of the environmental problems facing the planet. When people become too fearful, there is a tendency to react by denying or minimising the problems. Research studies on appeals to fear have shown that fear appeals are most likely to change people's behaviour under two conditions:

Vignette

Frances was the sustainability officer for a mental health organisation and was invited to give a guest lecture to a group of social work post-graduate students. She decided to give the students an overview of the scale of the looming climate catastrophe. She listed the dire climate threats anticipated around the world, citing IPCC expert conclusions that climate change affects everybody in every part of the world. She explained that we are on track for a further four to six degrees of warming by the end of the century, which will be catastrophic for humans and many other species on the planet.

Having thoroughly used a fear frame to begin her presentation, she then moved on to the solutions. She explained the good news, which is that we know what to do to reduce the threat (make a rapid transition to a zero carbon economy), we already have the technology to do this (rely on 100% renewables) and that there is much that we can all do, immediately, such as reducing our own carbon footprint (by switching to green energy, riding to work, buying local food, holidaying closer to home, planting trees), lobbying for good climate policy (ringing politicians to express support for renewable energy, public transport instead of freeways, protesting subsidies to coal and gas companies) and putting pressure on big business to develop sustainable practices (e.g. taking our money out of banks that fund fossil fuels, divesting our shares in polluting companies). She finished by asking the students to talk about what they were already doing to reduce environmental problems (helping them to see themselves as someone who cares about the environment) and then asked them to share with the group some additional action that they thought was important to begin doing. After each student named an activity they planned to do, Frances asked their course convenor to check in with them in a class in a few weeks' time to see how they were progressing.

(1) if people are aware of clear steps they can take to protect themselves, and (2) if these steps are conveniently available (e.g. Leventhal *et al.* 1980).

Intrinsic values frame

Tailoring the message to activate particular *values* is also a very useful communication technique.

Values are beliefs about what is important in life. They affect our behaviours, choices and feelings. Values are learned and can be shaped, cultivated, promoted and championed. Broadly, it is useful to distinguish between two classes of values: extrinsic and intrinsic. Intrinsic values are 'bigger-than-self' values, also referred to as self-transcendent values. They include things like empathy, cooperation, equality and justice (see Box 4.1). Extrinsic values, by contrast, are more self-interested, like valuing personal success, wealth, power and status. They are often contingent upon the perceptions of others and more aligned with conservative ways of thinking (Schwartz 2006).

Most people are significantly bi-conceptual and hold both types of values to be important. Research shows that humans are hard-wired both for self-interest *and* cooperation and empathy (Lakoff 2009). What is important is not whether the person holds extrinsic or intrinsic values per se, but rather the relative importance that he or she attaches to extrinsic as opposed to intrinsic values. In general, when people are focused on extrinsic values (e.g. financial success) experimental studies show that they are more likely to also show lower empathy, higher preference for social inequality and hierarchy, and less concern about environmental problems. Intrinsic values, by contrast, provide a better source of motivation for engaging bigger-than-self problems than extrinsic values (see Crompton & Kasser 2009).

Climate change is definitely a bigger-than-self sort of problem.

Unfortunately, many climate campaigns, or policy responses to climate change, are framed in ways that appeal to extrinsic values. They may highlight the economic self-interest of businesses investing in new green technologies, or in companies achieving status from being a 'first mover' in a sustainability industry. Householders may be encouraged to put solar panels on the roof on the grounds that these save money, 'buy a hybrid car because it looks cool' or 'insulate your house and save money'. Unfortunately, this sort of framing, while perhaps effective for increasing sales, may have a net negative impact on long-term environmental change. To realistically tackle climate change on the enormous scale that is required, it is only *intrinsic* values that can really help us to transform our society.

Research shows that different values can be engaged by the ways in which we communicate with people. To activate intrinsic values is to simultaneously inhibit extrinsic values (Crompton & Kasser 2009). Intrinsic values can be engaged by talking with people in a way that shows that these values are highly desirable (e.g. 'everyone has a human right to breathe clean air and drink clean water') or are the social norm (e.g. 'most Australians believe that they need to take action to protect the environment – you can see this in the huge numbers of homes with solar panels on their roofs'). When a communicator engages intrinsic values in a conversation, there is likely to be an increase in listeners' concerns about a range of bigger-than-self problems. At the same time, communicators need to be careful not to promote extrinsic values (like wealth and power) in their communications. Activating one class of values tends to inhibit the other class.

Box 4.1: Helpful intrinsic values

- Personal responsibility – taking care of yourself and others
- Valuing community spirit – working together to help each other
- Ethic of excellence – making the world a better place, including yourself
- Unity with nature
- Concern about equality
- Understanding, appreciation and tolerance for other people
- Concern for future generations
- Empathy towards others who are facing effects of humanitarian and environmental crises
- Recognition that human prosperity resides in relationships

Vignette

Vincent wanted to talk about sustainability within his workplace. He thought that it was time that his workplace divested from fossil fuels. He had some numbers and material facts he wanted to share, but knew he needed to frame them in terms of moral values so their overall significance could be understood. He began his conversation by talking about the values of his own organisation.

'I've been working at this organisation for seven years, and I admire the efforts that we all make to improve the health and living conditions of low-income people in our community. We are an organisation that really values early prevention of problems. It is not hard for us to see that climate change is a huge threat for the people who we care about – the disadvantaged people in our community. It fits perfectly with our overarching values to also take action on climate change. Right now, I'm concerned that we are contributing to the problems, rather than taking action to prevent the problems, by investing our money with institutions which fund fossil fuel expansion. The big four banks in Australia (NAB, Westpac, ANZ and Commonwealth) have invested millions of dollars in new fossil fuel projects since 2008. I think that the morally right thing to do is to divest our money and bank with institutions that do not fund fossil fuels which are driving climate change which will have such negative impacts on the people who we care about.'

Presenting information vividly

Climate change communication experts also emphasise the importance of presenting information vividly so that the message sticks and people are more likely to respond and take action. In the following sections we will look at the use of emotion, stories and vivid images to help create meaningful messages.

Cognitive psychologists explain that the effectiveness of vivid, emotional stories has got something to do with how the brain works, and the different information processing systems that humans use to understand information. The analytic processing system controls the intake of scientific information and uses logic, analysis, numbers and rules, when prompted, to take in information. The experiential system, on the other hand, operates automatically and is oriented towards survival. This system responds emotionally and instinctively to information that is presented in terms of concrete images, personal stories and vivid emotion (for a review of different human processing systems, see Evans 2008). Commonly, information about climate change is presented in terms of facts and figures, which relies on the analytic processing system. This might increase knowledge, but does not help to stimulate motivation. Evidence from the social sciences is that the experiential processing system is the stronger motivator for action. Personal stories and vivid images make communications more memorable, and therefore dominant in processing. The most effective communication, of course, targets both processing systems of the human brain (CRED 2009).

Using emotion and stories

Emotional arousal is a feature of stories that make them attract our attention and make listeners want to tell others about it. This is part of the reason that gossip is so compelling! Once feelings are involved, people are hooked to a story and are more likely to pass it on and respond personally to it.

The next issue is to find the right emotion to harness. Earlier, the pros and cons of using fear were considered. According to social psychologist and environmentalist Niki Harré, we can be even more successful about promoting sustainable behaviours when we can find emotionally-charged stories that elicit positive emotions and have a pro-social moral – stories that can generate emotions that steer us towards behaviours that are good for others and away from behaviours that damage others. These stories make us feel good and make us want to pass them on to others (Harré 2011).

Vignette

Lilly decided to run a series of film nights in her local community to share stories about people around the world working together to find sustainable solutions. She looked for films that told tales of joy – showing people doing amazing things on their farms so that they could provide good quality, local food for their community and reduce their dependence on food that is grown, processed and transported from far away. She advertised these film nights through her community house newsletter and around town. After the film people were invited to stay back for soup and bread (all made from local ingredients) and a discussion about the film, with a focus on how people could use the inspiration from the film to make changes in their own lives.

Harré adds that, to be really useful for conversations about climate mitigation and adaptation, 'stories need to provide the full motivational and detailed how-to-do-it package that we human imitators respond to so well' (Harré 2011, p. 54). What will pull us forward is living examples of how to be sustainable.

Creating sticky ideas

Chip and Dan Heath, communication experts from the United States, have written an excellent guide to communications that work. They call these 'sticky' ideas. They argue about the importance of making ideas simple, and evocative, to increase the chance that they will stick (Heath & Heath 2007).

Simple ideas are not necessarily short or simplistic, but are the essential core of the idea. It turns out that trying to say something simply can be quite complicated! Our own beliefs and previous knowledge can interfere with our capacity to talk about a concept in simple and understandable ways.

'Coal needs to stay in the ground' is a good example of an effective, simple message. To limit warming, we need to stop burning fossil fuels which release carbon into the atmosphere and significantly contribute to global warming. Keeping coal in the ground opens up a whole lot of possibilities about how to generate electricity to meet the world's needs, such as wind, solar and hydro. So the idea is simple, but world changing.

The RWJF research group agrees with the concept of simple ideas, and extends it to talking about the use of facts or data. They advise using one strong and compelling fact – a 'killer fact' – for maximum impact. This would be a surprising point that arouses interest, attention and emotion. Loading messages down with more than one or two facts tends to depress responses to them.

> *Regardless of how good or reliable the data is, this research showed us that less is more. If you can use two facts instead of three, use two. Or better yet, use just one great fact. (RWJF 2010, p. 10)*

The other aspect of simple ideas is to use understandable language, skipping jargon, acronyms and complicated scientific terms, and instead using words that make sense. The CRED guide to climate communication lists words and phrases commonly used to discuss climate change, and offers alternative words that get the same idea across more simply (CRED 2009). Chip and Dan Heath talk about the importance of concreteness in creating sticky ideas. They argue that concrete images – things that people can see, touch and easily understand – are much more successful in communicating clear ideas than using abstract language. Illustrating with examples like 'in the air we breathe and water we drink' makes the concept of environmental factors more tangible (Heath & Heath 2007).

Explaining climate adaptation in concrete terms would mean providing examples in terms of human actions rather than using complicated jargon, which means different things to different people. 'If we put up shade cloths in the garden, then the children will still be able to play outside during a heatwave' or 'If we buy a generator for our organisation, then we can turn it on if the power goes off during a heatwave/bushfire, and we will be able to stop the food and medicine that we store in our fridge for our clients from going bad'.

Cook and Lewandowsky (2011) also follow this principle of simplicity in their guide on debunking misinformation about climate change. Information that is easy to read, easy to understand and succinct is easier to process and is more likely to be accepted as true. Specifically, they advise using simple language, short sentences, subheadings and paragraphs, and ending on a strong and simple message that people will remember and be easily able to pass on to their friends, such as '97 out of 100 climate scientists agree that humans are causing global warming' (Cook & Lewandowsky 2011, p. 3).

Box 4.2: Anatomy of effective debunking

- Emphasise core facts you want people to know/remember
- Provide an explicit warning about myth if you have to repeat it
- State myth
- Provide alternative explanation
- Limit new information to about three facts
- Provide a graphic representation of the core facts

For more details go to www.sks.to/debunk

Additional evidence-based strategies for effective debunking of climate change myths are also important to know about. Cook and Lewandowsky (2011) warn that care needs to be taken to avoid the 'backfire effect', where a communicator inadvertently reinforces the very myths they seek to correct. The major elements of effective debunking, used to avoid these backfire effects, are outlined in Box 4.2.

Using social norms

A final, but very important, climate communication technique involves the use of norms to let the public know that people just like them are also heeding the climate warnings, and responding by reducing their own carbon footprints and adapting to the changed environment.

Social norms are group beliefs about how people should behave in a given situation. Some are created by formal rules, but many develop informally as a result of repeated use. Numerous studies have examined the power of social norms to change people's behaviour. People are very sensitive to cues about what is normal behaviour and like to follow suit. Doug McKenzie-Mohr, in his book *Fostering Sustainable Behaviour*, provides many examples of the power of norms to encourage sustainable farming practices, showering behaviour, littering behaviour, etc. (McKenzie-Mohr & Smith 1999).

Clearly, perceived norms can have a substantial impact on behaviour. People generally want to be like everyone else. They observe how others behave in order to determine how they should behave. So letting people know that many other people 'just like them' are talking about climate change, taking action to reduce their emissions and preparing for climate change impacts within their organisations and with their clients is a powerful way of encouraging pro-environmental behaviour change.

Information about social norms can be provided in two major ways. First, explicit statements can be used to state what people in the target context usually do. Descriptive norm messages that say 'everybody's doing it' to promote conservation-minded actions have been found in the research to be the most effective.

To be effective, though, norms also need to be visible. Psychologist Wes Schultz and associates discuss this in a 2007 review of environmental modelling. They explain that environments often show 'traces' of human behaviour that prompt people into following along, like paths worn through a forest showing us where to walk or website information on how many people have downloaded a particular music track (Schultz *et al.* 2007).

Vignette

Nicky wanted to use social norms to encourage other community organisations in her area to engage in climate adaptation. She thought that it was important to show others that her health organisation is addressing climate change, and that this is normal, sensible behaviour for any health organisation in this climate-changed world. Her community health centre decided that they needed to guarantee that they would continue to have a power supply in the event of power outages during heatwaves or extreme weather events. They decided to install solar panels with a battery bank to maintain power supply in a blackout. As well as enabling them to maintain refrigeration for necessary medicines, it also ensured that they would be able to use phone and internet during a heatwave, and continue to charge the mobile phones that were an important part of their outreach service.

Nicky worked up a few ways to let people know about her organisation's climate adaptation initiatives. She put stickers at the front of the building saying 'This organisation is powered by solar energy'. She wrote a press release for the local newspapers announcing what her organisation was doing and why, and she wrote a short article explaining the same in her industry newsletter.

To increase sustainability within her workplace, Nicky had a few other social norms to establish! Rather than leave her bike helmet on her bike, she brought it in to work and left in on her desk next to her keep cup to leave a behavioural trace of herself as a bike commuter who reuses coffee cups!

To encourage sustainability, be visibly sustainable yourself and leave behind as many behavioural traces as you can. A recent Canadian study found that 36% of restaurant diners who observed a pair of people using a compost bin and discussing with each other their decision to do so went on to compost themselves. Only 22% of those not exposed to a model did so (Sussman & Gifford 2013). This suggests that being sustainable and drawing attention to yourself, as awkward as it might feel, will win some people over.

Conclusion

Talking about climate change is not as hard as it first appears, and definitely not as hard as dealing with unmitigated climate change now and in the future! The strategies and skills explained in this chapter come from decades of social and behavioural science research into communication, risk perceptions, social norms, decision-making and other related fields. There are numerous climate-related conversations that need to be had, and these strategies, together with the personal learnings gained from each conversation you have, play a very important part in building better community understanding of climate change and how we can adapt and mitigate. Climate change is the single most important issue that is facing humanity. The challenge is huge, urgent and achievable. Now is not the time for denial or despair but action. Communicating about climate change among friends, among colleagues, within workplaces, with clients, within communities and beyond is an essential part of building the movement.

References

Albrecht G, Sartore G, Connor L, Higginbotham N, Freeman S, Kelly B, Pollard G (2007) Solastalgia: the distress caused by environmental change. *Australasian Psychiatry* **15**, S95–S98. doi:10.1080/10398560701701288

Buzzell L, Chalquist C (Eds) (2009) *Ecotherapy*. Sierra Club Press, San Francisco.

CRED (Center for Research on Environmental Decisions) (2009) *The Psychology of Climate Change Communication: A Guide for Scientists, Journalists, Educators, Political Aides, and the Interested Public*. Center for Research on Environmental Decisions, New York.

Cook J, Lewandowsky S (2011) *The Debunking Handbook*. University of Queensland, St Lucia.

Cook J, Nuccitelli D, Green SA, Richardson M, Winkler B, Painting R, Jacobs P, Skuce A (2013) Quantifying the consensus on anthropogenic global warming in the scientific literature. *Environmental Research Letters* **8**, 024024. doi:10.1088/1748-9326/8/2/024024

Costello A, Abbas M, Allen A, Ball S, Bell S, Bellamy R, Friel S, Groce N, Johnson A, Kett M, Lee M, Levy C, Maslin M, McCoy D, McGuire B, Montgomery H, Napier D, Pagel C, Patel J, de Oliveira JAP, Redclift N, Rees H, Rogger D, Scott J, Stephenson J, Twigg J, Wolff J, Patterson C (2009) Managing the health effects of climate change. *Lancet* **373**, 1693–1733. doi:10.1016/S0140-6736(09)60935-1

Crompton T, Kasser T (2009) *Meeting Environmental Challenges: The Role of Human Identity*. Greenbooks, London.

Doherty TJ, Clayton S (2011) The psychological impacts of climate change. *The American Psychologist* **66**(4), 265–276. doi:10.1037/a0023141

Evans JSBT (2008) Dual-processing accounts of reasoning, judgment, and social cognition. *Annual Review of Psychology* **59**, 255–278. doi:10.1146/annurev.psych.59.103006.093629

Fritze JG, Blashki GA, Burke S, Wiseman J (2008) Hope, despair and transformation: climate change and the promotion of mental health and wellbeing. *International Journal of Mental Health Systems* **2**(13). doi:10.1186/1752-4458-2-13

Futerra (2010) *New Rules New Game*. Futerra Sustainability Communications, London.

Harré N (2011) *Psychology for a Better World: Strategies to Inspire Sustainability*. University of Auckland, New Zealand, <http://psych.auckland.ac.nz/psychologyforabetterworld>

Heath C, Heath D (2007) *Made to Stick*. Random House, New York.

Hopkins R (2008) *The Transition Handbook: From Oil Dependency to Local Resilience*. Chelsea Green Publishing, White River Junction, Vermont.

Kidner D (2007) Depression and the natural world: towards a critical ecology of psychological distress. *International Journal of Critical Psychology* **19**, 123–146.

Lakoff G (2006) *Thinking Points: Communicating our American Values and Vision*. Farrar, Straus and Giroux, New York.

Lakoff G (2009) *The Political Mind: A Cognitive Scientist's Guide to your Brain and its Politics*. Penguin, London.

Leventhal H, Meyer D, Nerenz D (1980) The common sense representation of illness danger. In *Contributions to Medical Psychology*. (Ed. S Rachman) Vol. 2, pp. 7–30. Pergamon, New York.

Lewandowsky S, Gignac GE, Vaughan S (2013) The pivotal role of perceived scientific consensus in acceptance of science. *Nature Climate Change* **3**, 399–404. doi:10.1038/nclimate1720

Maibach EW, Roser-Renouf C, Leiserowitz A (2008) Communication and marketing as climate change-intervention assets: a public health perspective. *American Journal of Preventive Medicine* **35**, 488–500. doi:10.1016/j.amepre.2008.08.016

Mallon K, Hamilton E, Black M, Beem B, Abs J (2013) 'Adapting the community sector for climate extremes: extreme weather, climate change & the community sector – risks and adaptations'. National Climate Change Adaptation Research Facility, Gold Coast.

Marshall G (2014) *After the Floods. Communicating about Climate Change.* Climate Outreach and Information Network, UK. <http://www.climateoutreach.org.uk/portfolio-item/communicating-climate-change-around-recent-extreme-weather-events>

McKenzie-Mohr D, Smith W (1999) *Fostering Sustainable Behavior: An Introduction to Community-Based Social Marketing.* New Society Publishers, Gabriola Island, B.C. Canada.

Nicholsen SW (2002) *The Love of Nature and the End of the World: The Unspoken Dimensions of Environmental Concern.* MIT Press, Cambridge, USA.

Norgaard KM (2009) 'Cognitive and behavioral challenges in responding to climate change: background paper to the 2010 World Development Report'. World Bank Policy Research Working Paper 4940. The World Bank, Washington D.C.

Oskamp S (2000) Psychological contributions to achieving an ecologically sustainable future for humanity. *The Journal of Social Issues* **56**, 373–390. doi:10.1111/0022-4537.00173

Psychology for a Safe Climate (2013) *Let's Speak About Climate Change.* Psychology for a Safe Climate, Melbourne.

Peterson TC, Hoerling MP, Stott PA, Herring SC (2013) Explaining extreme events of 2013 from a climate perspective. *Bulletin of the American Meteorological Society* **94**, S1–S74.

Reser JP, Bradley GL, Glendon AI, Ellul MC, Callaghan R (2012) 'Public risk perceptions, understandings and responses to climate change in Australia and Great Britain'. National Climate Change Adaptation Research Facility, Gold Coast.

RWJF (Robert Wood Johnson Foundation) (2010) *A New Way to Talk about the Social Determinants of Health.* Robert Woods Johnson Foundation, New Jersey, <http://www.rwjf.org/en/research-publications/find-rwjf-research/2010/01/a-new-way-to-talk-about-the-social-determinants-of-health.html>

Schultz PW, Nolan JM, Cialdini RB, Goldstein NJ, Griskevicius V (2007) The constructive, destructive, and reconstructive power of social norms. *Psychological Science* **18**(5), 429–434. doi:10.1111/j.1467-9280.2007.01917.x

Schwartz SH (2006) *Basic human values: an overview.* The Hebrew University of Jerusalem. <http://segr-did2.fmag.unict.it/Allegati/convegno%207-8-10-05/Schwartzpaper.pdf>

Sheffield PE, Landrigan PJ (2010) Global climate change and children's health: threats and strategies for prevention. *Environmental Health Perspectives* **119**, 291–298. doi:10.1289/ehp.1002233

Stokols D, Misra S, Runnerstrom MG, Hipp JA (2009) Psychology in an age of ecological crisis: from personal angst to collective action. *The American Psychologist* **64**, 181–193. doi:10.1037/a0014717

Sussman R, Gifford R (2013) Be the change you want to see: modeling food composting in public places. *Environment & Behavior* **45**, 323–343. doi:10.1177/0013916511431274

Uzzell DL (2000) The psycho-spatial dimensions of global environmental problems. *Journal of Environmental Psychology* **20**(4), 307–318. doi:10.1006/jevp.2000.0175

PART 2:

VULNERABLE POPULATIONS AND APPROPRIATE ADAPTATIONS

5

People with disability and their carers

Rae Walker

Key points

- There is limited evidence from research for how climate change affects people with disability and their carers, but there is a significant amount of evidence from grey sources such as United Nations agencies, international non-government agencies and global institutions such as the World Bank.
- The wellbeing of people with disability and their carers is impacted more by climate change than people who do not experience the inequities of this population group.
- The two social justice issues for people with disability and their carers exacerbated most by climate change are housing and community resilience. Improving the energy efficiency of housing and strengthening the resilience of communities are both important social justice initiatives with benefits to people with disability and their carers under conditions of climate change.
- Climate emergencies are the one impact of climate change on people with disability and their carers for which there is a body of research and grey literature. This evidence, as well as its implications for people with disability and for how community-based organisations should respond, is an important part of this chapter.

Introduction

For people with a disability and their carers, three major impacts of climate change are clearly visible now: heatwaves, social and economic changes, and extreme weather-related emergencies. In this chapter, a disability approach to inequality and an inclusive approach to extreme weather emergencies are discussed. Other changes in our climate and related social environment that impact on some people with a disability are discussed elsewhere in this book. Some people with disability are vulnerable in heatwaves. The discussion of heatwaves in Chapter 6 is relevant to them. Many people with disability live on low incomes and have limited resources to adapt their housing to keep it cool in hot weather and make it more energy efficient to reduce the energy costs. Women and children, especially women and children with disabilities, become vulnerable in and after extreme weather emergencies and other disasters. The issues for women and children in emergencies are discussed in Chapter 7.

There is very little research evidence about the ways climate change impacts people with disability. The only exception is the experience people with disability have of extreme weather emergencies. On this issue there is a small body of research available. Most of the information about climate change and disability is found in reports prepared by organisations such as the United Nations agencies, the World Bank (GPDD & World Bank 2009) and non-government disability organisations.

This chapter begins with a discussion of the health risks climate change creates for people with disability and their carers, followed by a brief discussion of the population of people with disability and the issues climate change raises for them. The next two sections discuss a social justice approach to climate change and disability action then an inclusive approach to emergency planning, response and recovery. People familiar with the content of the first sections may want to go directly to the final two where actions appropriate for the community-based health and social services sector are discussed.

Climate change and health risks

The major categories of health risk for people with disability from climate change are summarised in Table 5.1. The evidence suggests that heatwaves, extreme weather events, social and economic change, and health and social inequalities are particularly significant for people with disability and their carers. There is a frequently expressed argument that, in general, 'the poorest and weakest groups, such as elderly people, persons with disabilities, children and minorities would be most exposed to climate change consequences' (Costello *et al.* 2009, p. 1721; Garnaut 2008). The logic of the argument is that 'because climate change acts mostly as an amplifier of existing risks to health, poor and disadvantaged

Table 5.1.   Immediate risks to the health of people with disability from climate change

Main categories of risks to health	Elaboration on the risks*
1. Health impacts of extreme weather events (floods, storms, cyclones, bushfires, etc.)	Extreme events cause injury to people, damage to infrastructure (e.g. power, homes, community facilities, businesses and water services) and economic activity, leading to contamination and disease, social and economic dislocation and the mental health effects of trauma. People with disability and their carers are a population disproportionately affected by extreme weather events.
2. Consequences of social, economic and demographic dislocations and inequalities	Disadvantaged populations, including people with disability, have limited capacity to adapt to the changing climate. The health effects are more diverse than simply mental health impacts. For example, when support networks are disrupted and employment, housing and transport, in particular, become less enabling, the impact on capacity for independent living of people with disability can be substantial.
3. Health impacts of temperature extremes, including heatwaves	Heatwaves are becoming more common leading to increased morbidity and mortality. Effects vary with duration, timing in the season and vulnerability of the population. People who are very old, very young, frail or have limited capacity to modify their environment are most at risk. This includes people with disability.

* Sources: Department of Human Services (2007); Russell *et al.* (2009); Garnaut (2008); Centre for Risk and Community Safety & Bushfire CRC (2010); Horton *et al.* (2008).

people will experience greater increments in the disease burden than rich, less vulnerable populations' (Costello *et al.* 2009, p. 1712). People with disability and their carers share many risks with other disadvantaged populations. However, there is evidence that risks to people with disability and their carers in the three climate impact categories in Table 5.1 are quite specific.

Definitions of disability

There are several definitions of disability in current use. The Australian Bureau of Statistics (ABS) collects data for decision-making by government and other institutions. Its definitions of disability focus on functional limitations in daily life. For the national *Survey of Disability, Ageing and Carers* disability is defined as 'any limitation, restriction or impairment which restricts everyday activities and has lasted or is likely to last for at least six months' (ABS 2010). Using this definition 18.5% of the population has a disability.

In the national census, where the intent is to measure the number of people requiring assistance with core daily activities, the concern is with the population with profound and severe disability. In this case disability is defined as 'those people needing help or assistance in one or more of the three core activity areas of self-care, mobility and communication, because of a long-term health condition (lasting six months or more), a disability (lasting six months or more), or old age' (ABS 2011). Using this definition 5.8% of the population has a disability. Although both definitions used by the ABS focus on function, they in fact measure different populations, making it important to identify which definition is used when referring to ABS statistics.

International organisations are more likely to focus on the human rights of people with disability and adopt a definition that reflects this. The World Health Organization (WHO 2011, p. 4) definition used in the *World Report on Disability* says: 'Disability is the umbrella term for impairments, activity limitations and participation restrictions, referring to the negative aspects of the interaction between an individual (with a health condition) and that individual's contextual factors (environmental and personal factors)'. All of these definitions share a focus on peoples' capacities to function in their context, but they differ in the ways they qualify that focus.

Population of people with disability and their carers

The national *Survey of Disability, Ageing and Carers* found that 18.5% of the population have a disability (ABS 2010). The disability creates specific limitations or restrictions for 16% of the population and no specific limitations or restrictions for 2.5% of the population. Of the 14.4% of the population that have limitations on core activities of self care, mobility and communication, 5.8% have severe to profound disability and 8.6% have moderate to mild disability (ABS 2010). People with profound or severe disability have the greatest need for assistance.

The proportion of the population with a disability increases with age. Of people aged four years and under, 3.4% have a disability. In the 65–69 years age group 40% have a disability, and of those over 90 years, 88% do. The causes of physical disability impacting core activities are diverse and include musculoskeletal conditions and disease of the neurological, respiratory or circulatory systems (ABS 2010). For many of these people heatwaves and other extreme weather events have significant impacts on wellbeing and capacity to cope with their environment.

Carers of people with disability, or of people aged 60 years or over, comprise 12% of the Australian population with slightly more females than males in the carer role (ABS 2010).

The most common forms of disability vary with age. Intellectual disability is most common (4%) for children under 15 years, while for people aged 15–64 years physical disability is most common (11%). For those aged 65 years and over the most common is a physical disability (40%), followed by sensory limitations and speech impairment (25%) (VicHealth 2012, p. 2). The forms of disability in a population have many implications for agencies including climate change communications and education of clients, carers and communities in regard to heat, emergencies and access to adaptation resources.

Issues for people with disability and their carers

People with disability and their carers are one of the most disadvantaged populations in Australia (VicHealth 2012, p. 4), making them especially vulnerable to the impacts of climate change (Polack 2008). The experience of disability follows from the 'dynamic interaction between health conditions and contextual factors both personal and environmental' (WHO 2011, p. 4) and varies enormously between individuals. While 'disability correlates with disadvantage, not all people with disability are equally disadvantaged' (WHO 2011, p. 8). Women, people with mental or intellectual impairments, or people with severe disabilities often experience greater disadvantage. People with 'wealth and status can … overcome activity limitations and participation restrictions' (WHO 2011, p. 8). Although we speak about people with disability as a population, we need to remain aware that its members have diverse experience and needs.

Of people with disability in Australia, 45% live in or near poverty and have an income that is only 70% of that of a comparable person without disability. The income of carers is a little over half that of people who are not in a caring role (VicHealth 2012, p. 5). People with disability have lower levels of education, are more likely to be unemployed and to work part time than people without disability, and the unemployment rates are higher for disabled women than for disabled men. The majority of people with disability (94%) live in households (74% with other people and 20% alone). They are more likely than the population at large to own their own house, a consequence of increasing prevalence of disability with age, but if they rent it can be difficult to find appropriate housing that is affordable, suitable and secure. People with disability with a low income are twice as likely to be in public housing as people without disability (ABS 2010). People with disability are more likely than others to experience violent crime, especially women with intellectual disability (VicHealth 2012). These disadvantages in daily life make it more likely that people with disability will be socially isolated, especially if they have an intellectual disability (VicHealth 2012). Over 20% of people with disability access the internet (in comparison to over two-thirds of non-disabled Australians) but this proportion declines for people with profound disability. The most common location of the computer is at home (ABS 2010).

The breadth and depth of social disadvantage reduces the adaptive capacity of people with disability and their carers in regard to climate change. For example, if they live in older houses income constraints are likely to make it hard to retrofit the building to conserve water and energy, and to maintain comfortable temperatures. Market-based adaptation policies (such as raising prices of utilities to reduce consumption) are likely to have a disproportionate negative impact on this population.

Of the 18.5% of the Australian population with a disability in 2009 (ABS 2010), over one-fifth required assistance (in descending order of frequency) with (1) property maintenance, (2) cognitive/emotional issues, (3) housework, (4) mobility, (5) transport and (6) health care. The need for assistance increased with the severity of the disability. Those with mild to moderate disability required most assistance with property maintenance and household chores, while those with severe and profound disability required most assistance with core activities of mobility (72–88%), self care (39–66%) and communication (13–29%) (ABS 2010). These needs are relevant to three major climate change issues currently impacting Australians (listed in Table 5.1). For example, household maintenance is an entry point for enhanced energy efficiency and the maintenance of temperature in very hot and very cold weather. Communication, transport and personal mobility are directly relevant to emergency planning, response and recovery from extreme weather events.

Disability advocates argue that people with a disability suffer a high rate of death and injury in emergencies but that the evidence to quantify this is rarely available. The inquiries that followed the catastrophic Victorian bushfires on 7 February 2009 that killed 173 people addressed this issue. In preparation for the coronial inquiry each death was investigated in detail. The file summaries were analysed and a report prepared for the subsequent Victorian Bushfires Royal Commission. O'Neill and Handmer (2012, p. 4) observe that:

This fatality dataset highlighted how many of the fatalities (44%) were particularly vulnerable due to age (either 70 or over, or under 12) and/or had a chronic and/or acute disability. Note that these vulnerabilities were sometimes compounded – 2% of fatalities had both a chronic and an acute disability; and a further 9% had a chronic disability and were 70 or over.

In the report to the Royal Commission the authors found that 24% of people who died had a chronic disability and 5% an acute disability. If the acute and chronic disability categories are added together, and those with both counted only once, 27% of people who died on Black Saturday had a disability at the time of the fire. However, the proportion of fatalities with acute or chronic disability was probably underestimated as not all case records contained detailed medical reports (Centre for Risk and Community Safety & Bushfire CRC 2010). In this bushfire, the proportion of people with disability who were fatalities exceeds what would be expected from their representation in the population as a whole. Also, the severity of peoples' disability is not reported, nor is the form they took, making it difficult to assess degrees of risk within the population of people with disability.

People with disability often use several sources of assistance but 87% use informal help (typically from family members) and 59% receive help from formal providers that might be government, not-for-profit or for-profit agencies (ABS 2010). Under conditions of climate change, the sources of support become critical when planning for, coping with and recovering from emergencies. For example, assistance providers need to be engaged in the preparation of personal and household emergency plans for them to be effective. If providers of assistance leave the neighbourhood of a person receiving help, or if the person with disability is evacuated leaving behind providers of assistance, their wellbeing and capacity to lead a dignified and independent life can be compromised. This problem becomes worse if response and recovery services are not designed to be inclusive of people with disabilities.

A social justice approach to climate change and disability

Action on climate change adaptation and mitigation has to be closely connected to issues of social justice. We know that, 'the most vulnerable groups, by lacking a voice and influence in climate change policymaking, are unlikely to receive the support they need as policies are less likely to account for their particular experiences' (Polack 2008, p. 17). People whose rights are poorly protected are likely to be less able to adapt to a changing climate and social environment (Von Doussa 2008; Preston *et al.* 2014), increasing social and health inequalities. People with disability and their carers are a group at risk. A social justice approach to climate change directs attention to the impacts on, and actions of, people and communities. '*Who* is likely to suffer *what* and *why*?' (Von Doussa 2008, p. 2). Appropriate action is about minimising climate risks and increasing the capacity of people to respond appropriately to, for example, rising temperatures and extreme weather.

The principles of capacity building approaches are explained in Sen's capability theory (Ruger 2004). From this perspective justice 'requires improvement of the conditions under which individuals are free to choose healthier life strategies and conditions for themselves and for future generations. A capability perspective emphasises the empowerment of individuals to be active agents of change in their own terms – both at the individual and collective level' (Ruger 2004, p. 1094). When people with disability have the capabilities to act themselves, individually or together, to improve their wellbeing and freedoms, their human rights and social justice can be realised (WHO 2011). The capability approach is particularly relevant to local-scale, community-based adaptation to climate change (Bell & Blashki 2014).

The capacity building or capability approach to justice requires organisations to create environments that enhance the capability of people with disability. For people with disability environments include accessible built environments and transport, signage and communications for people with sensory impairments, policies and practices that make services such as health, education and employment accessible, and changing knowledge and attitudes that exclude or diminish the self esteem of people with disability (WHO 2011). People with disability share issues and environments with other disadvantaged population groups. Initiatives that address the needs of disadvantaged people in general have benefits for many people with disability. People with disability also have specific needs, such as for enhanced communication, that may provide support to other segments of the community.

In an Australian study of climate change impacts on disadvantaged populations, Sevoyan *et al.* (2013) found:

- multiple disadvantage is common (e.g. disability plus low income)
- the greatest economic problem is the growing cost of utilities
- social networks do not compensate for lack of economic resources
- levels of social support and social participation are very low, especially in metropolitan areas
- extreme weather events are much harder for disadvantaged households to cope with than those which are not disadvantaged
- people in disadvantaged households were less aware of energy efficiency, to reduce household bills, than other households.

The two most important areas of action for disadvantaged populations related to climate change are housing improvements and strengthening community readiness.

Housing improvements can be achieved by strengthening regulations for rental properties, providing incentives for landlords to improve rental housing stock and retrofitting established housing stock. A high proportion of the population of people with disability are home owners, due to the increasing rate of disability with age. For this population, support for retrofitting is appropriate. Many renters in the disability population are in public housing, and so benefit from improvements in public housing stock.

Strengthening community readiness can occur through householder education and support for community service organisations to develop appropriate services to meet community need; engagement of disadvantaged people in adaptation planning; supporting community service organisations and local government to develop 'community connectedness, social capital and social support within their communities'; and supporting development of 'appropriate and affordable home and contents insurance products for low-income Australians' (Johnson 2012, p. 7).

The idea of mainstreaming many, but not all, initiatives to address needs of people with disability is also captured in the idea of universal design, which is discussed later in this chapter. The capability approach is often described in the emergency management literature when people with disability are included in emergency planning processes. Universal design principles are incorporated in, for example, emergency communication, and a dual track (sometimes called twin track) approach to recovery is used. A dual track approach includes accessible mainstream communication, facilities and services to meet basic needs and services provided by specialist disability organisations to meet specialist needs. Dual track approaches require partnerships between emergency planners and organisations serving people with disability.

There are social justice issues in the approaches communities take to emergency planning and response. People with disability and their carers may encounter problems in emergency situations such as 'unequal access to assistance; discrimination in aid provision; enforced relocation; sexual and gender based violence; loss of documentation; unsafe or involuntary return or resettlement; and, issues of property restitution' (HREOC 2008, p. 18–19). Emergencies also disproportionately affect a number of disadvantaged groups, and response efforts not recognising this increase their vulnerability. Disaster response, recovery and reconstruction should be inclusive of disadvantaged population groups, and value fairness and equity in actions. The high risk of mortality indicates that particular attention should be paid to people with disability.

A functional approach to emergencies and disability

From a functional perspective, people with disability are those with one or more activity limitations, such as 'reduced capacity or inability to see, lift, walk, speak, hear, learn, understand, remember, manipulate or reach controls, and/or respond quickly' (Kailes & Enders 2007, p. 233). A functional approach shifts the focus from diagnoses to people and their capacity to understand and act within their environment, and to modification of the environment to accommodate people with diverse capacities.

The climate change issue that has attracted most discussion in the climate change and disability area, in both the research and grey literature, is extreme weather events (and other emergencies) and their implications for people with disability and their carers. Such events are relatively common and life threatening. Some research on people with disability and their issues has been published, and overseas non-government organisations have made it a priority issue (e.g. NCD 2006; NOD 2009). The Australian Red Cross has prepared detailed guidance on emergency preparedness in the form of the *Emergency REDiPlan: Household preparedness for people with a disability, their families and carers*, which is available on the resources page of their website (www.redcross.org.au/emergency-resources.aspx).

Reasons disability organisations are concerned about emergencies

A consortium of international disability organisations summarised the experience people with disability have in emergencies:

> *Persons with disabilities are often literally and programmatically 'invisible' in the emergency response. They are excluded from or unable to access mainstream assistance programs as a result of attitudinal, physical, environmental and social barriers; they are often forgotten in the establishment of services specifically targeted for vulnerable groups; they are at risk of worsening their impairment or developing others due to lack of access to appropriate food, non-food items, health services, etc. which may even be the cause of death. (DPI et al. n.d., p. 2)*

A systematic literature review of research and grey literature on the Asian tsunami and Hurricane Katrina reached a similar conclusion. It showed that 'disabled people are at greater risk of injury, mortality, disease, destitution and displacement when compared with the general population' (Hemingway & Priestley 2006, p. 60). The barriers to survival and relief following emergencies reflect the barriers experienced by people with disability in everyday life (Hemingway & Priestley 2006, p. 60).

Two surveys of emergency planning personnel in the United States found little evidence of awareness of, or preparation for, meeting the needs of people with disability in emergencies. In one study no agencies had policies or guidelines designed to meet the needs of people with mobility impairments. All informants said they used the same policies to deal with everyone (they were not referring to universal design). Rural emergency personnel relied on advice from members of informal networks of people with disability to decide how they could be helped. Urban informants did not even know how many people with mobility impairments lived in their catchments. All informants expressed willingness to include people with disability in future planning but did not know how to do so (Rowland et al. 2007). In Australia emergency planners are required to take account of 'people with special needs' but the ways they are required to do this are limited. An analysis of submissions to the Queensland Floods Commission of Inquiry (into the 2010–11 floods) found that submissions from groups such as those involved in local disaster management failed to show substantial knowledge of flood management issues for vulnerable population groups, including people with disability (Bell & Blashki 2014).

A survey of emergency planners in 30 counties in the United States that had experienced disasters in the past five years assessed their preparedness for meeting the needs of people with disability. Only 6.9% of counties made changes post disaster to better meet the needs of people with disability. Of these changes, 60% were made in response to federal government requirements, while only 29% of changes were based on learning from the disaster. People with disability were included in planning in 13% of counties. Of county emergency managers, 57% did not know how many people with mobility impairments lived in their county, and of those who did 'know', most made broad estimates based on unreliable data or best guesses. Only 20% of counties had disaster guidelines for people with disability and operating procedures to follow the guidelines. The reasons for not having guidelines were cost and lack of trained personnel. As for prioritising guidelines for people with disability, 17% of counties said that public education was required before they would make guidelines for people with disability a priority and 25% would only do it if required by state level authorities (Fox *et al.* 2007). We do not have comparable evidence for Australia and we do need to ask how different, or similar, we are to the United States in this regard.

No Australian documents summarising the status of people with disability in emergency plans and planning have been found. But an online keyword search of a small number of Emergency Management Australia documents (using the term 'people with disability') found that in some documents there were no hits and in others the discussion was about populations with special needs, including people with disability, was very general. Some initiatives, such as registers of people at risk in emergency situations, have been established in some Australian jurisdictions. These initiatives, however, are very limited when compared to the responses to the needs and capabilities of people with a disability described in the literature.

Issues for people with disability in emergencies

Major issues for people with disability during emergencies are not always captured in the statistics. The experience of disability provides a deeper understanding of issues and required actions. For example, the National Organization on Disability (NOD 2009, p. 17) analysed the experience of people with disability during emergencies and developed a list of questions for organisations to consider when developing inclusive planning approaches. These questions are:

- What is it like to be a person with a disability during and after an emergency?
- Can one hear or understand the warnings?
- Can one quickly exit a home or workplace?
- Can one move about the community after evacuating?
- Are there necessary or even vital daily items (medicines, power supplies, medical devices) that are not likely to be available in emergency shelters?
- Are basic services, like rest rooms and showers, available and accessible to people with disability?
- Does the person require assistance from a caregiver?

There are nine categories of issues that create difficulties for people with disability and their carers in emergencies. Typically these issues are reported by disability organisations that have been involved in emergency response, in collating post-emergency evidence, or who report the limited research that is available. These categories are:

1. *General issues.* Movement and access to documentation is difficult. People become vulnerable to physical, sexual and emotional abuse, especially women and children with disabilities. Isolated people left behind have difficulty contacting tracing programs.
2. *Evacuation in an emergency.* Evacuation issues include inaccessible escape routes and loss of assistive technology. People providing physical assistance flee, leaving disabled people behind or carers stay, disrupting families. When social networks of people with disability are disrupted during an evacuation it has many impacts

including access to shelter, food, water and services. Where data has been collected it shows that disabled people's lives were put adversely at risk, not simply by individual limitations but by social and environmental factors. These included the 'vulnerability of buildings and facilities used by disabled people, an absence of specific evacuation plans, inaccessible warning information, lack of accessible evacuation transport, failure of backup systems (including power failures) and sometimes, the actions of neighbours, staff and rescue workers' (Hemingway & Priestley 2006, p. 61). Furthermore, 'emergency planning should include criteria for recognising the various forms of evacuations and for evaluating a proper response' (Christensen *et al.* 2007, p. 252).

3. *Information*. Because data on people with disability in communities is poor, people with disability are not included in registration systems. Furthermore, disaster relief personnel are poorly informed of options for people with disability. Communications are often inaccessible for people with sensory impairment (White 2006).

4. *Food aid and nutrition*. People with disability experience barriers to accessing food distribution points, long wait times, no special diets for people with special dietary needs, difficulty carrying home food rations, and food for work schemes that discriminate.

5. *Water and sanitation*. People with disability experience difficulty collecting and carrying water, and inaccessible toilet and bath areas.

6. *Shelter accessibility*. People most likely to experience shelter accessibility problems were wheelchair users, and 'people with visual, hearing and cognitive impairment' (Hemingway & Priestley 2006, p. 61). Sometimes people with disability were turned away from relief camps and encouraged to access specialist services which were not always able to respond appropriately. There are often few options for temporary accessible housing post emergency (Rooney & White 2007).

7. *Health care*. People with disability experience disruption to regular health care services and encounter services that are unable to meet disability-related needs.

8. *Livelihood*. People with disability experience loss of special tools needed to earn a living and obstacles to accessing new livelihood activities. They have problems returning to daily routines due to post disaster emotional trauma, lack of mobility and problems with the affordability and accessibility of temporary housing (Rooney & White 2007, p. 209).

9. *Participation*. People with disability also experience systematic exclusion from evaluations, assessments, interviews, community committees and leadership roles, thus reinforcing discrimination (DPI *et al.* n.d.; Hemingway & Priestley 2006, p. 61).

Inclusive planning to meet functional needs

Emergency planning needs to move away from lists of 'vulnerable persons' (i.e. persons with impairments) to 'vulnerable situations' that people move in and out of over time (Fjord & Manderson 2009, p. 67). From this point of view, an emergency is a situation in which the environment has changed to such a degree that the resources a person with disability uses may no longer be adequate to meet his or her needs.

People with disability can be considered those with one or more activity limitations (Kailes & Enders 2007, p. 233). Viewing disability in this way leads to the inclusion of people with functional limitations that are not normally labelled as 'having a disability', for example, people with heart disease, respiratory disease, emotional or psychiatric conditions, arthritis or asthma. It can also include temporary limitations on function resulting from, for example, accidents or injuries (sprains, broken bones) and people with functional limitations due to the disaster itself (Kailes & Enders 2007). 'In disaster management activities, it is important to think broadly about disability in terms of function and not in terms of an impairment or diagnosis. Traditional narrow definitions of disability are not appropriate' (Kailes & Enders 2007, p. 233). The emphasis in emergency planning should be on meeting function-based needs (NOD 2009).

Inclusive planning to meet functional needs of diverse populations is the most efficient and effective approach to emergency planning.

Adequately addressing functional support needs has a far greater impact on how well individuals survive than any specific diagnosis … By planning for people with functional needs, an operational set of predictable supports can be developed. A functional support framework provides for commonalities in planning among a large array of impairment types. This framework provides a way to operationalize support for functional needs and activity limitations that may be the same, even though the impairments may be very diverse. (Kailes & Enders 2007, p. 233)

A function-based framework for emergency planning has the following characteristics (NOD 2009):

- Communication is accessible to people who cannot see, hear or understand traditional communications, including people whose English is limited. Communication refers to 'languages, display of text, Braille, tactile communication, large print, accessible multi-media as well as written, audio, plain-language, human-reader and augmentative and alternative modes, means and formats of communication, including accessible information and communication technology' (United Nations 2007).
- Diverse medical needs are planned for including unstable medical conditions, infectious conditions, treatments and life support equipment.
- Planning assumes the maintenance of survivors' functional independence. Maintenance of support networks is critical for the continuity of independent living by people with disability. If an emergency disrupts the social networks, a disabled person may end up in inappropriate residential accommodation (Priestley & Hemingway 2007, p. 29). Early screening for functional needs allows them to be met, if necessary through specialist disability organisations. This enables people to 'maintain health, mobility and functional independence, as well as manage in mass shelters' (Kailes & Enders 2007, p. 235).
- Planning takes into account the needs of some people for supervision. A variety of reasons may lead to some people having difficulty functioning in a new or stressful environment. It may include some individuals with, for example, dementia, severe depression, schizophrenia or intellectual disability, and prisoners and unaccompanied children.
- Diverse transportation needs are included in planning. These needs include wheelchair-accessible vehicles, and affordable transport for people with little money and people who cannot drive for some reason.

Interagency partnerships are critical for good emergency planning. Organisations that specialise in the provision of services to people with disability need to be involved in emergency planning, preparedness, response, recovery and prevention activities. Local disability services are likely to respond differently, but more effectively, than many mainstream services to meet the needs of people with disability (Hemingway & Priestley 2006). Partnerships can be very effective at generating synergies between the diverse resources and capacities of organisations in the health and social services system, and of integrating the systems, but skilled management is necessary (Lasker *et al.* 2001; Lasker & Weiss 2003). In Victoria, the network of Primary Care Partnerships has developed high-level skills in this work and could take an enabling role in partnership development. In other states these skills sit with other institutions.

Following a major review of emergency planning, management and response, and the experience of people with disability in emergencies, the National Organization on Disability (NOD 2009, p. 23–24) defined the role of disability organisations in emergency planning and response as being to:

- identify those in the community who might have functional needs before, during, and after an emergency
- customise awareness and preparedness messages and materials for specific groups of people and put them in alternative and accessible formats, thereby increasing the ability of these individuals to plan and survive in the event of an emergency
- educate citizens with disabilities about realistic expectations of service during and after an emergency, even while demonstrating a serious commitment to their functional needs
- learn and gain from the knowledge, experiences and non-traditional resources that the disability community can bring to a partnership effort with emergency professionals
- work with institutional and industry-specific groups that are not typically considered emergency service resources but that can offer valuable and timely support to emergency professionals.

The National Organization on Disability defines a specific role for disability organisations that enables them to make a unique contribution to emergency planning. For their contribution to be effective and improve survival rates and wellbeing of people with disability, the mainstream planning process needs to operate in an inclusive way (Kailes & Enders 2007, p. 236) and:

- provide services that are inclusive and accessible to people with functional limitations
- employ some people with functional limitations in appropriate emergency roles (to capitalise on their knowledge and relationships with their communities)
- develop partnerships with community-based disability organisations that are connected to, and trusted by, their communities
- include community-based organisations in disaster planning and response, and develop agreements to include their strengths and skills in plans
- include a functional support coordinator in emergency shelters and recovery centres

- include screening questions in emergency shelter and recovery centre intake registration processes
- train emergency people in issues of functional need.

Planning that modifies environments to make them inclusive for people with diverse abilities is described in the universal design principles that offer an affordable way of increasing access to services, information and infrastructure. Universal design means 'the design of products, environments, programs and services usable by all people, to the greatest extent possible, without the need for adaptation or specialised design. Universal design shall not exclude assistive devices for particular groups of persons with disabilities where this is needed' (United Nations 2007). Universal design creates environments that are equally accessible for all people. The application of universal design can be critical for emergency evacuation of buildings and can enable communications during evacuations if new technologies are available that help people with sensory and cognitive impairments to keep informed (WHO 2011, p. 184). Disabled people's organisations are advocating for universal design principles to be included in information communication technologies developed by the emergency planning sector to facilitate communication for people with diverse impairments (WHO 2011, p. 191).

In the 1970s a group of architects, engineers, product designers and environmental designers from the Center for Universal Design at North Carolina State University developed a set of seven principles to guide the practice of inclusive design and planning (Universal Design n.d.). These principles are:

- *Equitable use.* The design is useful and marketable to people with diverse abilities.
- *Flexibility in use.* The design accommodates a wide range of individual preferences and abilities.
- *Simple and intuitive use.* Use of the design is easy to understand, regardless of the user's experience, knowledge, language skills or current concentration level.
- *Perceptible information.* The design communicates necessary information effectively to the user, regardless of ambient conditions or the user's sensory abilities.
- *Tolerance for error.* The design minimises hazards and the adverse consequences of accidental or unintended actions.
- *Low physical effort.* The design can be used efficiently and comfortably with minimum fatigue.
- *Size and space for approach and use.* Appropriate size and space is provided for approach, reach, manipulation and use, regardless of user's body size, posture or mobility.

The United States Federal Emergency Management Agency (FEMA) has adopted these principles and recommends that agencies use them to implement an inclusive approach to emergency planning to improve outcomes for the whole community. For a summary of FEMA's approach to inclusive emergency planning and links to resources visit the universaldesign.com website (www.universaldesign.com/2012-06-11-16-51-43/disability/1428-emergency-preparedness-for-people-with-disabilities-from-fema).

The application of universal design means that people with disability are able to use the same facilities and resources as the rest of the population before, during and after an emergency. In an accessible environment people with disability maintain their independence and are not directed to acute care facilities unnecessarily.

Key areas of action
Risk communication

An evaluation of general health promotion materials for people with disability (Williams-Piehota *et al.* 2010) found that the major sources of information in the United States were healthcare providers, friends and family, the internet and print materials. A service type not identified in this study, but one we know is most likely to be trusted by people with disability, is disability service providers. In the absence of evidence to the contrary, we should assume that these service providers are also likely sources of health promotion information in Australia and are appropriate sources of information on climate change and human health and wellbeing. The study authors recommend that the disability audiences be segmented by functional need, materials be tailored to each audience, language be simple and direct, materials delivered in multi-modal format, materials be pre-tested several times during development, and preferred communication channels used. Critically, people with disability needed to be involved in the production of materials (Williams-Piehota *et al.* 2010).

Alerts and emergency warnings

Pre-hurricane warnings in the United States were accessible to most people via television. When the hurricanes struck, emergency warnings were often inaccessible to people with sensory impairments due to the absence of sign interpreters or other communication aids. This is also a problem in everyday life. 'As barriers to the daily use of information and telecommunications technology decrease, so will the barriers to emergency communication' (NCD 2006, p. 4). People with hearing impairments could not comprehend evacuation instructions or other instructions in shelters. People with visual impairments could not comprehend information on television because audio descriptions of visual displays of maps or lists of affected areas were not provided. People who tried to use alternative communications such as cell phones were unable to access information when infrastructure failed. Radio was the only remaining communication technology but that failed people with hearing disability. It is recommended that daily media routinely adopt 'standardized methods, systems and services to identify, filter and present content in ways that are meaningful to people with disability leading up to, during and after emergencies' (NCD 2006, p. 6). Furthermore, a wide variety of media should be used for emergency information. The principles of universal design should inform risk communication to ensure that communications are accessible to as many people as possible.

Population movement
Evacuation procedures

Christensen *et al.* (2007) classify evacuations into four categories, each with different implications for people with disability:

1. *Protective evacuations.* Long-term, pre-impact responses to emergency situations (e.g. individual precautionary evacuation)

2. *Preventive evacuations.* Short-term, pre-impact responses (e.g. a bomb threat requiring movement to another part of a building)
3. *Rescue evacuations.* Short-term, immediately post-impact evacuation (e.g. evacuating a burning building)
4. *Reconstructive evacuations.* Long-term, post-impact in the context of 'impaired health and safety' (e.g. moving evacuees to camps or temporary housing).

Each kind of evacuation has implications for people with disability. For example, a person with a disability may be able to act appropriately in long-term evacuations but not in a short-term one due to their need for time to negotiate obstacles. Similarly, sheltering in place for people with disability (e.g. in a protected stairwell to await assistance from an emergency responder) may be appropriate for a preventive evacuation, but may create serious risks in a rescue evacuation. In general, reliance on planned systems for evacuation leaves people with disability vulnerable. More attention needs to be given to the built systems that modify environments and reduce obstacles to evacuation.

Emergency transportation

Before Hurricane Katrina, many people with disability in the United States were unable to evacuate because transport was inaccessible either in its location or design. Following an earthquake in Japan it was difficult to locate people with disability in order to provide transport. Registration that included home addresses could be found with some disability service provider organisations, private or municipal, but they could not be shared with emergency responders (Takahashi *et al.* 1997). The transportation barriers (related to Hurricane Katrina) were 'magnified versions of daily barriers to accessible transportation' (NCD 2006, p. 8). The emergency planners had either not considered the needs of people with disability or their arrangements did not work because they had not consulted people with disability to arrive at realistic options. Successful hurricane evacuation plans included:

- additional planned pick-up routes
- extra time to load and unload evacuation vehicles
- appropriate resources (e.g. buses with wheelchair lifts)
- clear articulation of methods for evacuating people with disability including the roles of schools, emergency managers and transport agencies
- encouragement of voluntary registration of people with disability
- practice of evacuation plans.

Emergency mass shelter and food

Some initial challenges faced by people with disability are inherent to any disaster response (e.g. confusion, shortage of trained people). However, many of the most significant problems could have been addressed with inclusive planning. Some people were refused entry to shelters and others could not access shelter services (e.g. medical care, restrooms, food, communication and transport services). People with intellectual disabilities found shelter environments distressing (Takahashi *et al.* 1997). People with disability who live independently in the community should be able to access general shelters. The most successful general shelters were not run by emergency personnel but by local people who practiced inclusiveness.

Continuity of services

For people with disability the requirements for daily living need to be maintained or restored as soon as possible after an emergency. Continuity of service is one way of describing the maintenance or restoration of requirements for daily living. The National Council on Disability (NCD 2006, p. 17) identified the requirements as being:

- affordable, appropriate housing
- accessible, affordable, reliable, safe transportation
- physical environment adjusted for inclusiveness and accessibility
- work, volunteer and education opportunities
- access to key health and support services
- access to civic, cultural, social and recreational activities.

Community-level action

Effective emergency planning necessarily occurs at multiple levels, including state, regional, municipal and local networks or communities of residents. The National Council on Disability (NCD 2006, p. 23) and Fox *et al.* (2007) recommend that local communities and municipalities:

- include people with disability in planning at all levels and as trainers in emergency planning
- ensure coordination of emergency plans
- develop a communication plan to inform people with disability about emergency plans
- establish an office or person with sole responsibility for disability issues
- raise awareness of people with disability of the need for personal and community-level disaster planning
- integrate emergency planning systems with healthcare providers and community-based organisations
- improve surveillance systems to identify people with disability by using registries and list of people known to disability organisations
- develop warning systems using multiple communication technologies
- work with disability organisations to promote personal disaster planning.

It is important that people with disability prepare personal emergency plans. Rooney and White (2007) identified the following set of self-help actions people with mobility impairments can take to improve their survival chances in an emergency:

- Prepare supplies, including general supplies, disability-related supplies, equipment and medication.
- Prepare evacuation plans, pre-register for emergency assistance and undertake emergency preparedness training.
- Prepare a supply of items that are useful to have after evacuation, including food, water, generators, cash and specialised mobility or medical equipment.
- Plan ahead, including self-assessments of risks, capacity to respond, emergency needs supplies and equipment.

- Build support networks of co-workers, family and friends that could provide assistance. The most highly rated support services in emergencies are family, friends, neighbours and police.

The Australian Red Cross has prepared personal emergency planning resources for people with disability, which is available on their website (www.redcross.org.au/emergency-resources.aspx).

In the recovery period post-emergency it is common for case management systems to be used to assist people with complex needs. Case management for people with disability needs to be qualitatively different from that with non-disabled people. It is more intense and with more contacts, of longer duration, and more complex. The complexity of the cases requires case managers to have expertise in the disability service area (Stough *et al.* 2010). People with disability share issues such as housing, transport and long recovery time with non-disabled people. However, the barriers for people with disability were much greater making recovery particularly difficult for many people with disability (Stough *et al.* 2010).

Conclusion

People with disability will almost certainly experience heightened disadvantage as climate change impacts strengthen and communities need to make increasingly substantial adaptations. Extreme weather events are an impact of climate change currently affecting populations across Australia. This is the one area where a small amount of research has been undertaken to understand the experience of people with disability. To address the needs of people with disability a social justice approach to climate change adaptation in general is needed. It also requires that mainstream emergency planning becomes more inclusive. Disability organisations have a role in aiding the adaptation of people with disability as well as working with mainstream emergency organisations to develop a dual track approach to emergency planning and response.

References

ABS (Australian Bureau of Statistics) (2010) *Disability, Ageing and Carers, Australia: Summary of Findings, 2009.* Cat. no 4430.0. Australian Bureau of Statistics, Canberra.

ABS (Australian Bureau of Statistics) (2011) *Census Dictionary, 2011.* Cat. no. 2901.0. Australian Bureau of Statistics, Canberra. <http://www.abs.gov.au/ausstats/abs@.nsf/mf/2901.0>

Bell E, Blashki G (2014) A method for assessing community flood management knowledge for vulnerable groups: Australia's 2010–2011 floods. *Community Development Journal* **49**, 85–110. doi:10.1093/cdj/bst002

Centre for Risk and Community Safety & Bushfire CRC (2010) 'Review of fatalities in the February 7, 2009, bushfires – final report'. Centre for Risk and Community Safety, RMIT University and Bushfire CRC, Melbourne.

Christensen KM, Blair ME, Holt JM (2007) The built environment, evacuations, and individuals with disabilities: a guiding framework for disaster policy and preparation. *Journal of Disability Policy Studies* **17**, 249–254. doi:10.1177/10442073070170040801

Costello A, Abbas M, Allen A, Ball S, Bell S, Bellamy R, Friel S, Groce N, Johnson A, Kett M, Lee M, Levy C, Maslin M, McCoy D, McGuire B, Montgomery H, Napier D, Pagel C, Patel J, de Oliveira JAP, Redclift N, Rees H, Rogger D, Scott J, Stephenson J, Twigg J, Wolff J,

Patterson C (2009) Managing the health effects of climate change. *Lancet* **373**, 1693–1733. doi:10.1016/S0140-6736(09)60935-1

Department of Human Services (2007) 'Climate change and health: an exploration of challenges for public health in Victoria'. Victorian Government Department of Human Services, Melbourne.

DPI, Handicap International, IDDC, Motivation, Women's Refugee Commission, World Vision (n.d.) 'For a UNHCR Executive Committee Conclusion on Disability. Joint submission'. International Disability and Development Consortium, Brussels, <http://iddcconsortium.net/sites/default/files/resources-tools/files/final_emergency-tg_lobby_paper_unhcr_disability.pdf>

Fjord L, Manderson L (2009) Anthropological perspectives on disasters and disability: an introduction. *Human Organization* **68**, 64–72.

Fox MH, White GW, Rooney C, Rowland JL (2007) Disaster preparedness and response for persons with mobility impairments: results from the University of Kansas Nobody Left Behind study. *Journal of Disability Policy Studies* **17**, 196–205. doi:10.1177/10442073070170040201

Garnaut R (2008) *The Garnaut Climate Change Review: Final Report*. Cambridge University Press, Port Melbourne.

GPDD (Global Partnership for Disability & Development), The World Bank (2009) 'The impact of climate change on people with disability. <https://www.ucl.ac.uk/lc-ccr/centrepublications/staffpublications/Impact_of_Climate_Change_on_Disability-Report-2010.pdf>

Hemingway L, Priestley M (2006) Natural hazards, human vulnerability and disabling societies: a disaster for disabled people. *The Review of Disability Studies* **II**, 57–68.

Horton G, McMichael T, Doctors for the Environment (2008) 'Climate change health check 2020'. The Climate Institute, Sydney.

HREOC (Human Rights and Equal Opportunity Commission) (2008) 'Human rights and climate change: background paper'. Human Rights and Equal Opportunity Commission, Sydney.

Johnson V (2012) 'Barriers to effective climate change adaptation: submission to the Productivity Commission Inquiry'. Brotherhood of St Laurence, Melbourne.

Kailes JI, Enders A (2007) Moving beyond 'special needs'. A function based framework for emergency management and planning. *Journal of Disability Policy Studies* **17**, 230–237. doi: 10.1177/10442073070170040601

Lasker RD, Weiss ES (2003) Creating partnership synergy: the critical role of community stakeholders. *Journal of Health and Human Services Administration* **26**, 119–130.

Lasker RD, Weiss ES, Miller R (2001) Partnership synergy: a practical framework for studying and strengthening the collaborative advantage. *The Milbank Quarterly* **79**, 179–205. doi:10.1111/1468-0009.00203

NCD (National Council on Disability) (2006) 'The impact of hurricanes Katrina and Rita on people with disability: a look back and remaining challenges'. National Council on Disability, Washington D.C.

NOD (National Organization on Disability) (2009) *Emergency preparedness initiative: Guide on the special needs of people with disability for emergency managers, planners and responders. Revised edition.* National Council on Disability, Washington D.C.

O'Neill SJ, Handmer F (2012) Responding to bushfire risk: the need for transformative adaptation. *Environmental Research Letters* **7**, 1–7.

Polack E (2008) A right to adaptation: securing the participation of marginalized groups. *IDS Bulletin* **39**, 16–23. doi:10.1111/j.1759-5436.2008.tb00472.x

Preston I, Nanks N, Hargreaves K, Kazmeirczak A, Lucas K, Mayne R, Downing C, Street R (2014) 'Climate justice and social justice: an evidence review'. Joseph Rowntree Foundation, York.

Priestley M, Hemingway L (2007) Disability and disaster recovery: a tale of two cities. *Journal of Social Work in Disability & Rehabilitation* **5**, 23–42. doi:10.1300/J198v05n03_02

Rooney C, White GW (2007) Narrative analysis of disaster preparedness and emergency response survey from persons with mobility impairments. *Journal of Disability Policy Studies* **17**, 206–215. doi:10.1177/10442073070170040301

Rowland JL, White GW, Fox MH, Rooney C (2007) Emergency response training practices for people with disability: analysis of some current practices and recommendations for future training programs. *Journal of Disability Policy Studies* **17**, 216–222. doi:10.1177/1044207307 0170040401

Ruger JP (2004) Ethics of the social determinants of health. *Lancet* **364**, 1092–1097. doi:10.1016/ S0140-6736(04)17067-0

Russell RC, Currie BJ, Lindsay MD, Mackenzie JS, Ritchie JS, Whelan P (2009) Dengue and climate change in Australia: predictions for the future should incorporate knowledge from the past. *The Medical Journal of Australia* **190**, 265–268.

Sevoyan A, Hugo G, Feist H, Tan G, McDougall K, Tan Y, Spoehr J (2013) 'Impact of climate change on disadvantaged groups: issues and interventions'. National Climate Change Adaptation Research Facility, Gold Coast.

Stough LM, Sharp AN, Decker C, Wilker N (2010) Disaster case management and individuals with disabilities. *Rehabilitation Psychology* **55**, 211–220. doi:10.1037/a0020079

Takahashi A, Watanabe K, Oshima M, Shimada H, Ozawa A (1997) The effect of the disaster caused by the great Hanshin earthquake on people with intellectual disability. *Journal of Intellectual Disability Research* **41**, 193–196. doi:10.1111/j.1365-2788.1997.tb00695.x

United Nations (2007) 'Convention on the rights of persons with disabilities'. United Nations, Geneva.

Universal Design (n.d.) *The Seven Principles of Universal Design.* <http://www.universaldesign. com/universal-design/1761-the-seven-principles-of-universal-design.html>

VicHealth (2012) 'Disability and health inequalities in Australia: research summary'. VicHealth, Melbourne.

Von Doussa J (2008) Climate change and human rights, *InSight*, June. Centre for Policy Development, Sydney.

White B (2006) Disaster relief for deaf persons: lessons from Hurricanes Katrina and Rita. *The Review of Disability Studies* **II**, 49–56.

Williams-Piehota P, Uhrig J, Doto JK, Anderson W, Williams P, Thierry JM (2010) An evaluation of health communication materials for individuals with disabilities developed by three state disability and health programs. *Disability and Health Journal* **3**, 146–154. doi:10.1016/j. dhjo.2009.08.002

WHO (World Health Organization) (2011) *World Report on Disability.* World Health Organization, Geneva.

6

People who are elderly or have chronic conditions

Margaret Loughnan and Matthew Carroll

Key points

- Older people and those living with chronic health conditions are more vulnerable to the impact of climate change.
- This vulnerability is influenced by a myriad of other factors including social isolation, lack of financial resources, inappropriate housing design and inadequate access to services.
- There is a pressing need to develop targeted public health strategies to support these vulnerable groups.
- Community service organisations play a lead role in supporting these groups at the local level and need to be provided with the skills and resources necessary to help them meet the demands of a climate changed future.

Introduction

There is growing recognition of the public health implications of global climate change (Costello *et al.* 2009). Particular attention needs to be paid to the impacts on the health and wellbeing of vulnerable groups, especially older people and those with chronic health conditions. The interaction between of health issues and reduced physiological and thermoregulatory reserve, often in combination with socio-economic factors, makes these groups more susceptible to the impacts of a changing climate, especially in light of population ageing. As a result, there is a critical need for improved public health policy and adaptation planning.

Significant risks are posed by extreme weather events such as extended heatwaves, bushfires and tropical cyclones or severe storms causing flooding. These events are predicted to increase in frequency and intensity in a climate-changed future (Alexander & Arblaster 2009). Those with chronic health conditions or disabilities, on low incomes, and who are socially isolated will be at an increased risk. These groups are over-represented among older people, compounding the problem and placing them in the high-risk category. Extreme weather events often require evacuation or confine people to their homes without sufficient access to necessities for extended periods of time. They also characteristically lead to the failure of social and technological infrastructure and support services, such as loss of electricity, communications and emergency response services. When all of this is taken into consideration, climate change induced stress and disruption will be detrimental to the health and wellbeing of older people and those with chronic conditions in Australia and around the globe.

Risk assessment, climate change adaptation, hazard mitigation, public health, and emergency preparedness and response must be coordinated. Community awareness, adaptation and preparation around climate change risks cannot proceed in isolation from one another. It is important that the special needs of vulnerable populations, including those of older people and those with chronic conditions, be represented and provided for. While climate change will have community-wide impacts, it will greater for these vulnerable groups. Older people are not without resources and resilience, but to stay safe they are going to need well-informed plans and targeted resources. To achieve this they will need to be informed and involved in policy planning and community response initiatives (Jennings 2011). Many of today's older people have grown up and lived through times of common struggle and hardship so have experience to contribute to adaptation strategy and policy development, taking into consideration the issues important to them as well as their concerns for their own welfare and that of their children and grandchildren (Jennings 2011; Loughnan *et al.* 2013b).

There are several questions surrounding the vulnerability to climate change of older people and those with chronic health conditions, such as: Who are older Australians? Where do they live? What is the prevalence of key chronic health conditions in Australia? What factors make older people and those with chronic health conditions more vulnerable? How can we reduce vulnerability? What is effective adaptation? The aim of this chapter is to outline the impacts of climate change on older people and people with chronic health conditions in Australia with specific reference to the impacts of extreme heat events, as this is the biggest threat in terms of mortality and morbidity within these groups (Nitschke & Tucker 2007; Vaneckova *et al.* 2008; Tong *et al.* 2010; Loughnan *et al.* 2014).

Who are older Australians and where do they live?

Projected changes in the climate over the coming decades will be overlaid on a changing age distribution with a growing proportion of older people (aged 65 years or over), especially the very old. The socio-economic circumstances and the geographic location of these adults will also change. Older adults are not a homogenous group of people; distinct subpopulations exist within this cohort based on ethnicity, age, socio-economic status, occupational and educational outcomes, social connectivity, family connectivity, general health, geographic location and access to services.

In 2011, 14% of Australians were aged over 65 years and this is expected to increase to 24.5% in 2061 (ABS 2013a). This is heightened in the very old, with the proportion of the population aged 84 years and over more than tripling from 0.5% in 1971 to 1.8% in 2011 (ABS 2012) and projected to increase to almost 6% by 2061 (ABS 2013a). This population ageing is due to a number of factors including the post-war 'baby boom', declining fertility rates and improved health care and other factors which have resulted in increased life expectancy. The likelihood of older people living alone also increases with age as a result of divorce (9.8%), separation (2.4%) and widowhood (26%) (ABS 2012), leaving them at greater risk of becoming socially isolated (Warburton & Lui 2007). This is often combined with lower socio-economic status and living in small and sub-standard housing, all of which are known to contribute to heat-related death (Kinney *et al.* 2008). Reduced financial reserve also decreases the capacity to respond to climate stressors in terms of adding insulation or other heat-protective measures such as the installation and operational costs of air-conditioning. Research has also shown that people living on low incomes may lack the confidence to approach and obtain disaster relief and that they are at a greater risk of responding to 'scams' during recovery phases after natural disasters (Harvison *et al.* 2011). The tendency for older people to live in older, sub-standard housing (Kinney *et al.* 2008) means that their homes are likely to be poorly designed to withstand climate impacts, lacking passive cooling, with little or no insulation, or natural or mechanical shading such as vegetation or awnings, pergolas etc., and limited ventilation or access to air-conditioning units. All of these factors increase the heat load within the house and the exposure of residents.

It must be recognised that Australia has a multicultural population, with 36% of all older Australians born overseas and coming from 120 different countries (ABS 2012). This includes 1.5% of older people who do not speak English at all, rising to 2.5% for those aged 85 years and over. Many of the languages other than English are spoken by relatively small numbers of people, making effective communication difficult outside the immediate family or language community. At this age, support needs are greater and poor language skills can impede the capacity to access health and social services and the ability to follow advice and instructions about general health and wellbeing, as well as responding to climate events. This connection between ageing, support needs and language is highlighted in the language reversion often seen in older people from non-English speaking backgrounds with dementia (Hansen *et al.* 2013).

It is also important to understand the geographic distribution of older Australians, the majority of whom live in Victoria, New South Wales and South Australia. The map shown in Figure 6.1 indicates the local government areas with the highest proportion of older residents. Many of these areas are already affected by increasing climate stresses such as heatwaves in south-eastern Australia and tropical cyclones and flooding in Queensland.

In the event that physiological and behavioural adaptation remains unchanged, and if current settlement patterns continue, the older population in these areas will continue to be at high and increasing risk, as these areas have already experienced significant impacts of climate change such as heatwaves, droughts, bushfires, floods and cyclones (Middelmann 2007).

The majority of the population resides in the state capitals, which are projected to continue ageing (depicted in Figure 6.2), increasing the risk of adverse health effects during extreme heat events (Loughnan *et al.* 2013a). This is due to multiple factors including the urban heat island effect (Morris & Simmonds 2000), higher housing density, poor housing design and negative perceptions of neighbourhood safety (Giridharan *et al.* 2007; Bambrick *et al.* 2011). In the coming decade the highest concentrations of older people will be in suburbs along the urban fringe in all capital cities except Brisbane. These are areas that already experience difficulty with service provision and are at higher risk of fire. These areas should be targeted for adaptation in terms of education and behavioural change, as well as urban reform and urban planning.

In addition to heightened risks associated with living in major metropolitan areas, there is also a growing trend towards older people moving to rural or coastal locales in order to enjoy the pastoral lifestyle. Known as counter-urbanisation or tree and sea changing (Ragusa 2010), the experience of living in a rural or coastal community may be initially positive but can become increasingly challenging with declining health and reduced access to services and facilities, and often with geographic separation from family and social networks (Morton & Weng 2013). While the research is scarce, there is evidence to suggest that rural communities face greater challenges because of the combined effects

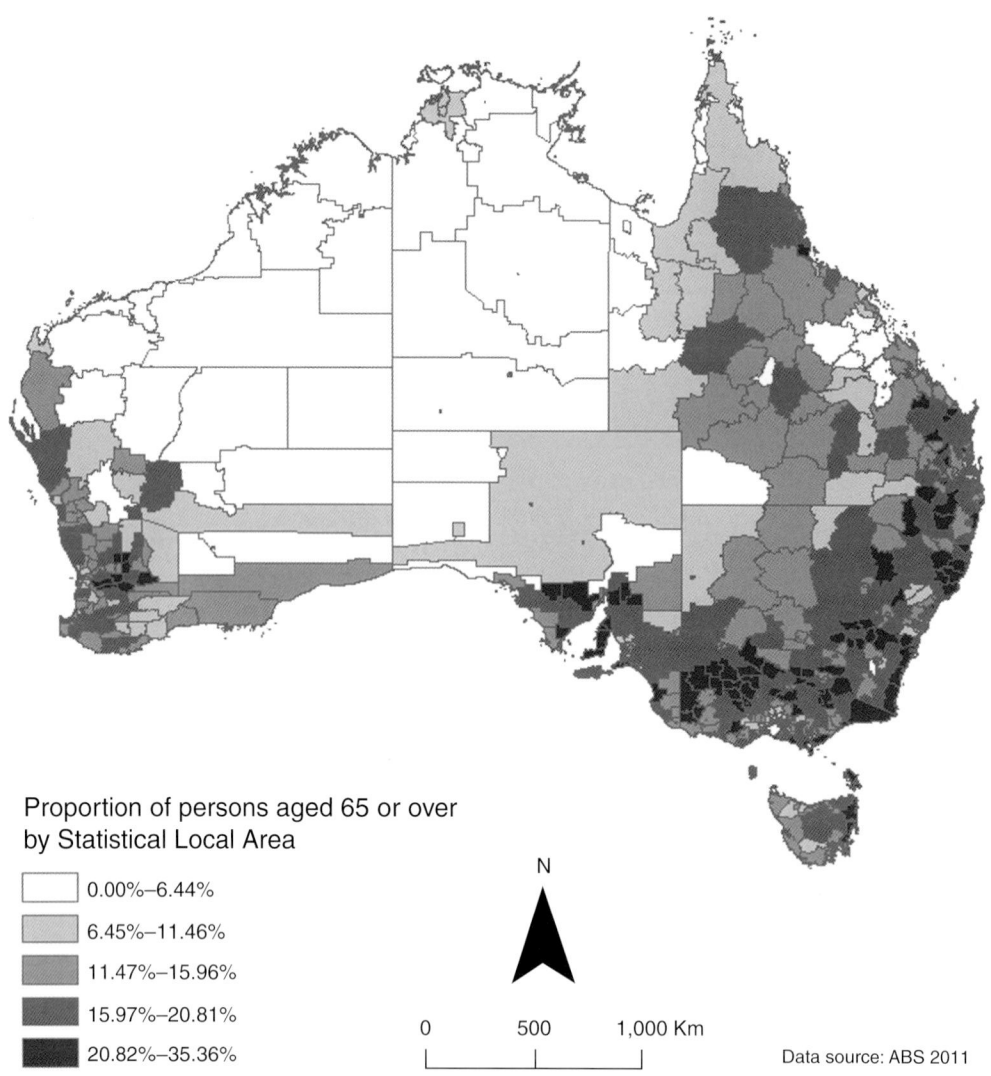

Figure 6.1.   The proportion of the population who are older people (aged 65+ years) in each local government area. The dark areas are high risk and have 16–34% of the resident population over the age of 65 years (ABS 2011).

of accelerated ageing of the population (due to the out-migration of younger people seeking greater work opportunities, and the in-migration of older people referred to previously), reduced access to services and facilities, environmental exposures (including the need to drive extended distances), lifestyle risk factors such as greater rates of smoking and drinking, and reduced likelihood of health service utilisation (Davis & Bartlett 2008; AIHW 2012a). These concerns can be expected to be magnified by the impacts of climate change, especially in low-lying coastal sea change locales that may be impacted by sea-level rise (Gurran *et al.* 2008).

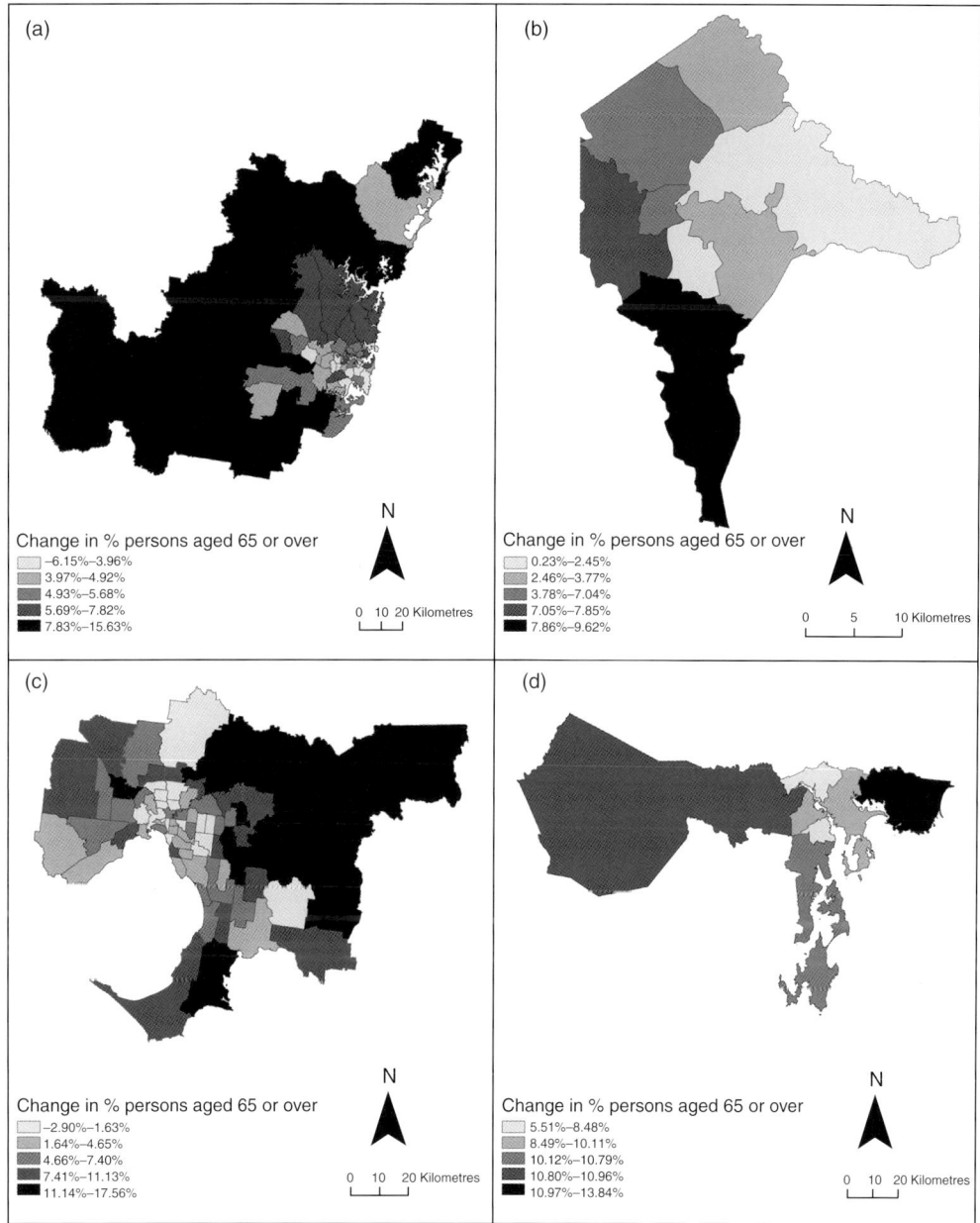

Figure 6.2.   Change in the proportion of the population over 65 years of age living in state capitals of Australia: (a) Sydney (2006–2026), (b) Canberra (2009–2021), (c) Melbourne (2006–2026), (d) Hobart (2007–2032).

(*Continued*)

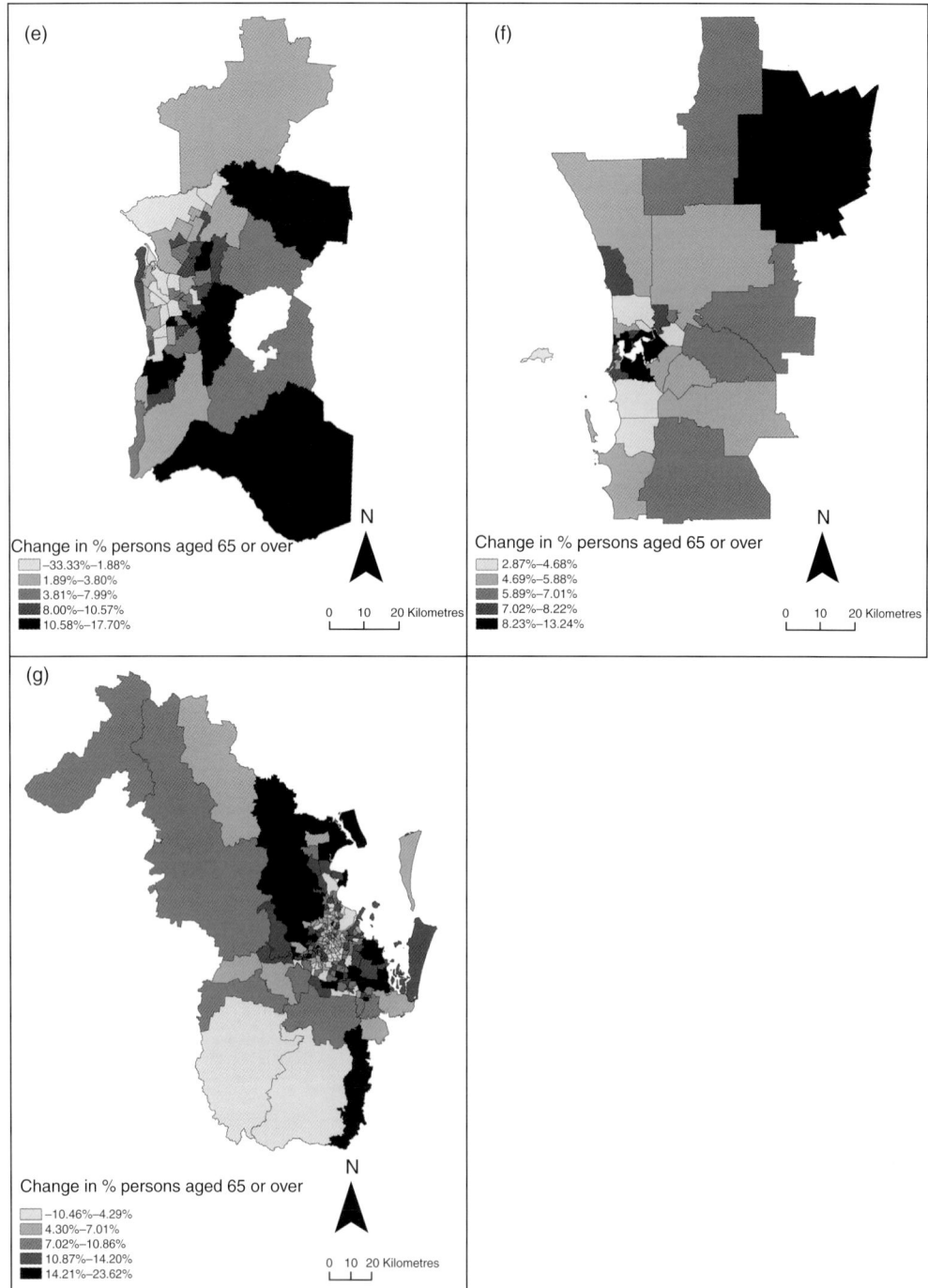

Figure 6.2. (*Continued*). (e) Adelaide (2006–2026), (f) Perth (2006–2026) and (g) Brisbane (2006–2031) (ABS 2013a).

What is the prevalence of chronic health conditions and increased vulnerability to climate impacts?

The most common chronic diseases (heart disease, stroke, cancer, type 2 diabetes and respiratory diseases) account for almost two-thirds of deaths globally and half of the global burden of disease in disability-adjusted life years (Kjellstrom *et al.* 2010). According to 2011–12 data from the Australian Bureau of Statistics, 5.1% of people aged 18 years or older had diabetes, 32.8% had high cholesterol, 10% had indicators of chronic kidney disease, and over 11% had indicators of impaired liver function (ABS 2013b). The relationship between climate change and chronic conditions is expected to be two way, with climate change predicted to increase the prevalence of cardiovascular and kidney disease, as well as people with existing conditions being more vulnerable to the impact of increased heat. The different climate change impacts (including heat, air pollution, extreme weather and water- or vector-borne diseases) are also expected to interact with chronic health conditions, multiplying the health effects (Balbus & Malina 2009), for example, the combined impact of air pollution and heat on the prevalence and severity of asthma and hay fever (Kjellstrom *et al.* 2010). While less attention has been paid to the impact of climate change on mental health, there is now growing recognition of significant direct impacts (such as anxiety, post-traumatic stress disorder, etc.) and indirect impacts through deteriorating physical health or community wellbeing (Berry *et al.* 2010). There is already evidence that heatwaves in Australia are associated with increased hospital admissions for mental and behavioural disorders (Nitschke & Tucker 2007; Hansen *et al.* 2008). These chronic health conditions are often in combination with advanced age, home-bound lifestyles and restricted access to transport and services, again multiplying the potential impact of climate change (Kenny *et al.* 2010).

Current policy that encourages people to age in place, while in line with the clear preferences of most older people and those living with chronic conditions, has the unintended consequence of creating a large, diverse and dispersed population. This may have implications in terms of a coordinated response to extreme weather events, creating problems for community service organisations that provide home-based services. These core services include supporting the health of individuals as well as wider community health and resilience, financial resilience, and enabling access to facilities and services, communication, transportation and water supplies (Mallon *et al.* 2013).

What are the physiological factors that increase vulnerability?

Many older people remain in good health for most of their lives, but in general there is an increase in the risk of chronic disease with advancing age. Climate change may aggravate some of the physiological limitations that are associated with older age. This includes respiratory distress due to increased air pollution, such as ground-level ozone concentration and increased airborne particulates found in bushfire smoke. Several studies have documented the exacerbation of chronic respiratory diseases such as asthma and chronic obstructive pulmonary disease and poor air quality (Jerrett & Finkelstein 2005; Maantay 2007; Hu *et al.* 2008). Cardiovascular disease and heat sensitivity in the elderly has also been well documented (Loughnan *et al.* 2008; Loughnan *et al.* 2010b; Reeves *et al.* 2010). In 2008, 16% of the Australian population had a disease of the circulatory system; the highest rates were in Tasmania, Victoria and South Australia (18.2–22.1%) (ABS 2013b). Some medications, in particular cardiac medications and some antipsychotics, can compromise thermoregulation making physiological adaptation extremely difficult for people using them (Peng *et al.* 2011).

Ageing and living with chronic conditions and disability are closely associated. A hazard of growing old is the increasing likelihood of disabilities, which will shape the types of lives people can live and the activities they can do. This may result in the need to have a carer living with them, have utilisation of community-based care, or move to non-private residential care (ABS 2012). In 2011, 537 300 older people were identified as having a profound or severe disability. Among people in the 65–74 years age group, approximately 1 in 10 people reported a profound or severe disability. This increased to 17% for the 75–79 years age group, rising to 68% for the 90 years and over age group. Older women generally had a higher rate of profound or severe disability (22%) than older men (16%) (ABS 2012). Therefore, a significant number of older people will rely on others for daily care and have limited ability to change their circumstances to accommodate changes in weather. This may mean anything from maintaining hydration to evacuating their home or aged care residence in response to an emergency. As the population continues to age there will be many more very old people in our community, increasing the need for support services and vulnerability to climate change. How best to accommodate this should be addressed as a matter of priority, with a pressing need for health and social services to come together to develop a coordinated approach and obtain the skills and resources needed to meet current and future demands.

What are the other factors that increase vulnerability to climate extremes?

Eric Klinenberg said heatwaves are 'silent invisible killers of a silenced invisible population' (Klinenberg 2002, p. 17). This suggests that the people most affected by extreme heat are those people unable to change their situation. Socio-economic characteristics, housing, adequate community infrastructure and the availability of social services will affect the degree to which older people are exposed to weather extremes and their ability to adapt (Sevoyan *et al.* 2013). Older adults living in poverty may experience greater exposure, for example, during heatwaves, because they may not have, or be unable or unwilling to bear the running costs of, air-conditioning (Hansen *et al.* 2011b). Lower socio-economic communities rely on implementation of infrastructure such as cooling centres and subsidised transport to minimise heat exposure. Low socio-economic status, ethnicity (including speaking a language other than English at home), having a disability and living alone have all been associated with increased mortality and morbidity among the elderly during heatwaves (Sevoyan *et al.* 2013; Loughnan *et al.* 2013a).

Depending on the condition, quality of construction and adaptive technologies, housing can either increase or decrease exposure to extreme weather and the associated adverse impacts on residents' health and wellbeing. Poorly maintained or portable housing are at particular risk during extreme storms. Lack of air-conditioning, insulation and small homes increase the risk of heat exposure during summer. Affordable housing in poorer neighbourhoods where safety and crime are an issue may leave home-bound people unable to open doors and windows in the evening to 'cool off' the house (Hansen *et al.* 2011b). Existing housing options do not always meet the changing needs of older people. Housing developments need to ensure an adequate and diverse housing stock that provides suitable choices for an ageing population. This will require appropriately designed home environments that can accommodate changing needs over time (APESPA 2011).

According to the Australian Council of Social Service (ACOSS), older people face a high risk of living in poverty, with over a third of people aged over 64 years living below the 60% poverty line (i.e. those people earning 60% or less of the median disposable income for all Australian households) (ACOSS 2012). Home ownership provides significant protection against poverty for many older people but the minority who rent face a higher poverty risk. Older people can also be more vulnerable to loss of property due to lack of insurance, limited personal finances and little ability to obtain credit, so they may not have the finances to prepare for or recover from climate events. Many older people live on a fixed income and struggle to pay for energy, food, rent and out-of-pocket medical expenses. Increased costs for utilities, housing, food and health care will have profound impacts on this group. People living alone or those with cognitive, vision, hearing or other sensory deficits or poor language skills are at risk of not receiving or misunderstanding emergency warnings and instructions, and may not have the capacity to respond appropriately (Hutton 2008).

An example of the impact poverty has on the lifestyle of older people is provided in ACOSS (2012, p. 17).

> *Mary is a 78-year-old pensioner who lives in Perth. Due to the cost of high rent and utility prices life has become a battle for survival. During winter Mary would turn her heating on for only one hour a day. She would spend much of the day in bed to keep warm. Her food budget each week is $40. This is a diet of baked beans, bread, and a small amount of fruit and vegetables. Once a month she buys a small piece of chicken. Once a week she eats a hot meal provided at a local church. Mary considers herself lucky as at least she has a roof over her head.*

Without adequate social service provision, people like Mary would not be able to withstand extreme weather events.

Limited electricity supply during heatwaves, including intentional 'brown outs' to compensate for excess demand (Maller & Strengers 2011), lack of potable water during floods and lack of transport to cooling centres or for evacuations, all increase exposure during extreme weather. Similarly, the lack of social service organisations, such as local community organisations, government organisations, religious groups, aid distribution centres, assisted living centres, hospitals, and home and community care providers, would lead to greater impacts and poorer outcomes for older people. Communication tools and the media are important for the transmission of information during weather extremes. The availability and dissemination of information is important, but it is also important to compile registries of vulnerable people to help target responses during emergency situations. One such example is the Vulnerable Persons Registers, which were established in Victoria following the 2009 Black Saturday bushfires (see www.dhs.vic.gov.au/facs/bdb/fmu/service-agreement/4.departmental-policies-procedures-and-initiatives/4.18-vulnerable-people-in-emergencies).

An active and diverse community can support its community wellbeing. Reports from the Chicago heatwaves in 1995 and 2006 indicated that older people living in vibrant residential and commercial areas were more proactive in seeking services, and areas with larger concentrations of older people had lower mortality rates, possibly due to shared social networks (Gamble *et al.* 2012). Worryingly, in Australia areas with larger numbers of older people have been found to have increased area-based vulnerability (Loughnan *et al.* 2013a), perhaps because they lack the strong social networks identified by Gamble *et al.* (2012) in the US. It may be possible that community-building initiatives that link people within communities would provide a buffer between vulnerable people and community service organisations during extreme weather events.

What do we already know about the impact of climate stressors on older people and those with chronic conditions?

Extreme heat

Climate models are predicting an increase in the frequency, duration and intensity of extreme heat events in Australia in the coming decades (Alexander & Arblaster 2009). This will be a major risk for older people (Loughnan *et al.* 2010a; Hansen *et al.* 2011b; Loughnan *et al.* 2013a). Health outcomes associated with extreme heat include heat stress, heat exhaustion and heat stroke, dehydration, acute renal failure, nephritis, cerebrovascular accidents or strokes, and acute myocardial infarction or heart attacks (Gamble *et al.* 2012). The 2009 heatwave in south-eastern Australia resulted in 374 excess deaths over a three-day period in Victoria, the greatest number of who were aged 75 years or older (Reeves *et al.* 2010). Nicholls *et al.* (2008) demonstrated that there was a 17% increase in mortality when the average daily temperature exceeded 30°C and a 21% increase in deaths when the overnight temperature exceeded 24°C. The impact of heat exposure has been outlined above as detrimental for the health of people suffering mental illnesses including schizophrenia and schizotypal and delusional disorders, as well as older people suffering from dementia (Hansen *et al.* 2008). Age-related changes in thermoregulation reduce the capacity of older people to physiologically adapt to heat. In addition, the mechanisms driving thirst are diminished in older people, increasing the likelihood of dehydration. Also, older people and those with chronic conditions are more likely to have mobility limitations that may impair their ability to reduce exposures.

Cyclones, severe storms and floods

A disaster may effect a largely urban environment, cause damage to an agricultural region, or both. In 1974 Cyclone Tracy caused devastation because it hit the city of Darwin. Cyclone Larry, in 2006, caused widespread devastation to agricultural crops and a number of towns in north Queensland. The effects of urbanisation and increasing population growth and density, most notably in the big cities and coastal regions, have led to greater demand for and concentration of infrastructure and a higher potential exposure to natural hazards (Middelmann 2007).

Tropical cyclones have caused 2100 fatalities in Australia since 1839 as a result of storm surges, high winds and extensive flooding (Middelmann 2007), all of which present a greater threat to older people and people with chronic conditions, who may not have the resources to evacuate or protect themselves. Apart from the immediate risk of injury and death during the event, extreme rainfall events have a range of secondary impacts such as respiratory effects from indoor moulds, impacts on food and water security, interruption to communications and utilities, and difficulty in the provision of health and support services. There is some uncertainty around the predicted change in tropical cyclone frequency and intensity, due to changes in the large-scale environment in which tropical cyclones form and evolve caused by greenhouse warming (Bureau of Meteorology 2014). Severe weather may also lead to mental or emotional stress before, during and after the event. Research from the US indicates that stress related to extreme weather resulted in decreased working memory function in older adults and residents and staff from aged care facilities, and had adverse effects on their mental health that required treatment up to five months after cyclones (Gamble *et al.* 2012). Similarly, the extent of adverse health effects among older people was positively related to the intensity and duration of the events (Gamble *et al.* 2012). Furthermore, the need to evacuate a region prior to an event presents several logistical problems for older residents or those with chronic conditions. For example, transportation may be unavailable or limited, and specialist temporary accommodation may be required. Hurricane Katrina in the US is a telling example of the increased vulnerability of older people, with 71% of those who died in Louisiana aged 60 years or older, with almost half of these aged over 77 years (McCann 2011).

While extreme weather events have been discussed here and elsewhere as separate entities (e.g. heatwaves or cyclones), it should be recognised that they often occur simultaneously. As such, vulnerable groups will be subjected to an ensemble of threats that stress the physiological response of people whose biophysical adaptive capacity has already been compromised by ageing and illness (Carnes *et al.* 2014). Many of the environmental changes associated with climate change, such as increases in air pollution and airborne allergens, increases in disease pathogens, increases in the frequency and severity of heatwaves, and degradation of water quality will all have profound health consequences for elderly adults (Carnes *et al.* 2014).

Will adaptation to warmer climate bring any benefits for older people or people with chronic disease?

It has been suggested that milder winters associated with a warming climate will change the seasonality of mortality and will reduce the number winter or cold-related deaths (McMichael *et al.* 2006). However, experience over the past few decades has suggested otherwise. In Britain and other temperate countries residents have received better housing, improved health care, higher incomes and greater awareness of the risks of cold. Socio-behavioural adaptation has indicated that the link between winter temperatures and increased mortality may no longer be as strong as before. Staddon *et al.* (2014) found that the association of year-to-year variation in winter mortality with the number of cold days in winter (less than 5°C) has disappeared, leaving only the incidence of influenza-like illnesses to explain most of the year-to-year variation in winter mortality over the past decade. Although winter mortality does exist, winter cold severity no longer predicts the number of people affected (Staddon *et al.* 2014).

Recent experience of climate change impacts on older Australians

Little research has been undertaken to determine how well older people understand the threats to their health and wellbeing presented by weather extremes associated with climate change. Loughnan *et al.* (2013b) explored this question in a study examining the response of older people to extreme heat during the summer of 2012. The research indicated that while the participants had good knowledge of heat avoidance behaviours, this was largely centred on perceived discomfort associated with hot weather rather than recognition that heat exposure has implicit health risks. There was also a lack of recognition of increasing vulnerability in a climate-changed future and a sense of self-reliance that made participants confident they could cope with hotter summers because they had coped during heatwaves in the past when they had limited or no access to air-conditioning. Overall, the group members were unaware of existing organisational responses to extreme heat, including the local heat health plan, but did identify possible locally-based responses, such as an at-risk register run by the district nurse and a Lions Club personal alert system, which could be expanded to include checking on vulnerable older residents during heat-waves. This highlights the need for greater connections between local government, community health and service organisations to ensure that there is a coordinated approach to meeting the needs of vulnerable community members.

The key theme from the focus group was the participants' reliance on personal experience rather than government opinion or prevailing science (Loughnan *et al.* 2013b). The need to respond to heat events may be clouded by current climate change debates, so future public health messages directed at older people may need to be presented in terms of climate variability, dealing with extremes and learning from past events. These findings regarding older people not perceiving heat as a threat and relying on personal or local support rather than state-led responses are backed up by earlier research in the UK (Abrahamson *et al.* 2009).

The main strategy to 'beat the heat' identified by participants in the Australian study was the use of air-conditioning. The interviews revealed that all households had at least one air-conditioning unit, with most also using fans to distribute cool air and supplement the use of the air-conditioning to minimise running costs. The fear of rising utility costs was raised on several occasions during the interviews (Loughnan *et al.* 2013b). This fear is not limited to older people in small rural towns. Productivity Commission figures show that costs have increased by 50% over the past five years (Strengers *et al.* 2012). With 70% of all Australians owning an air-conditioning unit and the recent increase in electricity costs, the use of air-conditioning as a primary means of cooling is becoming less affordable. As a result many people reserve the use of air-conditioning to periods of extreme heat, and in some low-income households people are skipping meals, selling possessions and cutting back on essential health care to pay their power bills. Sadly, some of these households are also likely to be vulnerable to heatwaves (Strengers *et al.* 2012). Environmental determinants of heatwave impacts relate to neighbourhoods as well as individual homes. Urban neighbourhoods need shade and vegetation cover, and streets oriented to provide ventilation to reduce the impacts of summertime heat (Barnett *et al.* 2013). Barnett *et al.* (2013) indicated that various studies estimate that properly-sited trees can save between 10% and 50% of annual energy use in conventional houses, compared with the same houses in the open.

Estimating the economic costs of climate change: drawing upon heat health effects

The economic costs of keeping cool during heatwaves are only a small part of the overall costs associated with the impact of climate change on vulnerable populations. There are also considerable costs associated with mortality and morbidity that result from extreme weather events. Heatwaves are responsible for more deaths than any other natural hazard in Australia (Coates 1996) and this data provides a useful insight into the potential economic impacts of climate change. Excess deaths are potentially avoidable deaths; heatwave preparedness planning can reduce the death toll and therefore can reduce the economic impact of extreme heat events. A difficulty arises when we try to assess what a life is worth, especially as the most vulnerable groups tend to be older and sicker. Previous

US studies have reported that the value of a statistical life for older Americans was between US$4–6.3 million (Krupnick *et al.* 2000; Ebi *et al.* 2004; Leonardi *et al.* 2006). In the absence of a similar Australian-based calculation, this US$4 million figure can be used as a broad indicator of the value of a statistical Australian life. During the summer of 2009, a three-day heatwave in Victoria resulted in an estimated 374 excess deaths. While heatwave deaths are potentially avoidable deaths, this may not always be possible and if only 50% of these deaths were avoided (187 deaths) than there would be a gross saving of A$748 million.

Another way to estimate the impacts is to look at the hospital and emergency costs from the same 2009 heatwave event. The National Hospital Cost Data Collection provides an estimate of the cost of acute hospital admissions and this data has been used to assess the extent of heatwave morbidity (AIHW 2010; Chikritzhs *et al.* 2010; AIHW 2012b). This report estimates the average cost data for acute admitted patients as A$4133 per patient. In addition, while the cost of emergency treatment and road transport by ambulance across Australia varies, in metropolitan Melbourne it is estimated to be $990 (Ambulance Victoria n.d.). During the 2009 heatwave there were approximately 1100 excess ambulance emergency calls in Melbourne resulting in increased healthcare costs of A$1.2 million. If 50% of these callouts resulted in emergency admission to hospital then this adds an estimated additional A$2.4 million per day of admission. For patients treated in emergency departments an additional cost would also be incurred. It is apparent from these figures that the economic impact of extreme heat events on the healthcare budget is huge.

What actions can be taken by government, health and social service agencies to ameliorate impacts and facilitate adaptation and mitigation?

Resilient communities

Building community capital and capability is important, as is empowering individuals to effect changes in their behaviour and environment that will increase their resilience. As a first step, it is important to clearly articulate the risks faced by communities and the possible adaptation strategies. This information must be targeted at high-risk groups such as the elderly, people with disabilities and those coming from non-English speaking backgrounds, especially those living in the urban fringes. These key indicators have been incorporated into a spatial vulnerability index for extreme heat events for all Australian capital cities, with corresponding maps of high-, medium- and low-risk areas developed to guide emergency management (Loughnan *et al.* 2014). An online tool has also been developed (see www.mappingvulnerabilityindex.com). Implementing surveillance systems for those at risk, as well as monitoring threats and understanding where these intersect, will help to reduce vulnerability and lessen the impacts of climate stressors.

Changes can be made to the physical design of environments to improve community resilience to climatic extremes, such as using climate-sensitive urban design principles or innovative urban reforms to ameliorate the risks of exposure for residents. Changes to neighbourhoods should directly optimise community resources. For example, early warning systems allow people time to plan evacuation or move to a safer location such as a cooling centre during heatwaves. People with special medical conditions could register with their local emergency service to receive information and additional support during extreme events. Emergency management checklists could be distributed to older residents with key contact information for public health issues, social networks, community care organisations and healthcare providers. These actions need to be pre-empted by rigorous education and information campaigns that involve community members rather than just a government directive. The key to establishing effective adaptation strategies is communication across and between sectors. There is clear evidence that older Australians will be at a greater risk during climate extremes than the general population. Given that the Australian population is ageing and that many older people live and will continue to live in areas that are experiencing the stresses associated with a changing climate, developing adaptation and mitigation strategies should be a priority.

Preparedness

Many cities around the world are designing and implementing strategic risk management action plans to lessen the impacts of climate change. These plans may have health and social co-benefits, such as reducing illnesses related to heat and air pollution, and other diseases associated with urban lifestyles (Harlan *et al.* 2006). The primary aims would be to:

- reduce greenhouse gas emissions by reducing energy consumption by industry, transport and households
- improve the built environment to include greener infrastructure and promote active transport and sustainable housing design
- promote adaptive behaviours for communities, individuals and organisations such as improved weather forecasting, heat warning systems, air quality alerts and emergency preparedness to deal with high-risk populations like the elderly.

Reducing fossil fuel consumption in cities has significant health co-benefits through improved air quality and a reduction in exposure-associated exacerbation of respiratory disease. Reducing emissions reduces anthropogenic heat releases and would reduce the urban heat island within cities. This, in turn, reduces exposure, particularly night-time exposure, thereby preventing heat-related exposures and deaths. Cities where mitigation strategies have resulted in cleaner living environments and restored ecosystems have the co-benefit of providing opportunities for a more active lifestyle for residents (Harlan *et al.* 2006; Younger *et al.* 2008). For example, reducing greenhouse gas emissions and providing greener, cooler environments promotes physical activity and reduces the risk of lifestyle diseases such as heart disease, respiratory disease, diabetes and obesity. As people age in such an environment they will remain healthier for longer. With good health there is an improvement in resilience to hazards such as heatwaves and better ability to respond and adapt. While the responsibility for mitigation efforts will largely lie with government, the community and the service organisations that support them will need to be engaged in these efforts for there to be any chance of success.

Heat-health alert warning systems targeting vulnerable groups

Adaptations plans such as improving housing standards and urban reform are important to address long-term changes to reduce risk from climate extremes; however, shorter-term adaptation measures, such as heat-health warning systems, must also be considered. Heat-health warnings systems were introduced to minimise harm to vulnerable populations during heatwaves. These plans are tailored to suit local meteorological and demographic conditions, and may include early alerts and advisories combined with emergency public health measures to minimise the impacts of heat exposures (Toloo *et al.* 2013). The issuing of a heat-health warning is usually accompanied by information about supportive measures that will provide a level of protection for vulnerable people. These are usually distributed by the media and are intended to increase awareness of the dangers of heat exposure, and to provide temporary measures to reduce impacts (e.g. advice to relocate to a cooling centre). Studies measuring the effects of heatwaves in regions before and after the introduction of heat-health alerts have shown these to be helpful in reducing mortality and morbidity (Weisskopf *et al.* 2002). However, further research is needed to determine how these warnings are understood and enacted, as the effectiveness of each intervention and the evidence behind it, albeit intuitively acceptable, remain unproven. This is particularly important for vulnerable groups such as older people and those with chronic conditions who may face barriers in accessing, comprehending or responding to warnings.

Toloo *et al.* (2013) suggested that while most people were aware of the risk, the ones who saw themselves as vulnerable were more likely to take actions such as using air-conditioners, maintaining hydration, staying indoors or in shade, dressing properly, avoiding strenuous activities and checking on elderly or disabled people. People who did not acknowledge they were at risk engage in fewer protective measures but were still aware of the need to protect others deemed more vulnerable than themselves. Protective behaviours are more likely to be employed if the costs and barriers to that action are seen as rational or the benefits of taking the action outweigh the costs (Sheridan 2007). The effectiveness of heat-health alert systems depends upon several factors, such as the interaction between human perception and behaviour within a social setting, and the extension of risk awareness to behavioural change with the most vulnerable groups (e.g. the elderly and those with chronic disease). The existing evidence supports the need for heat-health alerts from a public health and policy perspective; however, further research is needed to identify which programs are more effective and why, and how services are used by the vulnerable populations and groups during heatwaves.

Utility shut off protection for vulnerable groups

Regardless of age, health status, wealth or social status, everyone should have a fundamental right to sustainable, ongoing, secure and affordable access to energy. Electricity prices in Australia have risen on average by 32% between 2007 and 2010. Costs will continue to rise due to increasing network costs, rising gas prices and clean energy programs. These rising prices not only cause financial stress for older people, but will lead to negative impacts on the health and wellbeing of older Australians and people with chronic health conditions and disabilities. Many, but not all, older people are vulnerable to energy affordability and access issues. For example, older and chronically ill people have more pronounced heating and cooling needs to maintain health and wellbeing, particularly during extreme temperature events such as heatwaves and cold spells; they also have less ability to adapt to higher energy prices, generally due to being at home more often than other consumers (e.g. workers, families) and using more energy. Many older people are on low and fixed incomes and may have significant medical and other costs to meet. Due to the complexity of energy contracts, older people often do not take up market contracts that may save them money. Older people may be more vulnerable to door-to-door sales techniques and to signing up for energy contracts that do not meet their needs or budgets. Older people living in retirement villages, residential parks and other situations can be vulnerable to harmful price and access activities (COTA Australia 2012).

Appropriate planning is already underway in the US, with several states having developed utility protection plans for vulnerable people, for example, during the winter or summer disconnections are often limited as to when they can occur, as well as during extreme weather events and temperatures. The New York State Public Service Commission has programs for older people to protect the vulnerable from utility disconnections (NYSPSC 2012). These include waivers for up-front deposit payments, extended deadlines for payment and specific payment plans for older residents. In addition personal visits must be made three days prior to a disconnection and extra time must be allocated if a social service investigation is enacted. Disconnects must be postponed if a medical emergency exists in the household (including life-support systems). Michigan has similar guidelines (MPSC 2015). Clearly, adaptation is occurring in the US within the energy sector to reduce the impacts of extreme events on the health and wellbeing of vulnerable people. However, no documentation could be found to detail utility protection for vulnerable groups in Australia. It may be that Australia can learn from the US example.

What can be done at the local level to reduce the impact of climate change on vulnerable groups?

Supporting community service organisations

While the responsibility for climate change adaptation and mitigation is often seen as the responsibility of national and state government agencies, it is increasingly being recognised that the impact on communities will vary greatly, and local government (Harvison *et al.* 2011) and other community-level organisations will play a critical role (Ebi & Semenza 2008; Mallon *et al.* 2013). This was highlighted in the pilot study on the experiences of older people in a heat-exposed rural community where suggested solutions involved the local district nurse, Lions Club and Meals on Wheels (Loughnan *et al.* 2013b). These community-based organisations are embedded within their communities and already take the primary role in addressing the needs of isolated and vulnerable people, including older people and those with chronic conditions. It is expected that climate change will result in increased demand on their services (Girot *et al.* 2012; Mallon *et al.* 2013).

The need for community service organisations to work to address climate impacts and target vulnerable groups has been recognised in the *Liveable and Just Toolkit* which is an initiative of the Victorian Local Government Association in association with key stakeholders (Edwards & Wiseman 2009). The toolkit identifies a number of areas where climate change is expected to impact and those agencies that will be required to participate in the response. A key example is the provision of Home and Community Care (HACC) services, which provide essential services to older people and those with chronic health conditions, and which the toolkit sees as critical in addressing the impacts with respect to physical and mental health, food, access to health and community services, and social cohesion. HACC personnel can also play a role in helping vulnerable older people and those with chronic health conditions make their homes less susceptible to climate impacts. Environment Victoria is now providing training to HACC service personnel to enable them to undertake sustainability audits of their client's homes as part of their regular home visits (Environment Victoria 2014). These audits provide clients and their families with a list of simple measures to improve the comfort of the household and reduce electricity and water use.

While community service organisations will play an increasingly important role in responding to climate change, a recent review found that organisations may be just as vulnerable to climate change as the clients they serve (Mallon *et al.* 2013). The report estimates that up to 25% of community service organisations are likely to cease operations after a major climate event, leaving their vulnerable communities at considerably greater risk. The report calls for increased funding and support for these organisations to improve their physical infrastructure and skill base.

Development of local at-risk registers and alert functions

Another suggestion of the older participants in the Loughnan *et al.* (2013b) study was the development of personal alert services. This highlights the potential for community members to be unaware of existing services because the Victorian Department of Health already operates such a service, which uses a pendant worn on the wrist or neck that, when activated, alerts a HACC service response team (see www.health.vic.gov.au/agedcare/services/pav/info). This service currently targets people suffering from chronic conditions but could be expanded to cover the broader range of people vulnerable to the effects of climate change. Clearly more work is needed to publicise this service to communities, including more isolated rural communities such as the one targeted in Loughnan *et al.* (2013b).

The development of an at-risk register is another suggestion. This register would target vulnerable groups and be under the control of a local authority such as the local council or police. Such registers were one of the key recommendations of the 2009 Victorian Bushfires Royal Commission (Teague *et al.* 2010) and are now appearing as part of council climate change adaptation strategies across Australia, and as part of state and national bodies such as the Municipal Association of Victoria (see www.mav.asn.au/policy-services/environment/climate-change/impacts-adaptation).

One of the drivers for the Loughnan *et al.* (2013b) study investigating the experience of heat adaptation in older people was to gather the collective experience of these older people who have responded to climate events across their lifetimes and who have adapted to living in a heat-exposed community with limited access to support services. This awareness of the key role that older people and those with chronic conditions can play in sharing their experiences and advocating for their communities can be seen in a number of initiatives in Australia and internationally. This includes the 'Green Sages' initiative established by the Council on the Ageing (COTA Victoria 2014) which aims to empower older people to take action on environmental sustainability issues, develop education and training programs to build knowledge within the seniors community and enable seniors to play a lead role on these issues in their local communities. Similar initiatives targeting older people as peer mentors exist in the US (see www.greenseniors.org) and in Europe (see www.greenseniors.eu).

Conclusion

The vulnerability of older people and those living with chronic health conditions to the impacts of climate change means that it is important that adaptation and response strategies specifically target these groups, and take into consideration where they live and any barriers that they may face to receiving, comprehending and acting upon information regarding preparing for climate change or responding to climate events. While much of the responsibility lies with the state and federal governments, it is clear that climate change will impact very differently at the community level depending upon geographic, demographic and economic factors, so the responsibility for supporting the most vulnerable groups will fall heavily on community service organisations. These organisations will need considerable support to up-skill, to improve their infrastructure to protect them from climate events, and to build a collaborative approach bringing all the services together. Importantly, the role of older people and those with chronic health conditions will be critical as they have experiences to share and will play a key role in advocating for their peers and their communities.

Acknowledgement

The authors would like to acknowledge the assistance of Dr Thu Phan from the School of Earth Atmosphere and Environment at Monash University Australia for the preparation of the maps in Figures 6.1 and 6.2.

References

Abrahamson V, Wolf J, Lorenzoni I, Fenn B, Kovat S, Wilkinson P, Adger WN, Raine R (2009) Perceptions of heatwave risks to health: interview-based study of older people in London and Norwich, UK. *Journal of Public Health* **31**, 119–126. doi:10.1093/pubmed/fdn102

ABS (Australian Bureau of Statistics) (2011) *Australian Standard Geographical Classification (ASGC) Digital Boundaries, Australia, July 2011*. Cat. no. 1259.0.30.001. Australian Bureau of Statistics, Canberra.

ABS (Australian Bureau of Statistics) (2012) *Australian Demographic Statistics, March 2012*. Cat. no. 3101.0. Australian Bureau of Statistics, Canberra.

ABS (Australian Bureau of Statistics) (2013a) *2011 Census QuickStats*. 28/3/2013 ed. Australian Bureau of Statistics, Canberra.

ABS (Australian Bureau of Statistics) (2013b) *Australian Health Survey: Biomedical Results for Chronic Diseases, 2011-12*. Cat. no. 4364.0.55.005. Australian Bureau of Statistics, Canberra.

ACOSS (Australian Council of Social Service) (2012) 'Poverty in Australia' ACOSS Paper 194. Australian Council of Social Service, Strawberry Hills.

APEPSA (Advisory Panel on the Economic Potential of Senior Australians) (2011) 'Realising the economic potential of senior Australians: turning grey into gold'. Commonwealth of Australia, Canberra.

AIHW (Australian Institute of Health and Welfare) (2010) National Hospital Cost Data Collection, Appendix 3. In *Australian Hospital Statistics 2010-2011*. Australian Institute of Health and Welfare, Canberra.

AIHW (Australian Institute of Health and Welfare) (2012a) *Impact of Rurality on Health Status*. Australian Institute of Health and Welfare, Canberra, <http://www.aihw.gov.au/rural-health-impact-of-rurality>.

AIHW (Australian Institute of Health and Welfare) (2012b) 'Health expenditure Australia 2010-11'. Health and welfare expenditure series no. 47. Cat. no. HWE 56. Australian Institute of Health and Welfare, Canberra.

Alexander LV, Arblaster J (2009) Assessing trends in observed and modelled climate extremes over Australia in relation to future projections. *International Journal of Climatology* **29**, 417–435. doi:10.1002/joc.1730

Ambulance Victoria (n.d.) *Fees*. Ambulance Victoria, Doncaster, <http://www.ambulance.vic.gov.au/About-Us/Fees.html>.

Balbus JM, Malina C (2009) Identifying vulnerable subpopulations for climate change health effects in the United States. *Journal of Occupational and Environmental Medicine* **51**, 33–37. doi:10.1097/JOM.0b013e318193e12e

Bambrick HJ, Capon AG, Barnett GB, Beaty RM, Burton AJ (2011) Climate change and health in the urban environment: adaptation opportunities in Australian cities. *Asia-Pacific Journal of Public Health* **23**, 67S–79S. doi:10.1177/1010539510391774

Barnett G, Beaty RM, Chen D, McFallan S, Meyers J, Nguyen M, Ren Z, Spinks A, Wang X (2013) 'Pathways to climate adapted and healthy low income housing'. National Climate Adaptation Research Facility, Gold Coast.

Berry HL, Bowen K, Kjellstrom T (2010) Climate change and mental health: a causal pathways framework. *International Journal of Public Health* **55**, 123–132. doi:10.1007/s00038-009-0112-0

Bureau of Meteorology (2014) *Tropical Cyclone Trends*. Australian Bureau of Meteorology, Canberra, <http://www.bom.gov.au/cyclone/climatology/trends.shtml>.

Carnes BA, Staats D, Willcox BJ (2014) Impact of climate change on elder health. *The Journals of Gerontology. Series A, Biological Sciences and Medical Sciences* **69**(9), 1087–1091. doi:10.1093/gerona/glt159

Chikritzhs T, Whetton S, Daube M, Pascal R, Evans M (2010) Australia, the healthiest nation: death, hospital and cost savings of the Preventative Health Taskforce target reductions for alcohol, 2007 to 2020. *Australasian Medical Journal* **3**(8), 499–503. doi:10.4066/AMJ.2010.408

Coates L (1996) An overview of fatalities from some natural hazards in Australia. In *Proceedings of the NDR96 Conference of Natural Disaster Reduction*. 29 September–2 October, Surfers Paradise, Australia.

Costello A, Abbas M, Allen A, Ball S, Bell S, Bellamy R, Friel S, Groce N, Johnson A, Kett M, Lee M, Levy C, Maslin M, McCoy D, McGuire B, Montgomery H, Napier D, Pagel C, Patel J, de Oliveira JAP, Redclift N, Rees H, Rogger D, Scott J, Stephenson J, Twigg J, Wolff J, Patterson C (2009) Managing the health effects of climate change. *Lancet* **373**, 1693–1733. doi:10.1016/S0140-6736(09)60935-1

COTA (Council on the Ageing) Australia (2012) 'COTA Australia Policy & Position Statements, November 2012'. COTA Australia.

COTA (Council on the Ageing) Victoria (2014) *Green Sages*. COTA Victoria, Melbourne, <http://www.cotavic.org.au/action-advocacy/green-sages-2/>.

Davis S, Bartlett H (2008) Healthy ageing in rural Australia: issues and challenges. *Australasian Journal on Ageing* **27**, 56–60. doi:10.1111/j.1741-6612.2008.00296.x

Ebi KL, Semenza JC (2008) Community-based adaptation to the health impacts of climate change. *American Journal of Preventive Medicine* **35**, 501–507. doi:10.1016/j.amepre.2008.08.018

Ebi KL, Exuzides KA, Lau E, Kelsh M, Barnston A (2004) Weather changes and hospitalisations for cardiovascular diseases and stroke in California. *International Journal of Biometeorology* **49**, 48–58.

Edwards T, Wiseman J (2009) 'Liveable & Just: addressing social and equity impacts of climate change—the case for local government action'. McCaughey Centre, Melbourne.

Environment Victoria (2014) *Learn, Act, Give, Share*. Environment Victoria, Melbourne, <http://environmentvictoria.org.au/training-for-care-workers>.

Gamble J, Hurley B, Schultz P, Jaglom W, Krishnan N, Harris M (2012) Climate change and older Americans: state of the science. *Environmental Health Perspectives* **121**, 15–22. doi:10.1289/ehp.1205223

Giridharan R, Lau SSY, Ganesan S, Givoni B (2007) Urban design factors influencing heat island intensity in high-rise high-density environments of Hong Kong. *Building and Environment* **42**, 3669–3684. doi:10.1016/j.buildenv.2006.09.011

Girot P, Ehrhart C, Oglethorpe J, Reid H, Rossin T, Gambarelli G, Phillips J (2012) 'Integrating community and ecosystem-based approaches in climate change adaptation responses'. Ecosystems & Livelihoods Adaptation Network.

Gurran N, Hamin E, Norman B (2008) 'Planning for climate change: Leading practice principles and models for sea change communities in coastal Australia', University of Sydney, Faculty of Architecture Design & Planning, Sydney.

Hansen A, Bi P, Nitschke M, Ryan P, Pisaniello D, Tucker G (2008) The effect of heat waves on mental health in a temperate Australian city. *Environmental Health Perspectives* **116**, 1369–1375. doi:10.1289/ehp.11339

Hansen A, Bi P, Nitschke M, Pisaniello D, Newbury J, Kitson A (2011a) Older persons and heat-susceptibility: the role of health promotion in a changing climate. *Health Promotion Journal of Australia* **22**, S17–S20.

Hansen A, Bi P, Nitschke M, Pisaniello D, Newbury J, Kitson A (2011b) Perceptions of heat-susceptibility in older persons: barriers to adaptation. *International Journal of Environmental Research and Public Health* **8**, 4714–4728. doi:10.3390/ijerph8124714

Hansen A, Bi P, Saniotis A, Nitschke M, Benson J, Tan Y, Smyth V, Wilson L, Han G-S (2013) 'Extreme heat and climate change: adaptation in culturally and linguistically diverse (CALD) communities'. National Climate Change Adaptation Research Facility, Gold Coast.

Harlan S, Brazel AJ, Prashad L, Stefanov W, Larsen L (2006) Neighborhood microclimates and vulnerability to heat stress. *Social Science & Medicine* **63**, 2847–2863. doi:10.1016/j.socscimed.2006.07.030

Harvison T, Newman R, Judd B (2011) 'Ageing, the built environment and adaptation to climate change'. ACCARNSI Discussion Paper, Node 3 Built Environment, Innovation and Institutional Reform. University of NSW, Sydney.

Hu W, Mengersen K, McMichael A, Tong S (2008) Temperature, air pollution and total mortality during summers in Sydney, 1994–2004. *International Journal of Biometeorology* **52**(7), 689–696. doi:10.1007/s00484-008-0161-8

Hutton D (2008) 'Older people in emergencies: considerations for action and policy development'. World Health Organization, Geneva.

Jennings B (2011) 'Climate change and the elderly report in aging and public health'. American Public Health Association, Washington D.C.

Jerrett M, Finkelstein M (2005) Geographies of risk in studies linking chronic air pollution exposure to health outcomes. *Journal of Toxicology and Environmental Health* **68**, 1207–1242. doi:10.1080/15287390590936085

Kenny GP, Yardley J, Brown C, Sigal RJ, Jay O (2010) Heat stress in older individuals and patients with common chronic diseases. *Canadian Medical Association Journal* **182**, 1053–1060. doi:10.1503/cmaj.081050

Kinney PL, O'Neill MS, Bell ML, Schwartz J (2008) Approaches for estimating effects of climate change on heat-related deaths: challenges and opportunities. *Environmental Science & Policy* **11**, 87–96. doi:10.1016/j.envsci.2007.08.001

Kjellstrom T, Butler AJ, Lucas RM, Bonita R (2010) Public health impact of global heating due to climate change: potential effects on chronic non-communicable diseases. *International Journal of Public Health* **55**, 97–103. doi:10.1007/s00038-009-0090-2

Klinenberg E (2002) *Heat Wave: A Social Autopsy of Disaster in Chicago.* University of Chicago Press, Chicago.

Krupnick A, Alberini A, Cropper M, Simon N, O'Brien B, Goerce R, Heintzelman M (2000) 'Age, health and willingness to pay for mortality risk reductions: a contingent valuation survey of Ontario residents'. Discussion Paper 00-37, 51. Resources for the Future, Washington D.C.

Leonardi G, Haja S, Kovats R, Smith G, Cooper D, Gerard E (2006) Syndromic surveillance use to detect the early effects of heat-waves: an analysis of NHS direct data in England. *Sozial- und Praventivmedizin* **51**, 194–201. doi:10.1007/s00038-006-5039-0

Loughnan M, Nicholls N, Tapper NJ (2008) Demographic, seasonal, and spatial differences in acute myocardial infarction admissions to hospital in Melbourne Australia. *International Journal of Health Geographics* **7**, 42. doi:10.1186/1476-072X-7-42

Loughnan M, Nicholls N, Tapper N (2010a) The effects of summer temperature, age and socio-economic circumstance on acute myocardial infarction admissions in Melbourne, Australia. *International Journal of Health Geographics* **9**, 41. doi:10.1186/1476-072X-9-41

Loughnan ME, Nicholls N, Tapper NJ (2010b) When the heat is on: threshold temperatures for AMI admissions to hospital in Melbourne Australia. *Applied Geography* **30**, 63–69. doi:10.1016/j.apgeog.2009.08.003

Loughnan M, Tapper N, Phan T, Lynch K, McInnes J (2013a) 'A spatial vulnerability analysis of urban populations during extreme heat events in Australian capital cities'. National Climate Change Adaptation Research Facility, Gold Coast.

Loughnan ME, Carroll M, Tapper N (2013b) Learning from our older people: pilot study findings on responding to heat. *Australasian Journal on Ageing* **33**(4) 271–277. doi:10.1111/ajag.12050

Loughnan M, Tapper N, Phan T, McInnes J (2014) Can a spatial index of heat-related vulnerability predict emergency service demand in Australian capital cities? *International Journal of Emergency Services* **3**(1), 6–33. doi:10.1108/IJES-10-2012-0044

Maantay J (2007) Asthma and air pollution in the Bronx: methodological and data considerations in using GIS for environmental justice and health research. *Health and Place* **13**, 32–56. doi:10.1016/j.healthplace.2005.09.009

Maller CJ, Strengers Y (2011) Housing, heat stress and health in a changing climate: promoting the adaptive capacity of vulnerable households, a suggested way forward. *Health Promotion International* **26**, 492–498. doi:10.1093/heapro/dar003

Mallon K, Hamilton E, Black M, Beem B, Abs J (2013) 'Adapting the community sector for climate extremes: extreme weather, climate change & the community sector – risks and adaptations'. National Climate Change Adaptation Research Facility, Gold Coast.

McCann DGC (2011) A review of hurricane disaster planning for the elderly. *World Medical & Health Policy* **3**, 1–5. doi:10.2202/1948-4682.1189

McMichael AJ, Woodruff R, Hales S (2006) Climate change and human health: present and future risks. *Lancet* **367**, 859–869. doi:10.1016/S0140-6736(06)68079-3

Middelmann MH (2007) 'Natural hazards in Australia: identifying risk analysis requirements'. Geoscience Australia, Canberra.

Morris CJG, Simmonds I (2000) Associations between varying magnitudes of the urban heat island and the synoptic climatology in Melbourne, Australia. *International Journal of Climatology* **20**, 1931–1954. doi:10.1002/1097-0088(200012)20:15<1931::AID-JOC578>3.0.CO;2-D

Morton LW, Weng C-Y (2013) Health and healthcare among the rural aging. In *Rural Aging in 21st Century America*. (Eds N Glasgow and EH Berry) pp. 179–194. Springer, New York.

MPSC (Michigan Public Service Commission) (2015) 'Utility related programs and protections for senior citizens'. Consumer Tips, Alert 15-03, January. Michigan Public Service Commission, Lansing.

Nicholls N, Skinner C, Loughnan M, Tapper N (2008) A simple heat alert system for Melbourne Australia. *International Journal of Biometeorology* **52**, 375–384. doi:10.1007/s00484-007-0132-5

Nitschke M, Tucker G (2007) Morbidity and mortality during heatwaves in metropolitan Adelaide. *The Medical Journal of Australia* **187**, 662–665.

NYSPCS (New York State Public Service Commission) (2012) Consumer Guide: PSC Programs and Protections for Senior Citizens. New York State Public Service Commission, Albany, <http://www.dps.ny.gov/seniors>.

Peng B, William S, Loughnan M, Lloyd G, Hansen A, Kjellstrom T, Dea K, Saniotis A (2011) The effects of extreme heat on human mortality and morbidity in Australia: implications for public health. *Asia-Pacific Journal of Public Health* **23**, 27S–36S. doi:10.1177/1010539510391644

Ragusa AT (2010) Seeking trees or escaping traffic? Socio-cultural factors and 'tree-change' migration in Australia. In *Demographic Change in Australia's Rural Landscapes*. (Eds GW Luck, D Race and R Black) pp. 71–99. CSIRO Publishing and Springer, Melbourne.

Reeves J, Foelz C, Grace P, Best P, Marcussen T, Mushtaq S, Stone R, Loughnan M, McEvoy D, Ahmed I, Mullett J, Haynes K, Bird D, Coates L (2010) 'Impacts and adaptation response of infrastructure and communities to heatwaves: the southern Australian experience of 2009'. Queensland University Technology, Brisbane and National Climate Change Adaptation Research Facility, Gold Coast.

Sevoyan A, Hugo G, Feist H, Tan G, McDougall K, Tan Y, Spoehr J (2013) 'Impact of climate change on disadvantaged groups: issues and interventions'. National Climate Change Adaptation Research Facility, Gold Coast.

Sheridan SC (2007) A survey of public perception and response to heat warnings across four North American cities: an evaluation of municipal effectiveness. *International Journal of Biometeorology* **52**, 3–15. doi:10.1007/s00484-006-0052-9

Staddon P, Montgomery H, Depledge M (2014) Climate warming will not decrease winter mortality. *Nature Climate Change* **4**, 190–194. doi:10.1038/nclimate2121

Strengers Y, Maller CJ, Shove E, Walker G (2012) *Air conditioning is peaking out, time to rethink cool comfort*. The Conversation Media Group, Melbourne, <https://theconversation.com/air-conditioning-is-peaking-out-time-to-rethink-cool-comfort-10598>.

Teague B, McLeod R, Pascoe S (2010) '2009 Victorian Bushfires Royal Commission: final report'. Victorian Bushfires Royal Commission, Melbourne.

Toloo G, FitzGerald G, Aitken P, Verrall K, Tong S (2013) Evaluating the effectiveness of heat warning systems: systematic review of epidemiological evidence. *International Journal of Public Health* **58**, 667–681. doi:10.1007/s00038-013-0465-2

Tong S, Ren C, Becker N (2010) Excess deaths during the 2004 heatwave in Brisbane, Australia. *International Journal of Biometeorology* **54**, 393–400. doi:10.1007/s00484-009-0290-8

Vaneckova P, Hart MA, Beggs PJ, de Dear RJ (2008) Synoptic analysis of heat-related mortality in Sydney, Australia, 1993–2001. *International Journal of Biometeorology* **52**, 439–451. doi:10.1007/s00484-007-0138-z

Warburton J, Lui C (2007) 'Social isolation and loneliness among older people: a literature review'. Australasian Centre on Ageing, The University of Queensland, Brisbane.

Weisskopf M, Anderson H, Foldy S, Hanrahan L, Blair K, Torok T, Rumm P (2002) Heat wave morbidity and mortality, Milwaukee, Wis, 1999 vs 1995: an improved response. *American Journal of Public Health* **92**, 830–833. doi:10.2105/AJPH.92.5.830

Younger M, Morrow-Almeida HR, Vindigni SM, Dannenberg AL (2008) The built environment, climate change, and health: opportunities for co-benefits. *American Journal of Preventive Medicine* **35**, 517–526. doi:10.1016/j.amepre.2008.08.017

Women and children

Debra Parkinson, Brad Farrant and Alyssa Duncan

Key points

- Climate change crises deepen inequalities and act as a threat multiplier.
- Women's poverty and inequality, prescribed gender roles, violence against women and children, and the role of women in planning for climate change and disaster management are all implicated in climate change crises impacts.
- Internationally, mortality from disasters is higher for women and children.
- In Australia, approximately 40% of deaths from bushfires are females, while heatwaves affect more women than men.
- Women and children are vulnerable groups in society. Their fates are inextricably tied together through women's role as the primary caregiver to children.
- Climate change will disproportionately affect the physical and mental health of women and children, especially during disasters such as floods, bushfires and heatwaves. The primary source of this disadvantage is both the pre- and post-disaster social structure, including community and familial disruption and violence.
- Men dominate in emergency management and in disaster planning and recovery, resulting in limited gender analysis of these processes.
- Health and community professionals are in a unique position to alleviate human suffering from climate change.
- A range of short- and long-term actions by health and community service organisations will mitigate the challenges of climate change currently disproportionately affecting women and children.

Introduction

Climate change crises[1] deepen inequalities and act as a threat multiplier (GFDRR 2014; Rees et al. 2005; Fordham 2008). Poverty and inequality, not biology, determine women's vulnerability to climate change and ensure that women suffer more than men (Enarson 2012). The issue of gender inequality – marginalised in the everyday – will be further neglected as climate change impacts demand attention and resources (Rohr et al. 2009; Enarson, 2012).

As a 'climate change canary', Australia has already felt the impact of climate change, with recent disasters unprecedented in intensity and frequency (Hanna et al. 2011). Because much of Australia's arable land is already classified as marginal, and climate change is superimposed on top of a background of large climatic variability and extremes, our sensitivity to anthropogenic climate change sets us apart as arguably the most vulnerable of all developed countries (Hanna *et al.* 2011). The southward shifts in rainfall systems that are critical to farming practices add to Australia's vulnerability as both drought and weather-related disasters generate financial insecurity, social dislocation and loss of livelihoods in farming, peri-urban and regional communities (Strazdins *et al.* 2011).

Many health and community services professionals understand the damage that changes in climate bring to people's lives. Those working in regions of Australia affected by more frequent and intense weather events have seen the consequences for men, women and children (Department of Social Services 2008) and have frequently been personally affected. Lengthy periods of drought, for example, wear away at rural people's mental and physical health, and workers use their professional skills to offer support (Anderson 2009). Health professionals in urban areas also witness, and respond to, the effects of heatwaves on vulnerable populations. Indeed, the Canberra bushfires of 2003 and Brisbane floods of 2010–11 presented those urban populations and emergency managers with similar challenges (Camilleri *et al.* 2007; Houghton 2009) – although without the added complications commonly experienced in rural areas, including isolation from services, food, transport and medical assistance.

Extreme weather events can displace people and in some circumstances may inflame conflict in communities and families (Strazdins *et al.* 2011; Rohr *et al.* 2009; McCoy *et al.* 2014). In Australia, as in the rest of the world, women and children face increased interpersonal violence from men (e.g. Enarson 2012; Fothergill 2008; Parkinson & Zara 2013; Whittenbury 2013). The multiple and overlapping stressors that disasters bring result in both short- and long-term harm. Pre-existing social structures have been identified as the primary source of harm in the post-disaster period, as the poorest people suffer the most, and gender inequality becomes more salient (Alston 2013b; Austin 2008; Enarson & Chakrabarti 2009). However, health and community professionals are in a unique position to alleviate human suffering from the impacts of climate change (McCoy *et al.* 2014).

This chapter will conclude with key short- and long-term actions for community service organisations to help women and children face the challenges of climate change.

Disproportionate effect on women globally

Climate change scientists have documented rising sea levels, unprecedented heatwaves, decreasing rainfall and more frequent extreme weather events (e.g. Potsdam Institute 2012; Climate Institute 2013). Worldwide, this has resulted in an estimated number of excess deaths somewhere between 140 000 (WHO 2013) and 400 000 per year (DARA 2012) with another 4.5 million deaths annually due to a carbon-intensive energy system (largely due to air pollution) (DARA 2012). Many of the major killers are highly climate sensitive, such as heatwaves, and therefore expected to worsen with further climate change (WHO 2013). This is discussed in more detail in Chapter 1.

The focus of this chapter is the disproportionate effect of climate change on women and children, especially as it relates to disasters such as floods, bushfires and heatwaves, and their effect on societal institutions, community and family (Strazdins *et al.* 2011; Rohr *et al.* 2009). Women and children are vulnerable groups in society and their fates are inextricably tied with women generally being the primary caregivers to children. This greater burden on women is particularly apparent in countries where inequality and discrimination against women is high. In the developing countries where women are responsible for farming and water provision in the family, climate change brings them an increased workload (UNAIDS 2012). Decreasing supplies of water result in women and girls having to walk further to source water and gather fuel. School attendance is compromised as such tasks demand more time (UNAIDS 2012). In extreme circumstances, the valuing of boys and men over women manifests itself in the distribution of food. In the words of a local man during the 1971 Biafran famine (Biafra is now part of Nigeria): 'stop all this rubbish, it is we men who shall have the food, let the children die, we will make new children after the war' (Rivers 1982).

The lower nutrition for girls and women affects pregnancy and breastfeeding, and the survival of female babies and children (Henrici *et al*. 2010; Rylander *et al*. 2013). In disasters, some cultural practices have contributed to the death of women and girls. They may not have been taught to swim, or traditional clothing and modesty standards may compromise their survival in floods or storm surges; or they may have been expected to remain in the home (Lovekamp 2008; Tyler & Fairbrother 2013; Alston 2013a). Women's lower status often prevents their inclusion in decision-making in education or financial systems, even when they farm for the family (Lovekamp 2008; Alston 2013a).

Internationally, the risk during and after disasters for women and children is greater than for men, borne out by the higher death rate (True 2012; Enarson 2012; Dasgupta *et al*. 2010). In two tragic examples, during the Bangladesh cyclone in 1991, a man told an interviewer that he had been unable to hold both his children, and 'helplessly' chose to save his son to ensure continuance of the family line (Haider *et al*. 1991 cited in Fothergill 1998). In the second, a father told of being trapped with his son and five daughters. He released his daughters one by one, so his son could survive (Akhte 1992 cited in Lovekamp 2008). The ratio of deaths in the 2004 Asian tsunami was three females to one male (Henrici *et al*. 2010). Neumayer and Plümper (2007, p. 551) write that 'biological and physiological differences between the sexes are unlikely to explain large-scale gender differences in mortality', and instead social norms, role behaviours and, primarily, socio-economic status are responsible. This will be further discussed later in the chapter.

Disproportionate effect on women in Australia

In Australia, death tolls from bushfires and heatwaves are gendered. Statistics reveal a greater risk and mortality for women than men in heatwaves (Department of Human Services 2009; Hanrahan 2013). In the Victorian heatwaves of 2009, for example, 68% of heat-related presentations to health and emergency services were women; 50% of these were aged over 75 (Department of Human Services 2009). The higher mortality from heat stress is related to women's longer life expectancy and higher levels of poverty than men. In contrast, the mortality burden in bushfire is felt more heavily by men. On Black Saturday, 42% of the 173 people who died were female, a figure consistent with mortality data from bushfires over the previous 50 years where 40% of the 245 deaths were female (Haynes *et al*. 2010; Victorian Bushfires Royal Commission 2009). Until Black Saturday, the historic trend was that deaths of males and females were equalising, perhaps reflecting more effective bushfire education to men than women and the lack of firefighting skills among women (DeLaine *et al*. 2003; Haynes *et al*. 2010).

Climate change poses health risks for both men and women, especially those with limited personal and financial resources or existing physical and mental health conditions (e.g. UNAIDS 2012; Fritze *et al*. 2008; Brumby *et al*. 2011). There are particular risks for women in pregnancy for both maternal and infant health, as pre-term births are almost twice as likely in heatwaves (Department of Human Services 2009; Hanrahan 2013; Rylander *et al*. 2013). Suicide rates for men are higher than women in Australia, and suicide risk for farming men is particularly high (Carrington *et al*. 2013; Alston 2012). Yet, a gender 'lens' is rarely used when studying the sociological aspects of disaster and crises (Scanlon 1998; Eriksen *et al*. 2010).

In considering women and children, this chapter applies such a lens, recognising that gender shapes our world, and that in Australia as in the rest of the world, the consequences of climate change will hit women harder than men (Alston 2011). Five key issues will be further explored in relation to adaptation to climate change in this context: women's poverty and inequality; prescribed gender roles; violence against women; children; and the role of women in planning for climate change and disaster management. Actions for health and community service workers are then suggested.

Women's poverty and inequality

This section will describe current gender inequality in today's society. Later sections will then discuss how these inequalities are exacerbated by climate change disasters.[2] Climate change amplifies existing socio-economic and gender inequalities, *inter alia* (Costello *et al*. 2009). Women's higher levels of poverty increase women's exposure to climate change crises, and the lower assets and savings base before such crises means financial recovery is tougher for women than men (Henrici *et al*. 2010; Rohr *et al*. 2009; True 2012).

In Australia, the persistent wage differential of 17%[3] and discrimination against women in the workforce contribute to women's financial disadvantage (Pocock *et al*. 2013; Noble & Pease 2011; Summers 2013b). Women retire on just over half the superannuation balances of men and many move into old age precariously, at increasing risk of homelessness (Boetto & McKinnon 2013b; Sharam 2011). In Victoria, more than half the homeless people in 2011 aged 55–64 were women (51%). Sole parents, 82% of whom are women, often survive precariously, particularly if they are unable to find childcare and flexible employment (ABS 2011). An Australian Council of Social Service report in 2013 noted that sole parent families are far more likely to live in poverty than two parent households, and although comprising only 15.9% of all families, half of all children who live in poverty come from sole parent families (ABS 2011; ACOSS 2013). Cheaper housing, for both groups and for other women living in poverty, is usually found on the outskirts of cities or in rural areas. Both locales are more susceptible to bushfire and less likely to have accessible services (Essential Economics 2012; Armstrong 2013). Women have fewer financial resources to allow adaptation, for example, through relocating or retrofitting existing housing (Hansson 2007; DeLaine *et al*. 2003; Henrici *et al*. 2010).

Public transport, although preferred by women more than men, is rarely adequate in peri-urban and regional areas; women are less likely than men to own cars; and increasing petrol costs further limit women's ability to access services or places of employment (e.g. Dobbs 2007; Hamilton & Jenkins 2000; Johnsson-Latham 2007; Scanlon 1998). In times of disaster, women's relative lack of access to cars restricts their ability to escape or evacuate.

Extreme weather events also disturb relationships as a higher rate of marriage breakdown is evident after disasters and during prolonged drought (Alston & Whittenbury 2013; Phillips & Jenkins 2009; Shaw *et al*. 2012). The costs of marriage breakdown are greater for women than they are for men (Gray *et al*. 2010) and further complicated for women on farms. Women's contribution to the family farm, particularly when it is not financial, is frequently unacknowledged (Alston 2011). The value of the farm itself is often devalued by the impacts of climate change – with changing weather patterns that alter harvests, stock load, profits and land use; more common extreme weather events causing damage and loss, and increasing insurance costs; and the consequent volatility of farming reducing the asset value of the farm. Patrilineal inheritance, common in farming families, further disadvantages women, both as siblings of the incumbent farmer or inheritor, and as mothers and widows (Anderson 2009; Wendt & Hornosty 2010).

The highest financial penalty for women emerges from the social expectation that they take on the role of primary carer and domestic worker – recently estimated to cost Australian women approximately one million dollars over a lifetime (Summers 2013a; Broderick 2013). Climate change is known to amplify existing inequalities (Costello *et al.* 2009). As the effect of cumulative economic hardship on health is worse for women than men (Ahnquist *et al.* 2007), the long-term financial impost of climate change is clearly greater for women, with subsequent reductions in health (and access to health care) for women (Borrell 2011).

Prescribed gender roles: heroic men and protected women

The cultural conception of disaster is that men behave heroically to protect helpless women (Eriksen 2014b; Rivers 1982). However, after examining 18 disasters over three centuries, Elinder and Erixson (2012) instead conclude that in disasters, it is 'every man for himself'. Reporting of 'passive' women in disaster is equally misleading, resulting, as it does, from inaccurate reporting and cultural valuing of masculine traits and abilities (Scanlon 1997, 1998). Rather than being protected, Australian women are often alone or with children in actively escaping bushfires (Victorian Bushfires Royal Commission 2009; Eriksen 2014b; Parkinson & Zara 2011). Statements of women preferring to leave early to avoid bushfire are common (Bolin *et al.* 1998; DeLaine *et al.* 2003; Scanlon 1997), but this preference seems to result from socially constructed roles rather than biology. It appears that *parents* with caregiving roles – either male or female – are quicker to feel threatened and prefer to evacuate early, taking children out of harm's way (Fothergill 1998; Mulilis 1999; Tyler & Fairbrother 2013). In the United States after the 1998 Hurricane Bonnie, a cross-sectional survey of 1050 affected households in North Carolina supported a conclusion that gendered roles explain the different attitudes to evacuation:

> *Results from a series of bivariate and multivariate logistic regression analyses indicate that women are more likely to evacuate than men because of socially constructed gender differences in care-giving roles, access to evacuation incentives, exposure to risk, and perceived risk. (Bateman & Edwards 2002, p. 107)*

Australian research indicates that after participation in the 'Fiery Women' training program, women's confidence increased and they changed their preference from evacuation to staying and defending (DeLaine *et al.* 2003; Tyler & Fairbrother 2013). Lifelong conditioning in gender-appropriate activities, women's caring responsibilities and unwelcoming volunteer firefighting organisations combine to result in women having less access to survival skills in a bushfire (Eriksen 2014b). Yet, statistics show that women, skilled or not, are frequently at the fire front, with 18% of women (and 26% of men) in Eriksen's sample in New South Wales and California having fought fire on their property (Eriksen 2014b).

For women who fight fires by accident rather than design, their risk is heightened by a childhood that typically taught girls different skills to boys (Eriksen 2014b; DeLaine *et al.* 2003). In rural settings, this is particularly dangerous. Traditionally, bushfire prevention education has been structured for men by men without consideration of women's participation. The supports and structures that would promote an increase in participation, such as provision of child care, have been notably absent (DeLaine *et al.* 2003; Haynes *et al.* 2010). In the absence of their own knowledge and experience, women tend to rely on men's knowledge to survive. One woman who survived Black Saturday said of her partner, 'He was my fire plan' and risked her life in refusing others' help while she waited for him (Parkinson 2012). Women and children are often unintentionally in the path of fires (and no doubt other weather-related disasters), exposing their particular vulnerabilities. Christine Eriksen, a leading gender and disaster researcher in Australia, identified gendered differences in men's and women's perceptions of risk, desire to prepare for fire, their willingness to act, and a belief in their personal capacity to act. She writes:

When the gendered dimensions of wildfire are investigated in the context of hegemony, a paradox also emerges between women choosing not to take control of their own wildfire safety and women being denied the opportunity to take and be in control. Men, on the other hand, often take control and perform protective roles that many have neither the knowledge nor the ability to safely attempt to fulfil. (Eriksen 2014b, p. 39)

While a woman's role as primary carer of children or the elderly hampers escape in disasters, another barrier to women's autonomy in deciding whether to stay or go is that cultural norms have historically positioned the man of the family as its head, and often the decision-maker in important family matters (Haynes *et al.* 2010). This affects women's freedom to evacuate (Henrici *et al.* 2010; Tyler & Fairbrother 2013). One man interviewed after Black Saturday said:

I have first-hand knowledge that there are women, wives, on Black Saturday who wanted to leave town and their husband said, 'No, we're staying to fight this'. And they stayed to fight and they both died. (Zara & Parkinson 2013, p. 37)

In the community, too, men traditionally held an apparently natural authority, and this hegemonic rural masculinity persists, bolstered by the exclusion of women from firefighting or emergency management roles (Tyler & Fairbrother 2013; Eriksen 2014b). This same structure that is normalised within rural heterosexual couple families dominates in decision-making on drought and farming issues, when decisions largely sit with men about water licences, buy-back schemes and selling the family farm (Alston & Whittenbury 2013; Jones & Tonts 1995). Although largely excluded, women are critically and centrally affected by these decisions.

Influential rural researchers Margaret Alston and Kerri Whittenbury refer to the rural economic downturn reshaping the 'intimate sphere of the farm family' (Alston & Whittenbury 2013). In 2013, they reported that 73% of the women in their sample took off-farm work to supplement farm income compared to 17% of men (Alston & Whittenbury 2013). This work was in addition to bookkeeping, child care, domestic duties and emotional and physical care of husbands. Health costs inevitably emerge from the overwork implicit in the double- and triple-shift of work caused by stressors such as drought (Pini *et al.* 2007; Boetto & McKinnon 2013a; Anderson 2009). The additional burden of work for women is further exacerbated by loss of government services through disaster damage, budget cuts or rural economic decline (Alston 2004; Anderson 2009). The result for women is two-fold in limiting available time for paid work, and a shrinking rural job market as 'feminised' jobs are lost as banks, schools and health centres remove staffed services from rural areas (Alston 2004; Jones & Tonts 1995; Pease 2010).

Gendered roles rarely change in rural areas, and the lack of child care is a key barrier for women – but not for men. Research in 2009 into social impacts of drought found that women's health was affected by increased paid and unpaid work, and being 'overloaded' with demands and, in turn, this led to stress, depression and anxiety (Whittenbury 2013, p. 212). While men face increased workloads as a result of drought (Whittenbury 2013), there is not the expectation that they will be the primary carer of children and other family members as there is for women in most societies (Peek & Fothergill 2009). This is not limited to rural women. In urban areas, time use surveys by the Australian Bureau of Statistics demonstrate that gender roles for most couples are predetermined, with women doing most unpaid caring and domestic work (ABS 2008; AHRC 2013) but euphemisms, excuses and apologies can mask ongoing inequity in who does the double-shift of paid work and child care.

Cox (1998) writes of the invisibility of 'women's work' in caring for family and reconstructing communities following the 1983 Ash Wednesday bushfires, while 'men's work' in rebuilding was celebrated. In Australia and other countries, most post-crisis unpaid care work is still undertaken by women and the burden is heavier due to transport disruption, closed roads, service limitations, retail, school and child care closures (Anderson 2009; Peek & Fothergill 2009; Lovekamp 2008). It is predominantly women who are expected to buy food, provide meals, maintain the house, administer medication, transport family members and provide assistance with the clean up (Whittenbury 2013). This is all in depressed circumstances and with limited finances (Mallon *et al.* 2013).

Times of crisis indeed magnify the strengths and weaknesses in society, and discrimination against women increases as populations look to the past for security (Fothergill 1998; Mulilis 1999; Scanlon 1998). Hoffman (1998) writes that after the 1991 Oakland firestorm in California, there was a sharp return to the 1950s with strictly defined gender roles. Not only were individuals pressured into narrower gender roles but adaptation was managed in a way that reinforced stereotyping, reducing women to 'chief victim and care-taker' (Tschakert & Machado 2012). Tschakert and Machado (2012, p. 276) argue that:

> Undoubtedly, both men and women are impacted by droughts, floods, and heatwaves, and these impacts are experienced differently due to distinct roles determined by cultural norms, the gendered division of labour, and historically rooted practices and power structures.

They conclude that such differential vulnerabilities reflect power structures and that it is time to replace the notion of vulnerability with an examination of *the way* inequality, discrimination and injustice are reproduced through social, economic, political and cultural drivers (Tschakert & Machado 2012). Reproduction of both inequality and gender roles is facilitated and exacerbated in times of climate change induced crisis.

The impact of climate change on children

James Hansen, one of the world's leading climate scientists, has been arguing for many years that, for the sake of the children of today and tomorrow, we need to act quickly to prevent dangerous climate change. Hansen and colleagues argue that:

> *Exposure to media ensures that children cannot escape hearing that their future and that of other species is at stake, and that the window of opportunity to avoid dramatic climate impacts is closing. The psychological health of our children is a priority, but denial of the truth exposes our children to even greater risk. (Hansen* et al. *2013, p. 8)*

Even though children are innocent and non-consenting victims of climate change (Farrant *et al.* 2012) they are particularly vulnerable to its negative effects along with women, the elderly and the disadvantaged. As the predicted climate changes unfold it will be children and future generations who will be most impacted (Hanna *et al.* 2011). Indeed, some researchers estimate that children already suffer around 90% of the global disease burden from climate change (Sheffield & Landrigan 2011).

The time lag from cause to consequence makes climate change an inter-generational health equity issue (Strazdins *et al.* 2011). Understanding the effects of climate change on children requires us to develop new ways of understanding the interactions among child, family, community and the environment (MacCracken 2011).

The Intergovernmental Panel on Climate Change (IPCC) Working Group II report emphasised the lack of preparedness for climate change impacts they observed in many countries (IPCC 2014). Although more research at the regional and local levels is required to better understand, prepare for and adapt to the impacts of climate change, the challenges we face are becoming clearer with every passing day.

Climate change will affect children everywhere, but it will have different consequences in different regions (MacCracken 2011). Because of their immature physiological systems, young children are particularly vulnerable to overheating during heatwaves (Strazdins *et al.* 2011). Children living near rivers (more frequent and intense floods), seasonally arid areas (increased drought and bushfires) and areas prone to water scarcity will be most affected by the environmental changes (MacCracken 2011). Because the poor and the disadvantaged (who have fewer resources with which protect themselves) are and will continue to be among the first and most affected, climate change is expected to amplify the existing social gradient in health, leading to even greater health inequity (Strazdins *et al.* 2011).

At the global level the impact of climate change, including its effects on agricultural productivity and food security, will have a negative effect on child nutrition and health causing an estimated 20% increase in the number of malnourished children by 2050 (IFPRI 2009) and as many as 200 million additional 'environmental refugees' by 2050 (Myers 2002). Although wealthy countries like Australia are unlikely to be among the earliest or the worst affected, these changes along with the predicted rise in food prices are likely to increase the workload of health and community service professionals, particularly those who work with refugees and/or disadvantaged children and families.

Climate change will have a negative impact on public health because of increased average temperatures, extreme weather events and by altering the seasonality, range and incidence of existing diseases (Strazdins *et al.* 2011). Indeed, Australian children will face an estimated 30–100% increase across selected health risks by 2050 and a 3- to 15-fold increase in these health risks by 2100 (Strazdins *et al.* 2011). These include increases to the range and seasonality of vector-borne diseases such as dengue fever and Ross River virus as well as increased prevalence of mental health issues and suicide as a result of the loss and hardship associated with natural disasters including storms, floods, drought and desertification. Climate change will also increase the likelihood and severity of bushfires and therefore the health impacts of smoke (MacCracken 2011).

Increases in weeds, pollen, invasive species and other disease vectors are likely to lead to rises in the incidence of asthma, allergies, disease and other adverse health outcomes that disproportionately affect children. Climate change may also make serious epidemics more likely in communities that have been less affected in the past. It also has the potential to influence children's vulnerability to disease and stimulate the emergence of new diseases (UNICEF 2008).

Health and community service professionals need to be prepared for increased child illness and mental health problems caused by heatwaves, floods, storms and droughts. For example, a record number of days of extreme heat may impact the ability to provide counselling, parenting support, respite and other support services to families and children in out-of-home care (Mallon *et al.* 2013). The lack of support then has negative impacts on children's health and wellbeing, with abuse and neglect, and children being removed by child protection authorities (Mallon *et al.* 2013). These increased needs will often happen at times when the damage caused by the disaster has compromised the service providers' ability to meet the community's needs (Mallon *et al.* 2013). Indeed, the disruption of services caused by extreme weather events would also increase the risks of children being exposed to financial hardship, hunger and disease, and it is likely to make women and children more vulnerable to violence and homelessness (Mallon *et al.* 2013).

Children's immature neurobiology means that they are particularly vulnerable to toxic levels of stress (e.g. Shonkoff & Garner 2012). Exposure to elevated levels of trauma and stress in-utero and during childhood because of extreme weather and related disasters will lead to marked changes in brain development and longer-term cognitive and mental health impacts (Strazdins *et al.* 2011). For example, six months after the 2003 bushfires in Canberra, surveyed children showed much higher rates of emotional problems with nearly half having elevated symptoms of post-traumatic stress disorder (McDermott *et al.* 2005). Younger children were more likely to develop problems (McDermott *et al.* 2005).

An emerging area of research focuses on children and teenagers home alone during disaster or threatened disaster (Eriksen 2014b; Davie *et al.* 2014). Prolonged exposure to extreme weather (e.g. drought) has also been found to be associated with increased child and adolescent psychological distress over time (Stain *et al.* 2011). Many children will suffer the adverse mental, physical and social health impacts caused by climate-change-related forced migration (Britton & Howden-Chapman 2011). There are already six islands in the Torres Strait that are facing inundation. There is also evidence that anxiety about future climate change is already affecting children's mental health (UNICEF 2008).

Violence against women and children increases after disasters (Bartlett 2008; Sety 2012) and stressed parents are more likely to have family conflict and engage in maladaptive parenting. This means that climate change will put children at greater risk for the adverse developmental outcomes associated with maladaptive parenting (Strazdins *et al.* 2011). These adverse outcomes are physical, psychological and behavioural. For example, research in the United States in the six months after a hurricane found a five-fold increase in inflicted traumatic brain injuries in children under two linked to child abuse (Keenan *et al.* 2004). In these situations older children often turn to people and organisations they know for support (e.g. schools, sporting clubs) so it is important that they are alert to the likely issues and know what support is available for these children. Children and adults alike will also experience place-based distress (solastalgia) at the unwelcome changes to home environments at the local and regional levels as a result of climate change (Farrant *et al.* 2012; Albrecht *et al.* 2007). It is therefore important that health and community service professionals are adequately trained and resourced to help identify and ameliorate these effects.

The protection of children is society's highest calling. If we do not act fast to do the mitigation and adaptation required to protect them, our children and future generations will be the first to have poorer health and mental health than the generation before them. Indeed, as Health Care Without Harm pointed out in their latest report, the responsibility to lead mitigation and adaptation efforts is deeply connected to the mission of health workers and the sector more generally (HCWH 2014). This includes leading the switch to renewable energy and focusing on energy and water efficiency (HCWH 2014).

We must include the voices of women, young people and children in the debate about climate change and encourage them to participate in planning our mitigation and adaptation strategies. In conversations about the urgent mitigation that is necessary to avoid dangerous global warming, it is also important to acknowledge the many health promoting co-benefits that are associated with reducing our greenhouse gas emissions related to energy, food, buildings and transport (IPCC 2014). These include the benefits from active transport, reduced air pollution, and eating less red and processed meat. We, too, must demand more action at the personal, local, national and international levels.

Climate change and violence against women

Climate change and extreme weather events are associated with increased violence against women by male intimate partners around the world – including in developed countries such as the United States, the United Kingdom and New Zealand (e.g. Enarson & Chakrabarti 2009; Fothergill 1999, 2008; Austin 2008). In Australia, violence against women increases after disasters such as bushfires (Parkinson *et al.* 2011; Parkinson & Zara 2013) and slow-onset disasters such as drought (Whittenbury 2013). The link between extreme heat and domestic violence is not clear; however, Anderson *et al.* (2000) found increased domestic violence and sexual assault during short heatwaves. Other studies looking at wider community crime was inconclusive (Walker 2012). Several explanations for links between violence and extreme weather events have been proposed. One theory is that men's feelings of inadequacy in meeting gender-based expectations in either catastrophic disaster or the failure of the family farm result in some men reverting to a form of hyper-masculinity and using violence against women (Enarson 1999; Whittenbury 2013; Austin 2008). Other pressures – for both men and women – include the trauma of the disaster itself; loss of control; loss of place and solastalgia (Albrecht *et al.* 2007); loss of people through death or relocation; new reliance on 'charity'; red tape and bureaucracy; and perceived unfairness in grants, insurance payouts, awards and allocation of paid work in reconstruction. The post-disaster period is fraught, too, with practical pressures of unemployment, financial problems, temporary crowded accommodation and homelessness.

However, only some men choose to be violent and whether such stressors 'cause' violence is controversial and complex. Delaying action while waiting for ever more proof is counter-productive and leaves women and children unsafe in times of climate crises (Bain 2014). Traditionally women have been reluctant to speak of intimate partner violence as domestic violence and sexual assault remain shameful to women experiencing assaults by husbands and male partners (Carrington *et al.* 2013; Lievore 2003). This reluctance is magnified after disasters as women are even less likely to be heard by family and even by professionals. A woman interviewed after Black Saturday said:

> I rang [his counsellor] and said, 'Listen, you need to know it's not all rosy here, he needs help, he's angry, he's scaring me, this is not healthy for a baby, not for a [child] to be around, it's not right.' And then as soon as she started talking to him in the next session he comes home and goes, 'That was my final session, she says I'm doing really well'. (Angela) (Cited in Parkinson & Zara 2013, p. 24)

Health professionals and community members alike can assist women by understanding that disaster is no excuse for domestic violence, and by responding effectively when women try to speak of their partner's violence.[4] In normal circumstances, health professionals working with women affected by such crises may be unlikely to hear about the women's experiences of violence by their intimate partner unless they ask. Rurality increases women's reluctance to report due to lack of confidentiality, high conservatism and patriarchy in rural areas (Anderson 2009; Carrington *et al.* 2013). Disaster and drought further diminish any likelihood that women will seek help.

Disaster intensifies the everyday excusing of men's violence in Australia. A 2010 report revealed that one in five Australians would excuse domestic violence if it follows 'temporary loss of control' or is followed by regret by the perpetrator (VicHealth 2010, p. 8). In a climate change crisis, sympathies tend to lie with suffering and heroic men, while women face increasing pressure to sacrifice their health and wellbeing for the good of the family and community. Tolerance of violence is observed to be high in the 'mate' cultures in emergency organisations (e.g. Ainsworth *et al.* 2014; Pacholok 2013; Eriksen 2014b), and disaster researchers in the United States write that experience of disaster may be seen to justify violent behaviour (Anastario *et al.* 2009; Austin 2008; Fothergill 1999).

Research with a sample of 30 women after Black Saturday revealed that service providers and community members excused men's violence, advising some women to 'give it some time', that 'he's not himself', and that 'things will settle down'. Some women spoke of police, health professionals, case managers, and even trauma counsellors turning away from sound domestic violence responses as a way of respecting the men and the trauma they had experienced (Parkinson 2012).

In other crises, too, women's reporting of violence is silenced before it is voiced (Austin 2008; Enarson & Phillips 2008; Bradshaw 2004). Post-disaster and drought services are rarely established with attention to domestic violence, resulting in inappropriate responses to women seeking help (e.g. Enarson 1998; Phillips & Jenkins 2009; Sety 2012). After Black Saturday, no formal body had oversight of domestic violence or of the recording or collation of accurate data (Parkinson *et al.* 2011). For example, no case managers were trained to identify family violence by the Victorian bushfire case-management system established immediately after Black Saturday. In a constructive move, the Victorian Department of Human Services Emergency Management Unit has since taken action to ensure an improved response in future disasters. Data gathering forms have been amended to include domestic violence, and part funding was made available for a downloadable training package in identifying and responding to domestic violence after disaster (see www. whealth.com.au/environmentaljustice/family-violence-and-disaster.html).

Knowledge of increased violence after disaster must inform future disaster planning. In evacuation and recovery facilities, and meetings, awareness of the vulnerability of women and children to violent men can minimise risk (Phillips & Jenkins 2009; Wilson *et al.* 1998; Enarson 2012), and provision of housing options will mean women are not faced with a choice between homelessness or returning to abusive partners (Phillips & Jenkins 2009).

Women in climate change planning and emergency management

This section begins with an overview of women's role in climate change planning at the global level as this is critical in determining the contribution women can make in decision-making. It then considers women in firefighting, and is followed by recommendations for what actions can be taken so that women and children gain the knowledge, skills and opportunities they need. For example, a gendered review of policies and practices by emergency service organisations and relevant bodies, such as local government, will be prerequisite to structural change.

In addition to progressing women's rights, women's equal leadership in climate change at senior and global levels will change the outcomes for the future. Research indicates that women take climate change more seriously and are more likely to implement new strategies to reduce climate change risk (Peek & Fothergill 2009; Rohr *et al.* 2009; Armstrong 2013). Men, on the other hand, tend to trust that a higher level or technological solution will be found (Rohr *et al.* 2009).

This kind of gender analysis is often absent from climate change planning and management, probably as men dominate the decision-making. Since the Earth Summit[5] in 1992, the United Nations has facilitated an international approach to mitigating climate change. Each year many representatives have met to monitor progress at events known as COPs (Conferences of the Parties). A consistent and central role for women – in equal numbers to men – in this endeavour over the past two decades may have seen greater progress. Although the first COP in 1995 included a women's forum with 200 women, inclusion of women at this high level has been inconsistent over the 19 COPs to 2013 (Rohr *et al.* 2009). Instead, worldwide, governments have prevaricated on this critical issue. This same issue of male dominance is replicated in national and state government bodies, where women's voices at senior levels are few. UN Women, an arm of the United Nations, stresses that women's contributions to addressing climate change – whether in communities or globally – are essential, stating that it:

> ... continues to advocate for the adoption of a comprehensive, universal and legally binding climate change agreement ... one that is necessarily gender-responsive, and therefore transformative ... Whether in global discussions or in actions on the ground, women's contributions, participation and leadership in this area are of paramount importance. (UN Women 2013)

In order to effect change at every level, it is recommended that at least a third of those in decision-making positions should be women (Armstrong 2013). Women, like men, are not homogenous. A critical mass is needed to reflect the concerns that research consistently shows women have in regard to the health of the planet (Armstrong 2013). Socially constructed roles influence gendered responses to disaster (Bateman & Edwards 2002; Scanlon 1998). To prioritise the health of future generations decision-making bodies must include people experienced in primary caring roles.

As in political crises, leaders addressing or managing climate crises commonly dismiss gender as a legitimate consideration, nominating the urgency of the task at hand as reason for not 'diverting' attention and resources (Enarson 2012; Rohr *et al.* 2009). In the 50 years since scientists first alerted the world to the greenhouse effect, men in decision-making roles have failed to resolve the climate change crisis. As Rohr *et al.* (2009, p. 298) ask in relation to gender-sensitive policy, 'But can this wait until tomorrow if we are not able to integrate the needs, livelihoods and innovative ideas of more than half of the population today?' Further, in countries where women participate more equally in politics, carbon dioxide emissions per capita are lower (Armstrong 2013).

Women in firefighting

Men, mateship and heroism dominate disaster imagery (e.g. Eriksen *et al.* 2010; Livingston 2011; Phillips & Morrow 2008). Firefighting has historically been characterised by hegemonic masculinity and, as such, brings prestige (e.g. Pease 2011; Connell 2003, 2005). Both action on the fire front and the high physical and mental costs make firefighting one of the most dangerous occupations (Cook & Mitchell 2013).

Several recent studies in Australia and the United States refer to the masculine culture of firefighting and subsequent barriers to women attempting to assume positions either on the front line or in senior roles (e.g. Pacholok 2013; DeLaine *et al.* 2003; Eriksen 2014b). In 2011, only one-fifth (21%) of operational New South Wales Rural Fire Service volunteers were women, and negligible proportions of women reached the level of captain (2%) and deputy captain (4%) (Eriksen 2014a). In Victoria women comprised only 14% of volunteer firefighters in the Country Fire Authority (CFA) in 2011 (Ainsworth *et al.* 2014).

By way of explanation, a 2014 study found the Victorian CFA was perceived by female firefighters as sexist and discriminatory, with some men using offensive language and threatening behaviour, even watching pornography (Ainsworth *et al.* 2014). The culture was described as 'blokey', with lack of attention to basic needs of women such as having uniforms that fit and that allow urination. This led some women to avoid drinking water to minimise the need to remove the uniform – a potentially harmful practice in the heat of fires (Ainsworth *et al.* 2014). Some men used this to tease and humiliate women. According to this study, organisational change was overdue to address discriminatory practices and professionalise the system of promotions to ensure its basis in merit and ability (Ainsworth *et al.* 2014). Eriksen's (2014b) triangulated research comparing and contrasting experiences of gender and wildfire in New South Wales and California found the same alarming discrimination against women in firefighting environments.

> *The challenges many women face when striving to gain recognition for their fire fighting competencies are often a result of subliminal behaviour by men who, in theory, condone equal opportunities in the workplace but have never questioned the ways in which their own behaviour reproduces inequalities and sexism ... [W]omen within the male-dominated sphere of fire fighting are continually reminded of how their gender is a source of discrimination through the habituated and unconscious practices of many male colleagues. (Eriksen 2014b, p. 130)*

A study from the United States reported that firefighters considered competence as being calm in a crisis, using aggressive tactics, controlling emotions and taking risks (Pacholok 2009), none of which are likely to halt climate change. If women are considered incapable – either blatantly as in the firefighting example, or by omission as in the UN COPs – and management of climate change and disasters remain the purview of men, it seems little will change after decades of inaction or inadequate action (Ferris *et al.* 2013; Rohr *et al.* 2009).

Short- and long-term actions

Systems for monitoring weather predictions along with service delivery plans for extreme weather events are vitally important (Mallon *et al.* 2013). Accurate data collection reflecting the status and experience of diverse groups within the community, while at present a low priority in the aftermath of a major disaster, must become a high priority. Sex-disaggregated data will help prevent inequality in disaster recovery. The general failure worldwide to collect this data has been noted as critical (Martin 2010). In the short term, it is important to continue to identify vulnerable groups, adding other groups to those usually noted (e.g. the frail and elderly, people with disabilities and children) such as women in domestic violence situations, sole parents with small children, newly arrived immigrant people, adolescents at home alone, and others who may self-identify as having particular needs (Enarson & Fordham 2001; Martin 2010).

Health and community professionals and their organisations are well placed to address the harmful effects of strict adherence to gender roles for both men and women by:

- promoting non-stereotyped gender roles in the workplace and with clients, and reducing stereotypes of women's lesser competence in crises
- hiring into non-traditional gender roles
- employing women at the highest levels, using temporary special measures until parity is reached
- educating all staff on gender equity
- using language that recognises women's unpaid work and encourages men to do their share
- ensuring a workplace culture that encourages men to equally share parenting and caring roles
- training women as spokespeople for climate change crises
- regularly conducting gender audits of staff and clients and policies
- educating women and children in emergency skills, such as the 'Weathering the Storm' (www.weatherthestorm.net.au) or 'Fiery Women' training (DeLaine *et al.* 2003).

To address children's issues, strategies could include lobbying for a safer world through:

- ensuring mitigation practices are followed in the workplace and at home
- educating school children on climate change mitigation and adaptation
- developing a register of at-risk children and families along with response plans for before and after extreme weather events (Mallon *et al.* 2013)
- advocating for adequate reductions in greenhouse gas emissions at local, national and international levels
- fostering links and knowledge exchange between support services and the schools, sporting and other organisations to which children turn to for support in times of need.

Health and community professionals and organisations could address men's violence against women by:

- ensuring all staff members undertake training on identifying domestic violence after disaster or in other climate change crises, such as prolonged drought or exposure to harmful chemicals such as lead poisoning and water contamination or asthma from air pollution (www.whealth.com.au/work_family_violence_after_disaster.html)
- including information about increased violence against women in the aftermath of crises in brochures on available services and in community hubs, meetings and events
- encouraging staff and others to ask women if they feel safe at home
- ensuring information on referrals to assist women and children with domestic violence issues is available to staff, clients and community members
- involving domestic violence professionals in climate change crises planning, recovery and reconstruction.

Conclusion

The recently released IPCC Working Group II report clearly identifies the significant worldwide impact of climate change on human health and wellbeing (IPCC 2014). These health impacts will dramatically increase in the absence of intensive mitigation and adaptation efforts (IPCC 2014). Our clear responsibility is to push for reduced greenhouse gas emissions (IPCC 2014) and to ensure that women's and children's voices are heard in the climate change debate.

Any crisis offers a window of opportunity for change, and the change can be positive or negative in terms of equity (Birkmann *et al.* 2010; Seib 2008; Hawe 2009). Renegotiating gender roles and adopting wider cultural change will contribute towards greater gender equality. A significant focus must be women and children's right to live free from violence (Council of Australian Governments 2011; OHCHR 1993). In this shift, both men and women would benefit from reduced gendered expectations and the future for the planet itself enhanced by women's equal participation in urgently addressing climate change (Ferris *et al.* 2013). The alternative – as observed after past disasters – is reinforcement of male privilege to the detriment of women, children and other marginalised groups (Enarson & Phillips 2008; Hoffman 1998; Shaw *et al.* 2012). A lot is a stake.

If we are to avoid catastrophic climate change and bequeath a sustainable planet worth living on, we must push, as individuals and as a profession, for a transformed, sustainable, and fair world. (McCoy et al. 2014, p. 2)

Acknowledgements

Thanks to Claire Zara for editing, and to Women's Health In the North, Women's Health Goulburn North East, Monash University and Monash Injury Research Institute. And thank you to Claire Zara and Cathy Weiss for long-standing collaboration.

Endnotes

1. The term 'climate change crisis' is used in this chapter to refer to the felt impacts of climate change, and as such, natural disasters will be an important focus.

2. For detailed discussions and case studies on the gendered impacts of climate change see 'The Impact on Women's Health of Climatic and Economic Disaster: Position Paper 2014' published by the Australian Women's Health Network, Melbourne. ISBN: 978-0-9578645-4-2.

3. Calculation of the gender pay gap excludes part-time, casual earnings and overtime payments. See 'Behind the gender pay gap' published by the Workplace Gender Equality Agency (www.wgea.gov.au/learn/about-pay-equity).

4. A postcard resource produced by Women's Health Goulburn North East states four steps to help: Ask 'Are you safe at home?'; name it as violence; respond with contact details of the Domestic Violence Service and police; and follow up (www.whealth.com.au/environmentaljustice/family-violence-and-disaster.html).

5. The Earth Summit is officially known as United Nations Conference on Environment and Development (UNCED).

References

ABS (Australian Bureau of Statistics) (2008) *How Australians use their time, 2006.* Cat. no. 4153.0. ABS, Canberra, <http://www.abs.gov.au/AUSSTATS/abs@.nsf/Lookup/4153.0Main +Features12006>.

ABS (Australian Bureau of Statistics) (2011) *2011 Census QuickStats.* ABS, Canberra, <http://www.censusdata.abs.gov.au/census_services/getproduct/census/2011/quickstat/0>.

ACOSS (Australian Council of Social Service) (2013) 'Back to basics: simplifying Australia's family payments system to tackle child poverty'. ACOSS, Strawberry Hills.

Ahnquist J, Fredlund P, Wamala SP (2007) Is cumulative exposure to economic hardships more hazardous to women's health than men's? A 16-year followup study of the Swedish Survey of Living Conditions. *Journal of Epidemiology and Community Health* **61**, 331–336. doi:10.1136/jech.2006.049395

AHRC (Australian Human Rights Commission) (2013) 'Investing in care: recognising and valuing those who care, Vol. 1: research report'. Australian Human Rights Commission, Sydney.

Ainsworth S, Batty A, Burchielli R (2014) Women constructing masculinity in voluntary fire-fighting. *Gender, Work and Organization* **21**, 37–56. doi:10.1111/gwao.12010

Albrecht G, Sartore GM, Connor L, Higginbotham N, Freeman S, Kelly B, Stain H, Tonna A, Pollard G (2007) Solastalgia: the distress caused by environmental change. *Australasian Psychiatry* **15**, S95–S98. doi:10.1080/10398560701701288

Alston M (2004) Who is down on the farm? Social aspects of Australian agriculture in the 21st century. *Agriculture and Human Values* **21**, 37–46. doi:10.1023/B:AHUM.0000014019. 84085.59

Alston M (2011) Gender and climate change in Australia. *Journal of Sociology* **47**, 53–70. doi:10.1177/1440783310376848

Alston M (2012) Rural male suicide in Australia. *Social Science & Medicine* **74**, 515–522. doi:10.1016/j.socscimed.2010.04.036

Alston M (2013a) Environmental social work: accounting for gender in climate disasters. *Australian Social Work* **66**, 218–233. doi:10.1080/0312407X.2012.738366

Alston M (2013b) Introducing gender and climate change: research, policy and action. In *Research, Action and Policy: Addressing the Gendered Impacts of Climate Change.* (Eds M Alston and K Whittenbury) pp. 207–222. Springer, Victoria.

Alston M, Whittenbury J (2013) Does climatic crisis in Australia's food bowl create a basis for change in agricultural gender relations? *Agricultural Human Values* **30**, 115–128. doi:10.1007/s10460-012-9382-x

Anastario M, Shehab N, Lawry L (2009) Increased gender-based violence among women internally displaced in Mississippi two years post-Hurricane Katrina. *Disaster Medicine and Public Health Preparedness* **3**, 18–26. doi:10.1097/DMP.0b013e3181979c32

Anderson D (2009) Enduring drought then coping with climate change: lived experience and local resolve in rural mental health. *Rural Society* **19**, 340–352. doi:10.5172/rsj.351.19.4.340

Anderson CA, Anderson KB, Dorr N, DeNeve KM, Flanagan M (2000). Temperature and aggression. In *Advances in Experimental Social Psychology*. (Ed. M Zanna) pp. 63–133. Academic Press, New York.

Armstrong F (2013) Climate change: How it will put women's health at risk globally. *Women's Agenda*, <http://www.womensagenda.com.au/talking-about/opinions/climate-change-how-it-will-put-women-s-health-at-risk-globally/201305092120>.

Austin DW (2008) Hyper-masculinity and disaster: gender role construction in the wake of Hurricane Katrina. In *American Sociological Association Annual Meeting, Sheraton Boston and the Boston Marriott Copley Place*, 31 July, Boston. American Sociological Association, <http://research.allacademic.com/meta/p_mla_apa_research_citation/2/4/1/5/3/p241530_index.html>.

Bain A (2014) Gender-based violence: stop looking for 'proof' and put survivors first. *The Guardian*, 19 February 2014, <http://www.theguardian.com/global-development-professionals-network/2014/feb/18/gender-based-violence-service-based-data>.

Bartlett S (2008) 'Climate change and urban children: Impacts and implications for adaptation in low- and middle-income countries'. International Institute for Environmental Development, London.

Bateman J, Edwards B (2002) Gender and evacuation: a closer look at why women are more likely to evacuate for hurricanes. *Natural Hazards Review* **3**(3), 107–117. doi:10.1061/(ASCE)1527-6988(2002)3:3(107)

Birkmann J, Buckle P, Jaeger J, Pelling M, Setiadi N, Garschagen M, Fernando N, Kropp J (2010) Extreme events and disasters: a window of opportunity for change? Analysis of organizational, institutional and political changes, formal and informal responses after mega-disasters. *Natural Hazards* **55**, 637–655. doi:10.1007/s11069-008-9319-2

Boetto H, McKinnon J (2013*a*), Gender and climate change in rural Australia: a review of differences. *Critical Social Work* **14**(1), 15–31.

Boetto H, McKinnon J (2013*b*), Rural women and climate change: a gender-inclusive perspective. *Australian Social Work* **66**, 234–247. doi:10.1080/0312407X.2013.780630

Bolin R, Jackson M, Crist A (1998) Gender inequality, vulnerability, and disaster: issues in theory and research. In *The Gendered Terrain of Disaster: Through Women's Eyes*. (Eds E Enarson and BH Morrow) pp. 17–32. Praeger Publishers, Westport, Connecticut.

Borrell J (2011) Rupture, loss, identity and place following the 2009 Victorian bushfires: a theoretical exploration. *New Community Quarterly* **9**, 14–22.

Bradshaw S (2004) 'Socio-economic impacts of natural disasters: a gender analysis'. Sustainable Development and Human Settlements División, Women and Development Unit, United Nations Publications, Santiago, Chile.

Britton E, Howden-Chapman P (2011) The effect of climate change on children living on Pacific islands. *International Public Health Journal* **2**, 459–468.

Broderick E (2013). Caring for the carers. *The Australian*, <http://www.theaustralian.com.au/national-affairs/opinion/caring-for-the-carers/story-e6frgd0x-1226585430408>.

Brumby S, Chandrasekara A, McCoombe S, Kremer P, Lewandowski P (2011) Farming fit? Dispelling the Australian agrarian myth. *BMC Research Notes* **4**, 89. doi:10.1186/1756-0500-4-89

Camilleri P, Healy C, Macdonald E, Sykes J, Winkworth G, Woodward M (2007) 'Recovering from the 2003 Canberra Bushfire: a work in progress'. Australian Catholic University, Canberra.

Carrington K, Mcintosh A, Hogg R, Scott J (2013) Rural masculinities and the internalisation of violence. *International Journal of Rural Criminology* **2**(1), 3–24.

Connell RW (2003) 'The role of men and boys in achieving gender equality'. United Nations Division for the Advancement of Women in collaboration with International Labour Organization, Joint United Nations Programmes on HIV/AIDS, United Nations Development Programme, Brasilia, Brazil.

Connell RW (2005) *Masculinities*. 2nd edn. University of California Press, Berkeley and Los Angeles, California.

Cook B, Mitchell W (2013) 'Occupational health effects for firefighters: the extent and implications of physical and psychological injuries'. United Firefighters Union of Australia, Victorian branch, Melbourne.

Costello A, Abbas M, Allen A, Ball S, Bell S, Bellamy R, Friel S, Groce N, Johnson A, Kett M, Lee M, Levy C, Maslin M, McCoy D, McGuire B, Montgomery H, Napier D, Pagel C, Patel J, de Oliveira JAP, Redclift N, Rees H, Rogger D, Scott J, Stephenson J, Twigg J, Wolff J, Patterson C (2009) Managing the health effects of climate change. *Lancet* **373**(9676), 1693–1733. doi:10.1016/S0140-6736(09)60935-1

Council of Australian Governments (2011) 'National Plan to Reduce Violence Against Women and their Children'. Commonwealth of Australia, Canberra.

Cox H (1998) Women in bushfire territory. In *The Gendered Terrain of Disaster: Through Women's Eyes*. (Eds E Enarson and BH Morrow) pp. 165–179. Praeger Publishers, Westport, Connecticut.

DARA (2012) *Climate Vulnerability Monitor: A Guide to the Cold Calculus of a Hot Planet*. 2nd edn. Fundacion DARA Internacional, Madrid.

Dasgupta S, Siriner I, Partha SD (Eds) (2010) *Women's Encounter with Disaster*. Frontpage Publications, London.

Davie S, Stuart M, Williams F, Erwin E (2014) Child friendly spaces: protecting and supporting children in emergency response and recovery. *Australian Journal of Emergency Management* **29**(1), 25–30.

DeLaine D, Probert J, Pedler T, Goodman H, Rowe C (2003) Fiery women: consulting, designing, delivering and evaluating pilot women's bushfire safety skills workshops. In *The International Bushfire Research Conference 2008 - incorporating the 15th annual AFAC Conference*, September, Adelaide, South Australia.

Department of Human Services (2009) 'January 2009 heatwave in Victoria: an assessment of health impacts'. Victorian Government, Melbourne.

Dobbs L (2007) Stuck in the slow lane: reconceptualizing the links between gender, transport and employment. *Gender, Work and Organization* **14**, 85–108. doi:10.1111/j.1468-0432.2007.00334.x

Department of Social Services (2008) 'Report from National Rural Women's Summit'. Department of Social Services, Canberra.

Elinder M, Erixson O (2012) Gender, social norms, and survival in maritime disasters. *Proceedings of the National Academy of Sciences of the United States of America* **109**, 13220–13224. doi:10.1073/pnas.1207156109

Enarson E (1998) Through women's eyes: a gendered research agenda for disaster social science. *Disasters* **22**, 157–173. doi:10.1111/1467-7717.00083

Enarson E (1999) Violence against women in disasters: a study of domestic violence programs in the United States and Canada. *Violence Against Women* **5**, 742–768. doi:10.1177/10778019922181464

Enarson E (2012) *Women Confronting Natural Disaster: From Vulnerability to Resilience*. Lynne Rienner Publishers, Boulder, Colorado.

Enarson E, Chakrabarti P (Eds) (2009) *Women, Gender and Disaster: Global Issues and Initiatives*. Sage Publications, New Delhi.

Enarson E, Fordham M (2001) Lines that divide, ties that bind: race, class, and gender in women's flood recovery in the US and UK. *The Australian Journal of Emergency Management* **15**, 43–52.

Enarson E, Phillips B (2008) Invitation to a new feminist disaster sociology: integrating feminist theory and methods. In *Women and Disasters: From Theory to Practice*. (Eds BD Phillips and BH Morrow) pp. 41–74. Xlibris, USA.

Eriksen C (2014a) Gendered risk engagement: challenging the embedded vulnerability, social norms and power relations in conventional Australian bushfire education. *Geographical Research* **52**(1), 23–33. doi:10.1111/1745-5871.12046

Eriksen C (2014b) *Gender and Wildfire: Landscapes of Uncertainty*. Routledge, New York.

Eriksen C, Gill N, Head L (2010) The gendered dimensions of bushfire in changing rural environments. *Journal of Rural Studies* **26**, 332–342. doi:10.1016/j.jrurstud.2010.06.001

Essential Economics (2012) 'One Melbourne or Two? Final Report'. Essential Economics, Carlton, Victoria.

Farrant BM, Armstrong F, Albrecht G (2012) Future under threat: climate change and children's health. *The Conversation*, Australia, <https://theconversation.edu.au/future-under-threat-climate-change-and-childrens-health-9750>.

Ferris E, Petz D, Stark C (2013) 'The year of recurring disasters: a review of natural disasters in 2012'. The Brookings Institute, London.

Fordham M (2008) The intersection of gender and social class in disaster: Balancing resilience and vulnerability. *International Journal of Mass Emergencies and Disasters* **17**(1), 15–36.

Fothergill A (1998) The neglect of gender in disaster work: an overview of the literature. In *The Gendered Terrain of Disaster: Through Women's Eyes*. (Eds E Enarson and BH Morrow) pp. 11–25. Praeger Publishers, Westport, Connecticut.

Fothergill A (1999) An exploratory study of woman battering in the Grand Forks flood disaster: implications for community responses and policies. *International Journal of Mass Emergencies and Disasters* **17**, 79–98.

Fothergill A (2008) Domestic violence after disaster: voices from the 1997 Grand Forks flood. In *Women and Disasters: From Theory to Practice*. (Eds BD Phillips and BH Morrow) pp. 131–154. Xlibris, USA.

Fritze J, Blashki GA, Burke S, Wiseman J (2008) Hope, despair and transformation: climate change and the promotion of mental health and wellbeing *International Journal of Mental Health Systems* **2**(17 September), 2–13.

GFDRR (Global Facility For Disaster Reduction and Recovery) (2014) *Climate Change Adaptation*, Washington D.C., <https://www.gfdrr.org/cca>.

Gray M, De Vaus D, Qu L, Stanton D (2010) 'Divorce and the wellbeing of older Australians'. Research paper No. 46. Australian Institute of Family Studies, Melbourne.

Hamilton K, Jenkins L (2000) A gender audit for public transport: a new policy tool in the tackling of social exclusion. *Urban Studies (Routledge)* **37**, 1793–1800. doi:10.1080/00420980020080411

Hanna EG, Mccubbin J, Strazdins L, Horton G (2011) Australia, lucky country or climate change canary: what future for her rural children? *International Public Health Journal* **2**, 501–512.

Hanrahan C (2013) Heatwaves linked to preterm births. *Medical Observer*, Sydney, <http://www.medicalobserver.com.au/news/heatwaves-linked-to-preterm-births>.

Hansen J, Kharecha P, Sato M, Masson-Delmotte V, Ackerman F, Beerling DJ, Hearty PJ, Hoegh-Guldberg O, Hsu SL, Parmesan C, Rockstrom J, Rohling EJ, Sachs J, Smith P, Steffen K, Van Susteren L, Von Schuckmann K, Zachos JC (2013) Assessing 'dangerous climate change': required reduction of carbon emissions to protect young people, future generations and nature. *PLoS ONE* **8**(12), e81648. doi:10.1371/journal.pone.0081648

Hansson SO (2007) 'Gender issues in climate adaptation'. Swedish Defence Research Agency, Ministry of Defence, Stockholm.

Hawe P (2009) 'Community recovery after the February 2009 Victorian bushfires: a rapid review'. The Sax Institute for the Victorian Government Department Of Health, Melbourne.

Haynes K, Tibbits A, Coates L, Ganewatta G, Handmer J, Mcanerney J (2010) 100 years of Australian civilian bushfire fatalities: exploring the trends in relation to the 'stay or go policy'. *Environmental Science & Policy* **13**(3), 185–194. doi:10.1016/j.envsci.2010.03.002

HCWH (Health Care Without Harm (2014) 'Health care and climate change: an opportunity for transformative leadership'. Health Care Without Harm, <http://noharm-uscanada.org/documents/health-care-climate-change-opportunity-transformative-leadership>.

Henrici JM, Helmuth AS, Braun J (2010) 'Women, disasters, and Hurricane Katrina'. Institute for Women's Policy Research, Washington, D.C.

Hoffman S (1998) Eve and Adam among the embers: gender patterns after the Oakland Berkeley firestorm. In *The Gendered Terrain of Disaster: Through Women's Eyes*. (Eds E Enarson and BH Morrow) pp. 55–61. Praeger Publishers, Westport, Connecticut.

Houghton R (2009) 'Everything became a struggle, absolute struggle': post-flood increases in domestic violence in New Zealand. In *Women, Gender and Disaster: Global Issues and Initiatives*. (Eds E Enarson and P Chakrabarti) pp. 99–111. Sage, New Dehli.

IFPRI (International Food Policy Research Institute (2009) 'Climate change: impact on agriculture and costs of adaptation'. The International Food Policy Research Institute, Washington D.C.

IPCC (2014) *Climate Change 2014: Impacts, Adaptation, and Vulnerability*. Intergovernmental Panel on Climate Change, <http://www.ipcc.ch/report/ar5/wg2/>.

Johnsson-Latham G (2007) 'A study on gender equality as a prerequisite for sustainable development'. Environment Advisory Council, Sweden.

Jones R, Tonts M (1995) Rural restructuring and social sustainability: some reflections on the Western Australian wheatbelt. *The Australian Geographer* **26**, 133–140. doi:10.1080/00049189508703142

Keenan HT, Marshall SW, Nocera MA, Runyan DK (2004) Increased incidence of inflicted traumatic brain injury in children after a natural disaster. *American Journal of Preventive Medicine* **26**, 189–193. doi:10.1016/j.amepre.2003.10.023

Lievore D (2003) 'Non-reporting and hidden recording of sexual assault: an international literature review'. Department of the Prime Minister and Cabinet, Barton, ACT.

Livingston M (2011) A longitudinal analysis of alcohol outlet density and domestic violence. *Addiction (Abingdon, England)* **106**, 919–925. doi:10.1111/j.1360-0443.2010.03333.x

Lovekamp W (2008) Gender and disaster: a synthesis of flood research in Bangladesh. In *Women and Disasters: From Theory to Practice.* (Eds BD Phillips and BH Morrow) pp. 99–116. Xlibris, USA.

MacCracken MC (2011) Climate change impacts will be pervasive. *International Public Health Journal* **2**, 371–376.

Mallon K, Hamilton E, Black M, Beem B, Abs J (2013) 'Adapting the community sector for climate extremes: extreme weather, climate change and the community sector – risks and adaptations'. National Climate Change Adaptation Research Facility, Gold Coast.

Martin J (2010) Disaster planning and gender mainstreaming: Black Saturday bushfires. *New Community Quarterly* **8**, 3–9.

McCoy D, Montgomery H, Arulkumaran S, Godlee F (2014) Climate change and human survival. *British Medical Journal* **348**, g2351. doi:10.1136/bmj.g2351

McDermott BM, Lee EM, Judd M, Gibbon P (2005) Posttraumatic stress disorder and general psychopathology in children and adolescents following a wildfire disaster. *Canadian Journal of Psychiatry* **50**, 137–143.

Mulilis JP (1999) Gender and earthquake preparedness: a research study of gender issues in disaster management: differences in earthquake preparedness due to traditional stereotyping or cognitive appraisal of threat? *The Australian Journal of Emergency Management* **14**(1), 41–50.

Myers N (2002) Environmental refugees: a growing phenomenon of the 21st century. *Philosophical Transactions of the Royal Society of London. Series B, Biological Sciences* **357**, 609–613. doi:10.1098/rstb.2001.0953

Neumayer E, Plümper T (2007) The gendered nature of natural disasters: the impact of catastrophic events on the gender gap in life expectancy, 1981–2002. *Annals of the Association of American Geographers. Association of American Geographers* **97**, 551–566. doi:10.1111/j.1467-8306.2007.00563.x

Noble C, Pease B (2011) Interrogating male privilege in the human services and social work education. *Women in Welfare Education* **10**, 29–38.

OHCHR (Office of the United Nations High Commissioner for Human Rights) (1993) 'Declaration on the elimination of violence against women'. United Nations, Geneva.

Pacholok S (2009) Gendered strategies of self: navigating hierarchy and contesting masculinities. *Gender, Work and Organization* **16**(4), 471–500. doi:10.1111/j.1468-0432.2009.00452.x

Pacholok S (2013) *Into the Fire: Disaster and the Remaking of Gender.* University of Toronto Press, Toronto.

Parkinson D (2012) 'The way he tells it – Vol. 1 Relationships after Black Saturday'. Women's Health Goulburn North East, Wangaratta.

Parkinson D, Zara C (Eds) (2011) *Beating the Flames.* Women's Health Goulburn North East, Wangaratta, <www.whealth.com.au/environmentaljustice>.

Parkinson D, Zara C (2013) The hidden disaster: domestic violence in the aftermath of natural disaster. *Australian Journal of Emergency Management* **28**, 28–35.

Parkinson D, Lancaster C, Stewart A (2011) A numbers game: women and disaster. *Health Promotion Journal of Australia* **22**(4), 42–45.

Pease B (2010) Reconstructing violent rural masculinities: responding to fractures in the rural gender order in Australia. *Culture, Society and Masculinity* **2**, 154–164. doi:10.3149/CSM.0202.154

Pease B (2011) Governing men's violence against women in Australia. In *Men and Masculinities Around the World: Transforming Men's Practices.* (Eds K Pringle, J Hearn, E Ruspini and B Pease) pp. 177–189. Palgrave Macmillan, New York.

Peek L, Fothergill A (2009) Parenting in the wake of disaster: mothers and fathers respond to Hurricane Katrina. In *Women, Gender and Disaster: Global Issues and Initiatives.* (Eds E Enarson and PG Dhar Chakrabarti) pp. 112–130. Sage, New Delhi.

Phillips B, Jenkins P (2009) Violence and disaster vulnerability. In *Social Vulnerability to Disasters.* (Eds, B Phillips, D Thomas, A Fothergill and L Blinn-Pike) pp. 279–306. CRC Press, Boca Raton.

Phillips BD, Morrow BH (Eds) (2008) *Women and Disasters: From Theory to Practice.* Xlibris, USA.

Pini B, Panelli R, Dale-Hallett L (2007) The Victorian women on farm gatherings: A case study of the Australian 'Women in Agriculture' movement. *The Australian Journal of Politics and History* **53**, 569–580. doi:10.1111/j.1467-8497.2007.00475.x

Pocock B, Charlesworth S, Chapman J (2013) Work-family and work-life pressures in Australia: advancing gender equality in 'good times'? *The International Journal of Sociology and Social Policy* **33**, 594–612. doi:10.1108/IJSSP-11-2012-0100

Potsdam Institute (2012) *Turn Down the Heat: Why a 4°C Warmer World Must Be Avoided.* World Bank, Washington D.C., <http://documents.worldbank.org/curated/en/2012/11/170 97815/turn-down-heat-4%C2%B0c-warmer-world-must-avoided>.

Rees S, Pittaway E, Bartolomei L (2005) Waves of violence – women in post-tsunami Sri Lanka. *Australasian Journal of Disaster and Trauma Studies* **2**, 1–6.

Rivers J (1982) Women and children last: an essay on sex discrimination in disasters. *Disasters* **6**, 256–267. doi:10.1111/j.1467-7717.1982.tb00548.x

Rohr U, Hemmati M, Lambrou Y (2009) Towards gender equality in climate change policy: challenges and perspectives for the future. In *Women, Gender and Disaster: Global Issues and Initiatives.* (Eds E Enarson and PG Dhar Chakrabarti) pp. 289–202. Sage, New Delhi.

Rylander C, Odland JØ, Sandanger TM (2013) Climate change and the potential effects on maternal and pregnancy outcomes: an assessment of the most vulnerable; the mother, fetus, and newborn child. *Global Health Action* **6**, 1–9. doi:10.3402/gha.v6i0.19538

Scanlon J (1997) Human behaviour in disaster: the relevance of gender. *Australian Journal of Emergency Management* **11**, 2–7.

Scanlon J (1998) The perspective of gender: a missing element in disaster response. In *The Gendered Terrain of Disaster: Through Women's Eyes.* (Eds E Enarson and BH Morrow) pp. 45–55. Praeger Publishers, Westport, Connecticut.

Seib GF (2008) In crisis, opportunity for Obama. *The Wall Street Journal*, 21 November 2008, <http://online.wsj.com/news/articles/SB122721278056345271>.

Sety M (2012) Domestic violence and natural disasters. *Australian Domestic and Family Violence Clearinghouse Thematic Review* **3**, 1–10.

Sharam A (2011) 'No home at the end of the road?' Swinburne Institute and Salvation Army Australia Southern Territory, Melbourne.

Shaw C, Van Unen J, Lang V (2012) 'Women's voices from the floodplain'. Economic Security4Women, North Sydney.

Sheffield PE, Landrigan PJ (2011) Global climate change and children's health: threats and strategies for prevention. *Environmental Health Perspectives* **119**, 291–298. doi:10.1289/ehp.1002233

Shonkoff JP, Garner AS (2012) The lifelong effects of early childhood adversity and toxic stress. *Pediatrics* **129**, e232–e246. doi:10.1542/peds.2011-2663

Stain HJ, Dean J, Kelly B, Blinkhorn S, Carnie T (2011) Climate adversity: yet another stressor for rural adolescents. *International Public Health Journal* **2**, 513–519.

Strazdins L, Friel S, Mcmichael A, Butler SW, Hanna E (2011) Climate change and child health in Australia: likely futures, new inequities? *International Public Health Journal* **2**, 493–500.

Summers A (2013a) Gender pay gap still a disgrace. *The Sydney Morning Herald*, 5 January 2013, <http://www.smh.com.au/federal-politics/gender-pay-gap-still-a-disgrace-20130104-2c8o6.html>.

Summers A (2013b) *The Misogyny Factor.* New South Publishing, Sydney.

Climate Institute (2013) *Climate Risks Around Australia: Implications from the Latest IPCC Report Media Brief* The Climate Institute, Melbourne, <http://www.climateinstitute.org.au/articles/media-briefs/climate-risks-around-australia.html>.

True J (2012) *The Political Economy of Violence Against Women.* Oxford University Press, New York.

Tschakert P, Machado M (2012) Gender justice and rights in climate change adaptation: opportunities and pitfalls. *Ethics and Social Welfare* **6**, 275–289. doi:10.1080/17496535.2012.704929

Tyler M, Fairbrother P (2013) Gender, masculinity and bushfire: Australia in an international context *Australian Journal of Emergency Management* **28**, 20–25.

UNAIDS (Joint United Nations Programme on HIV/AIDS) (2012) 'Impact of the global economic crisis on women, girls and gender equality'. Joint United Nations Programme on HIV/AIDS, Geneva.

UNICEF (United Nations International Children's Emergency Fund) (2008) 'Our climate, our children, our responsibility: the implications of climate change for the world's children'. United Nations International Children's Emergency Fund, UK.

VicHealth (2010) 'National Survey on Community Attitudes to Violence Against Women 2009: changing cultures, changing attitudes – preventing violence against women. A summary of findings'. Victorian Health Promotion Foundation, Carlton.

Victorian Bushfires Royal Commission (2009) 'Final report: Volume 1 – The fires and the fire-related deaths'. Victorian Bushfires Royal Commission, Melbourne. <http://www.royal-commission.vic.gov.au/Commission-Reports.html>.

Walker R (2012) 'The relationship between climate change and violence: a literature review'. La Trobe University, Melbourne.

Wendt S, Hornosty J (2010) Understanding contexts of family violence in rural, farming communities: implications for rural women's health. *Rural Society* **20**, 51–63. doi:10.5172/rsj.20.1.51

Whittenbury K (2013) Climate change, women's health, wellbeing and experiences of gender based violence in Australia. In *Research, Action and Policy: Addressing the Gendered Impacts of Climate Change.* (Eds M Alston and K Whittenbury) pp. 207–222. Springer, Dordrecht.

WHO (World Health Organization) (2013) *Climate Change and Health.* World Health Organization <http://www.who.int/mediacentre/factsheets/fs266/en/>.

Wilson J, Phillips BD, Neal DM (1998) Domestic violence after disaster. In *The Gendered Terrain of Disaster: Through Women's Eyes.* (Eds E Enarson and BH Morrow) pp. 115–124. Praeger Publishers, Westport, Connecticut.

Women UN (2013) *Women for Results: Climate Change Initiatives Highlighted at COP-19.* United Nations, <http://www.unwomen.org/en/news/stories/2013/11/gender-days-at-cop19>.

Zara C, Parkinson D (2013) 'Men on Black Saturday: Risks and Opportunities for Change'. WHGNE, Wangaratta.

8

Climate change: impact on country and Aboriginal and Torres Strait Islander culture

Kerry Arabena and Jonathan 'Yotti' Kingsley

Key points

- The chapter aims to provide better mechanisms for incorporating Aboriginal and Torres Strait Islander perspectives into climate change discussions and dialogue.
- Aboriginal and Torres Strait Islander understandings of country are explored as an important contribution to the climate change debate.
- Strength-based approaches to effectively tackle climate change will require humans overcoming disconnections from our ecosystems and moving to approaches that embrace cultural and ecosystem diversity.
- Key strategies mentioned for ensuring Aboriginal and Torres Strait Islander peoples' perspectives are incorporated revolve around respect, collaboration, reciprocity (two-way learning), resilience, community control, better understanding of Indigenous models of health and the environment.
- Clear recommendations have been provided for social/cultural, economic and spiritual action to assist services and individuals to better integrate diverse Aboriginal and Torres Strait Islander knowledge systems.

Introduction

Prior to colonisation in 1788 the Aboriginal and Torres Strait Islander population had the primary responsibility for managing 100% of Australian ecosystems. Now that the number of Aboriginal and Torres Strait Islander people totals less than 4% of the entire Australian population, traditional methods of ecosystem management have been severely disrupted and are excluded from current climate change debates. Throughout this chapter Aboriginal and Torres Strait Islander peoples is the preferred terminology to represent the diverse and distinct populations across the Australian continent that are collectively known as the First Peoples of Australia. The term 'Aboriginal and Torres Strait Islander people' recognises the two different Indigenous populations of Australia: Aboriginal people on mainland Australia and in the island state of Tasmania, and Torres Strait Islander peoples who have occupied islands between mainland Australia and Papua New Guinea for thousands of years. The term 'indigenous' is an international term and refers to populations of peoples that retain historical, social, economic, political and cultural ties to territories other than Australia (AHRC 2013). In this chapter we seek to enhance our understanding of how ecosystems are conceived in Australia, and challenge our current societal obligations to ecosystem management. This is done by incorporating Aboriginal and Torres Strait Islander peoples' perspectives, practices and knowledge in climate change debates.

Ecological scholars have described the urgency to move humanity away from non-sustainable behaviours towards developing new social realities and a more holistic understanding of public health discourse, putting ecosystem health at the forefront (Albrecht *et al.* 2008; Parkes & Horwitz 2009; Gislason 2013). Aboriginal and Torres Strait Islander peoples in Australia have often asserted their 'earth caring' philosophy germane to their identity, social structures and culture, which is often challenged by governments, policy-makers and people who consider themselves to be members of mainstream society (Arabena 2006). The view that both ecological philosophers and Indigenous peoples have knowledge and practices that offer an alternative way to think about environmental and social engagement, outside of the confines of 'modern Australian society', is emphasised in this chapter.

Increasingly, there is congruence between the voices of Indigenous peoples and of ecological philosophers in highlighting how our achievements in science, technology, industry, commerce and finance have brought us into a new age at the expense of the diversity of life and life-enhancing processes. In an attempt to reflect this congruence, the first author of this chapter, a Torres Strait Islander ecological social worker, and the second author, an ecosystem advocate, merge a range of thematic areas, some which we come to differently. The findings advocated herein derive from personal experiences of living and working in both Aboriginal and Torres Strait Islander communities and in 'modern Australian society' and a shared concern for the environment, and from within each of the author's worldviews which is founded in different knowledge, philosophies, life experiences and cultural backgrounds.

This task was undertaken in the knowledge that we were challenging conventional wisdom by offering an alternative view to that of modern Australian society, by holding land and sea as the enduring reality, binding together people, culture and livelihood. For Aboriginal and Torres Strait Islander people, land and sea constitute 'country', a term that encapsulates the geographical landscape and ancestral and social connections to a place, to which people have a reciprocal relationship and responsibility. Ecological philosophers focus on the utter dependence of human beings on the vitality of life-sustaining services (air, water, soil), and on their capacity to recognise and draw on faith-based traditions and an established evidence of connectivity. Aboriginal and Torres Strait Islander laws and customary practices have shaped the environments of Australia for thousands of generations. Ecological evidence points to the extent of human activities and their significant global impact on the earth's ecosystems.

Frameworks, knowledge and practices incorporating both the holistic and reciprocal notion of country can considerably strengthen our ability to tackle the environmental concerns of the future (Kingsley *et al.* 2013b; Rose 2013). Scholars have gone to great lengths to explain the importance of local and Indigenous perspectives of the land in improving current ecosystem and human health problems (Johnston *et al.* 2007; Burgess *et al.* 2008; Parkes 2010). However, settler society has a commitment to a financial system that has provided comfort and security to a great many people at the expense of the ecospheres in which we live. Strang (2009, p. 22) highlighted this by arguing that:

Australia has been committed to a positive vision of competitive growth for so long that it has become widely normalized … However, the ecological crisis has brought simmering doubts about the guiding principles to the fore. Australian society is thus being forced to consider whether it is functional … to ignore pressing issues such as climate change … or to cling to the mantra that technical advances and efficiencies will deal effectively with the ecology problems that will inevitably accompany further intensification in resource use. Most people demonstrated a perennial human capacity to contain conflicting ideas simultaneously, arguing for most sustainable environmental management while maintaining lifestyles that – replicated throughout the population – make this impossible to achieve. Such capacity for denial is a source of frustration to those who see a need for real change …

From our perspective, real change requires a disengagement from processes that are not intrinsic to the ecosystems in which we live.

Climate change, Indigenous adaptation, Indigenous knowledge

While the ways in which Indigenous peoples can contribute to climate change adaptation are little understood, studies around the world have shown that indigenous adaptation and environmental knowledge systems will be extremely useful to those forced to relocate due to the consequences of climate change (Devkota *et al.* 2013; Zander *et al.* 2013). Therefore, it is not surprising that a decade ago Rose (2005) recommended that to truly tackle this dangerous climate change we must challenge ourselves to have cross-cultural discussion between Aboriginal and non-Indigenous Australians. Some academics, however, identify that Indigenous peoples have been excluded or have had a lack of input into climate change policy, research and practice that impacts on their regions (Gerrard 2008; Petheram *et al.* 2010; Bradsley & Wiseman 2012; Ford 2012; Green *et al.* 2012; Choy *et al.* 2013; Cochran *et al.* 2013).

Advocating for a change in this dialogue is justified because Indigenous peoples' deep historical and holistic understanding, management and knowledge of the ecosystem (involving social, cultural and environmental elements) will contribute greatly to environmental sustainability through appropriate climate change adaptation measures that can complement current processes (Turner & Clifton 2009; Alexander *et al.* 2011; Wiseman & Bradsley 2013). Unfortunately, McIntyre-Tamwoy *et al.* (2013, p. 105) found that currently in Australia:

Indigenous communities have had little input or influence in the shaping of government policies and practice relating to climate change … This is in contrast to trends elsewhere in the western world where public risk perception plays a key role in shaping natural hazards policy and management response systems.

Collaborative work in this space in Canada, for example, has developed to the point where indigenous knowledge has transformed mainstream policy. Research with Inuvialuit people in the Western Arctic region looking at short- and long-term climate adaptation responses led to the recommendation of co-management institutes and learning at all levels of government and community (Berkes & Jolly 2001).

Despite such responses, the extent of the health impacts of climate change may be greater even than figures suggest because of Indigenous peoples' deep connection to the environment (Strand *et al.* 2010). This is evident when looking at Aboriginal peoples and the impact climate change has had, is having and will have, on these populations in exacerbating their existing disadvantage (Campbell *et al.* 2008). The traditional lifestyles and practices of indigenous peoples are becoming increasingly vulnerable to the impacts of climate change, although through traditional storytelling people have demonstrated a capacity to adapt and maintain connection to these 'disappearing places' (Sakakibara 2008). Sakakibara (2008, p. 473) explains that the 'sea may be eroding … but storytelling weaves old and new homes into a viable place of cultural survival'. Climate change, therefore, has been identified as a human rights issue in indigenous populations for two reasons: its direct impact on the health, economic and social conditions of indigenous peoples; and its destruction of traditional places of spiritual significance causing flow-on psychosocial health issues (Heinamaki 2009; Lanyi 2012).

Spartz and Shaw (2011) identified that these health consequences occur when modifications are made to places and/or displacement occurs from one's environments, which can cause a serious form of psychological disorder. Read (1996, p. 197) explained that loss of places is often not explored and shared with individuals, who tend to have to manage their own psychological grief in isolation, stating that:

> … *in Australia, environmental impact assessments had considered the usual criteria of dust, noise, vibration and environmental damage. They have not assessed the impact of loss of home, community and countryside.*

This is similar to the idea of solastalgia, the psychological distress felt when a local environment to which there is personal attachment is affected (Connor *et al.* 2004; Albrecht 2005; Albrecht *et al.* 2007). Albrecht *et al.* (2007) acknowledged that the psychological consequences felt by Aboriginal and Torres Strait Islander peoples who have viewed the destruction of their lands are often expressed as a loss of identity and increasing social isolation. Indigenous communities have been disproportionately affected by environmental change and, until recently, have had little or no involvement in the management of their lands due to social, historical, economic and political factors, a situation that has caused a decline in ecological knowledge (Pilgrim *et al.* 2008). McNamara and Westoby's (2011) study of Torres Strait Islander women's perspectives of climate change used solastalgia as a theoretical framework, and found that destruction and environmental change led to sadness, fear and distress.

Climate change does not only impact on human health but also on animals, marine life, wetlands, plant life and community/health infrastructure, which in turn impacts on traditional land management practices (Green *et al.* 2009b). With these changes Aboriginal and Torres Strait Islander peoples have observed shifting movement and migration patterns on animal behaviour (Alexander *et al.* 2011; Tam *et al.* 2013). CSIRO (2011) and Green *et al.* (2010) identified these changes through examining shifts in Indigenous seasonal calendars, which are the barometer of local knowledge about flora, fauna, climate, wind, fire, plants and ecosystem movements, growth and periods of transformation. This is also evident in marine animal populations: for example, in northern parts of Australia Elders have identified that the turtle and barramundi population migration times have shifted, with community members claiming that even their biology has changed (NITV 2014). This is similar to the distress felt by indigenous elders in Canada, Russia and Sweden who are seeing shifts in reindeer populations with which they have a deep connection (Vitebsky 2005; Löf & Carriere 2011). On a global scale the intersection of these shifts with environmental biodiversity, and its consequent impact on human health, is evident when looking at declining bee populations, because we are dependent on their pollination role for the food we eat (Greenpeace International 2013).

Environmental health workers are starting to identify the stresses and risks associated with climate change and are documenting their impacts, especially on those living in the tropics where the problems are more prevalent. Evidence is emerging that due to an increased frequency of cyclones in tropical Aboriginal and Torres Strait Islander communities there is a likelihood of a rise in vector-, water- and mosquito-borne diseases (Green et al. 2009a; Davis 2013) like dengue fever and Ross River virus (Green et al. 2009b). Further, with the increase in temperature due to climate change there will be more cases of heat stress, dehydration and respiratory illness like asthma (Green et al. 2009b). Both of these factors will be associated with a rise in communicable diseases such as bacterial diarrhoea (Green et al. 2009a).

There are two fundamental health concerns for Aboriginal and Torres Strait Islander peoples, no matter their geographical location, that will develop from the impacts of climate change: displacement from their country (and the psychological consequences) and a greater stress placed on urban settings, with an increased migration to our cities. Frumkin (2002) noted that with increased urban density/sprawls there are public health consequences – including increased air pollution and temperature, mental health and social capital issues, as well as decreased water quality and quantity – so community and health services will need to be prepared. Couple this with the extreme weather events expected in urban, regional and remote areas, such as increased flooding, heatwaves and bushfires, and all Aboriginal and Torres Strait communities will be put under considerable stress and risk, as has already happened elsewhere in Australia (Hughes & Steffen 2013; Steffen et al. 2014).

Currently in Australia, Aboriginal and Torres Strait Islander people face inequalities that are exacerbated by the impacts of dangerous climate change, no matter where they live. We acknowledge that Aboriginal and Torres Strait Islander people who live in urban and non-urban regions have a similar connection to country, although there is diversity in how they express it. Often discussions around the difference between this urban/remote divide can lead to prejudices because it stereotypes Aboriginal people living in remote areas as being more connected to country. Kingsley et al. (2013a) have disproved this by identifying Aboriginal Victorian peoples in urban and regional areas as having similar connection to their country as Aboriginal and Torres Strait Islander communities elsewhere. Hence, Aboriginal and Torres Strait Islander peoples across Australia – living in urban, regional and remote areas – should be recognised for their diverse knowledge of their country and respected accordingly. The same respect should be given to all communities when tackling climate change and responses tailored through collaboration.

Overcoming the disconnect

Prior to colonisation it was estimated that Australia's Aboriginal and Torres Strait Islander population spoke more than 200 languages (Charlesworth 2005). If, as Shiva (1995, p. 1) claims, 'linguistic diversity corresponds to a living diversity of cultures and ecosystems', then there were more than 200 ecosystems managed by Aboriginal and Torres Strait Islander peoples through collective management, inter-generational practices and knowledge. However, over the past century the knowledge and languages that maintained these ecosystems have been eroded through interactions with modern Western, homogenised knowledge and political systems. This homogeneity goes against Aboriginal and Torres Strait Islander peoples' relationships with country (Strang 2008; Kingsley *et al.* 2013b). As Dockery (2010, p. 321) noted, 'while we sometimes speak of 'Indigenous culture' as if it were one homogenous culture, there is in reality considerable diversity among Indigenous peoples from different tribes and regions'. Milton (2006, p. 352) explained that an 'understanding of cultural diversity can be a source of ecological wisdom, but nowhere is this wisdom ready-made'.

Currently in Australia, it appears that natural and cultural diversity is not viewed as a source of wealth or as a resource for current modern and political models. However, it is important to note there have been some positive steps to incorporate increased cultural diversity, such as support for collaborative and joint approaches to the management of national parks (Layton 2001; Strang 2008) and in the primary healthcare system with increased support for Aboriginal Community Controlled Health Organisations (Kingsley *et al.* 2013a).

Sustainability and diversity, then, are crucial themes in discussing climate change. What is required is a disengagement from the harmful exploitation of the earth's resources and an acknowledgment of the diversity of the relationships between humans and other members of the earth community. Without these two principles, reunification between human beings and the ecosystems of which we are a part cannot be achieved. No one group can develop a context of diversity, nor can this be achieved by addressing only one aspect of the problem of exploitative engagement. This achievement requires alternative answers, from diverse sources, and amalgams not readily accessible or acknowledged in monocultured societies. Monocultural society is a feature of Western local knowledge systems that were spread throughout the world by intellectual colonisation (Shiva 1995).

Monocultures are neither tolerant of other systems nor able to reproduce themselves sustainably. Aboriginal and Torres Strait Islander peoples, more so than others in the Australian context, appreciate the capacity for ecosystems to establish what sustainable living is: that the health and wellbeing of an ecosystem establishes the health and wellbeing of all entities that live within it. This Indigenous-led knowledge system uses sustainability and diversity as core themes that redress crucial negative characteristics of monocultures creating a human-land ethic, critical to destabilising monocultures. In Australia, for example, commercial farming techniques such as overgrazing and intensive agriculture have systematically eroded both the biological productivity of the land and the knowledge systems that have, over the course of millennia, managed ecosystem diversity and productivity (see, for example, Mary White's book *Listen… Our Land Is Crying – Australia's Environment: Problems and Solutions*, published in 1997). This erosion has, in turn, prevented the development of a human land ethic arising from within 'place' in which we all recognise and contribute to maintaining an ecosystem's health and wellbeing (Read 1996).

A core proposition of this chapter is that to address issues of climate change adaptation properly, particularly from a strengths-based point of view, all people will need to disengage from monocultured thinking that does not allow for or recognise biological, cultural, social, linguistic and spiritual diversity as a necessary feature of a sustainable future. We need to move away from monocultured thinking and action, and promote contextual strategies that protect the biological productivity of ecosystems as well as the health and wellbeing of all who need to thrive within them.

Climate change adaptation, therefore, creates opportunities to consider the development of frameworks that reduce the influence of monocultured modes of thinking, speaking and acting, along with narrow definitions of nature, to increase the influence of those that accommodate diversity. Experienced from a non-Western and deep ecology cultural viewpoint, monocultures of the mind erase local knowledge systems, especially those that have been based on the life-support capacities of ecosystems. This is still the case for many Aboriginal and Torres Strait Islander communities and deep ecological thinkers.

In order to write this chapter we had to find the centre – to assume the integrity of the domain of reliable and systematic knowledge of Aboriginal and Torres Strait Islander people – and appreciate that many ecological agendas have been marginalised in the pursuit of a global modernity. This forms part of the growing research field broadly defined as 'ecohealth'. Ecohealth research is carried out at a local community level, combining ecology and public health together in a transdisciplinary way to identify how the environment impacts on people's lives and to promote the sustainable health of individuals and ecosystems (Wilcox & Kueffer 2008; Albrecht et al. 2008; Charron 2012). Albrecht et al. (2008) noted that ecohealth acknowledges that health and wellbeing is linked to the complex interaction between ecosystems and the sociocultural and economic factors that affect them.

Reciprocity acknowledges the need to displace the prevailing concept of 'modernity' as it is based in monoculture thinking and action. Such a move would create opportunities to align ecologically sensitive and humane social orders within systems founded on the principles of Western modernity (e.g. social utopia movements within Western democracies), and at other times to replace these systems altogether with entirely new social formations (e.g. socialist countries). We recommend that all systems need to incorporate and acknowledge Aboriginal and Torres Strait Islander knowledge (from Elders) in a way that puts caring for country principles at the forefront.

For Western society to find compatibility with the meaning and values contained in indigenous and ecological cultures and literatures we need to create a joined up narrative about climate change adaptation. This will need to describe both an unfolding systematic framework and a narration of a learning process. The following section is an attempt to think about how climate change adaptation could be applied to such a framework and process. Certainly, the first component of a systematic framework would be to structure discourses of building on strength and resilience, and the second to identify areas for action in health and community services.

Structuring discourse of strength and resilience

Of all those in Australia who are becoming increasingly susceptible to the social, environmental, cultural and economic impacts of climate change, it is Aboriginal and Torres Strait Islander people living in regions where extreme weather events are increasing in their frequency and severity who are most under threat (Zander et al. 2013). Flooding, cyclone activity, coastal erosion, heatwaves, fires and drought have contributed to a loss of community and environmental assets, exacerbated conditions that facilitate the spread of plant and animal disease, and increased the likelihood of vector-borne illnesses (Davis 2013; Choy et al. 2013). These factors, when combined with the tangible social and economic impacts of climate change, are expected to affect adversely the quality of life of all people in Australia but particularly Aboriginal and Torres Strait Islander people living in regions most vulnerable to climate change impacts.

While this discourse brings attention to the needs of Aboriginal and Torres Strait Islander people, we need to avoid making the collective story of climate change adaptation one in which the experience of vulnerability, climate change and Indigenous peoples are intertwined. It is important not to frame Aboriginal and Torres Strait Islander peoples as only being vulnerable to climate change, without also exploring their potential contribution to responding to and influencing climate change adaptation strategies in Australia. There is much to learn from the strength and resilience of populations who have consistently experienced disruption and upheaval, but there is also a risk of resourcing people's relocation from country rather than directing resources towards applying Indigenous ways of knowing to climate change adaptation more broadly.

Although we should be concerned about current and future relationships between human health and wellbeing and climate change, research in Australia has also identified the real contribution of Indigenous knowledge and systems of 'country-based governance' to ensure a holistic approach to climate change monitoring and observation. Further, these studies have measured the impact of Indigenous community organisations in developing adaptive capacity through environmental justice, 'Caring for Country' initiatives and the participation of Indigenous peoples in abatement and climate change economies (Tran *et al.* 2013). As previously explained, 'Caring for Country' models can be applied in urban, regional and remote contexts (Kingsley *et al.* 2013a,b).

This strengths-based approach to determining adaptive strategies is also imperative, as Aboriginal and Torres Strait Islander people through Native Title claims are responsible for more than 20% of Australia's total land mass, covering extensive areas of the coast and interior. Therefore, Aboriginal and Torres Strait Islander institutions have cultural, legal, social and economic rights and responsibilities to manage land and respond to emergent issues including moving from country because of climate change impacts, and cannot be left out of the climate change debate.

Aboriginal community-controlled models of health and connection to country

In reference to services' ability to adapt to climate change we first need to understand the importance of Aboriginal community control, or organisations governed, managed and incorporated by Aboriginal and Torres Strait Islander boards, to service Aboriginal and Torres Strait Islander people. Aboriginal community-controlled organisations have been instrumental in incorporating holistic and culturally appropriate models into practice in healthcare settings and environmental management because they exist to meet the holistic needs of the community they service. The ability of community-controlled organisations to meet these needs is well recognised, with over 4000 Aboriginal corporations now registered with the Office of the Registrar of Aboriginal Corporations (www.oric.gov.au). For over 40 years, these organisations have provided a range of services to people who were often alienated within Australia, and today they inform and teach others in mainstream society about community knowledge and practices. This is often referred to as two-way learning, or bi-cultural learning. Aboriginal Community Controlled Health Organisations have as their core a set of ideas and principles founded in the view that land and seas are concentric to culture, health and wellbeing. Country is central to wellbeing, as according to Swan and Raphael (1995, p. 13) this holistic concept:

> ... is steeped in the harmonized inter-relations which constitute cultural wellbeing. These inter-relating factors can be categorized largely as spiritual, environmental, ideological, political, social, economic, mental and physical ... when the harmony of these inter-relations is disrupted, Aboriginal ill health will persist.

Because of this, many organisations take a rights-based and ecological approach to health care and so in practice are well placed to respond to the health effects of climate change.

Custodianship, country and stewardship

Tse *et al.* (2005) identified that, although traditional owner groups differ in 'cultural geography', the Aboriginal concept of health is holistic, encompassing mental, physical, cultural and spiritual health (Lutschini 2005). It is a 'whole-of-life view' incorporating the cyclical concept of 'life–death–life' and connection to country (NAHSWP 1989). In this model, the past, present and future intertwine to impact on Aboriginal people's health and the focus is on the whole community rather than the individual. Wahbe *et al.* (2007) noted that greater impacts could be made in Aboriginal and Torres Strait Islander health by strengthening partnerships between local communities, universities, governments and the non-government sector. However, often the Indigenous model of health is disregarded by non-Indigenous academics, policymakers and populations and the whole is reduced to measurable indicators that suit their agendas, rather than looking at the positive determinants of Aboriginal culture (Kingsley *et al.* 2013a).

Research indicates that more meaningful measures of Aboriginal people's 'voices' are required in health and wellbeing literature (Taylor 2008; Prout 2012), such as incorporating concepts like country. Country is a template for understanding elements of Aboriginal culture, a way of teaching and maintaining identity, and is essential to creating projects that ensure the health of both people and country (Jampijinpa *et al.* 2008; Holmes & Jampijinpa 2013). Jampijinpa *et al.* (2008) recommend that in Western discourse we re-imagine caring for country because it actually refers to looking after one's home. This is due to the transferable values of fostering ecosystem stewardship, of which cultural connections between people and place are integral components.

However, as Holmes and Jampijinpa (2013, p. 7) note, country is still:

[A]ssumed ... to mean the material landscape, the flora and fauna, and landforms. However ... country is defined by its linkages with the other elements ... Country is the biophysical world as understood through its associated law, skin, ceremony, and language ... country, encompassing the physical environment but also the various social, spiritual, and cultural relationships that transform an ecological landscape into a socio-cultural one ... country is so deeply ingrained in the fabric of Aboriginal culture ... and all activities are underpinned by a broad concept of spiritual and physical unity with the land ... the result is more akin to the land possessing people than people possessing the land ...

This custodianship is the capacity to harmonise activities within the requirements of the natural environment and familial obligations. Often this world view is displaced by institutions that require land to be 'developed' and the relationships to others are enacted through financial, commercial and industrial pathways, and not through country. These shifts from earth-centred meaning and value can lead to our relationships, our economies and our identities being displaced. Some academics are incorporating holistic knowledge and adopting transformational sciences to work between and across different cultures and domains of knowledge. This ability to be transdisciplinary is the way of the future.

The United Nations (2007) Declaration on the Rights of Indigenous Peoples states that it is a human right for indigenous people to have access to and full enjoyment of land and seas, and to cultural practices that connect people to country. Collins and Murtha (2009, p. 961) identified that 'Indigenous people … around the globe … bear a disproportionate share of environmental burden compared to their non-Indigenous counterparts, a trend that has been described as environmental racism'. Scientists, governments and communities make value judgments on what they believe should be kept with reference to the environment (flora and fauna) and to health (prevention and hospital care), and what no longer strengthens the ecology and society. They do this by using a lens of personal, economic, cultural and social values predominantly taking on Western viewpoints, moving away from World Health Organization definitions of health which request more holistic approaches to these matters. We believe that incorporating Indigenous values, knowledge and practices into services may have considerable benefits.

Actions to connect to country and culture in health and community services

How do we support diverse organisations and the services they provide to respond to the climate change adaptation needs of the Aboriginal and Torres Strait Islander population? This is no easy task as Indigenous knowledge systems are diverse and holistic containing cultural, economic, social and spiritual domains. The points of action are, therefore, aligned with the complexity and connections between these domains. However, if agencies harmonise their delivery of services through these interconnected domains, so that the relevant elements of their programs are associated with these knowledge systems, there is an ability to move from monoculture thinking to adopting diversity as a principle. This is well demonstrated in community-controlled organisations, particularly those that represent traditional owner groups (e.g. Prescribed Bodies Corporate, Land Councils, Aboriginal Medical Services). Located in urban, regional and remote settings, these services are based on self-determination and holistic approaches – the model that will be required to tackle extreme weather events in a manner appropriate for Aboriginal and Torres Strait Islander populations. One concrete examples of this is the Victorian community-controlled Lake Condah Sustainable Development Project.

Social and cultural actions

This chapter highlights the importance of incorporating social and cultural understanding of Aboriginal and Torres Strait climate change adaptation into the mainstream discourse, and makes the following recommendations for health and community services to integrate into their everyday actions in urban, regional and remote settings:

1. Incorporate this information into Reconciliation Action Plans in consultation with traditional owners. This allows organisations and staff to build reciprocal relationships, respect and opportunities as a mechanism for knowledge exchange. Actions that can be taken to strengthen this incorporation include:
   a. developing partnerships with organisations such as Indigenous Community Volunteers (www.icv.com.au) where staff are orientated with and understand working on country with Aboriginal and Torres Strait Islander people, applying skills transfer and mutual learning

     b. conducting training, development and orientation programs with traditional owners for staff, and holding community events on country to learn about local Indigenous culture (e.g. Charles Darwin University's Mawul Rom Program – see www.cdu.edu.au/enews/versions/040411/cduandmawulromlaunch.html – a Masters program that has been developed through mediation, negotiation and partnership with Aboriginal communities)

     c. delivering seminars and workshops linking culture, country, health and wellbeing to strengthen the conversation and to privilege Aboriginal peoples' knowledge.

2. Incorporate Indigenous values of climate change adaptation into health and community services by integrating understandings of country, such as is evident in organisations like the National Climate Change Adaptation Research Facility.

3. Ensure that organisations and staff are guided by Aboriginal and Torres Strait Islander research protocols that are then transferable to health and community services to strengthen relationships with Aboriginal and Torres Strait Islander communities. Ethical guidelines can be found through the National Health and Medical Research Council (www.nhmrc.gov.au/guidelines/publications/e52), Australian Institute of Aboriginal and Torres Strait Islander Studies (www.aiatsis.gov.au/research/ethics/gerais.html) and The Lowitja Institute (www.lowitja.org.au/lowitja-publishing/L009).

We recognise that different communities will have a diversity of connections to their country, but that these connections should be respected as all communities will have valuable and unique contributions to the climate change adaptation debate.

Economic actions

The knowledge identified within this chapter is part of an ancient economic system that should be recognised. If health and community services are to take climate change adaptation seriously they will need Aboriginal and Torres Strait Islander representative groups advising on this work, and these groups will need to be financially compensated. Because there are more than 350 different traditional owner groups throughout Australia, this will require considerable effort to ensure appropriate collaboration.

A health and community issue that could be applied to climate change adaptation is food security, as climate change impacts are reducing fresh food and its availability in Aboriginal and Torres Strait Islander communities (Davis 2013). This situation has also been exacerbated by changing hunting and gathering methods (Davis 2013) and a reduction in bush food (Green *et al.* 2009b), which has led to less of the fresh food that used to be in abundance on country. These changes in regional food yields, combined with a disruption to fishing activities due to climatic shifts, will impact greatly on economic activities (Green *et al.* 2009b). Therefore, services are encouraged to take on advocacy roles to sustain both the productivity of country and sustainable hunting practices. This could be invigorated by services taking youth and Elders out on country as a collaboration with, for example, local schools (e.g. www.ehp.qld.gov.au/ecosystems/community-role/ranger/junior).

Arabena (2009) has highlighted the cultural differences between modern food systems (leading to increased food insecurity) and traditional food production (being significant in connecting communities to country). Her findings on food security in Aboriginal communities identified the critical impacts of climate change and noted several issues that will occur in the future due to increases in temperature and extreme weather events: that the transportation and storage of food will become increasingly hard as a hotter climate will lead to a rise in bacterial diseases, poor food hygiene practices and the cost of food; that the quality of food will be reduced because of an increase in processed rather than traditional foods; and that, on top of the other issues, rising fuel costs will diminish the foods available. Arabena (2009) has provided recommendations that could be applied by health and community services to tackle such issues:

1. Determine which sustainable food products and supply activities contribute to our food/environmental security, and engage local community and agencies in the supply chain.
2. Consider stores as supplementing the food production capacity of the ecosystem in which people live to reduce the 'ecological footprint' of stock.
3. Ascertain whether different distribution networks, especially those incorporating traditional food production, could be developed at a local level.
4. Consider reconfiguring Aboriginal communities into sustainability ecohealth academies where individuals can learn about climate change adaptation.
5. Dedicate resources to research into the concepts of sustainability stores and communities.
6. Assess ecosystem services and localised food production or supply as possible sources of employment and income.

Spiritual actions

To understand the spiritual notion of country we recommend that services incorporate knowledge around caring for country into their programs. This is critical as caring for country is a vital component of Indigenous ecological knowledge and a useful mechanism for improving the health status of Aboriginal people, while at the same time improving Australia's ability to adapt to climate change. Indigenous ecological knowledge derives from frequent interaction with the local environment and being able to read signs of the land rather than taking control of it. Pilgrim *et al.* (2008, p. 1008) emphasised that 'time and money could be spared if the knowledge, experience and capacity of local people were protected and used in resource management efforts today'.

Along with caring for country, ranger programs have also been fundamental in successfully adapting to climate change and to improving the wellbeing of Aboriginal and Torres Strait Islander peoples (Green *et al.* 2009b; Berry *et al.* 2010; McIntyre-Tamwoy *et al.* 2013). For example, controlled burning has proven to be a good method in reducing the extreme effects of bushfires. We believe if ranger programs can be brought together with services to improve health and environmental outcomes, this will have a great influence on Australia's ability to tackle climate change across all settings and geographies. Further, it will allow for climate change adaptation programs to be integrated into health and community services.

Although engagement in caring for country may be a mechanism for reducing the health inequalities affecting many Aboriginal people, it will not necessarily have universal application. Practical ways that services could incorporate caring for country into their programs may include:

1. Developing organisational policies that reinforce access to country. This could include car policies that allow for hunting and bush tucker activities, or cultural leave policies (which could be called care for country leave) and incorporating these activities within an organisation. Pearson (2006) acknowledges that Aboriginal people have to 'orbit' between two worlds – their own cultural identity and 'modern society' in Australia. He suggests that Aboriginal people should be able to determine their own direction so they can effectively integrate into global economies, for example, by not needing to be on country to have a connection with it. Therefore, we acknowledge that these policies will need to be flexible.
2. Undertaking clinical treatment for long-term health issues on country when there is no harm to the patient in doing so. An example of this is the Purple House project by the Western Desert Nganampa Walytja Palyantjaku Tjutaku Aboriginal Corporation mobile renal dialysis unit (www.westerndesertdialysis.com/the-purple-house).
3. Using spaces in large compounds under Aboriginal management. This may include services running programs around Indigenous food and bush tucker gardens, but also transferring skills and incorporating new technologies (managing desertification) to tackle ecosystem issues.

Bringing these actions together

In this chapter we recommend that services start moving beyond Western health approaches and start taking a planetary view of the earth. We need to involve ourselves in taking into account the intra-species and ecological effects of climate change in all sectors. For example, in the Torres Strait turtle nests are now 'drowning' due to increased higher tides, which is everyone's business. We need to work together – with business, local councils, Elders, health and community services, park rangers   not necessarily by focusing on health, but tackling health consequences. Creating a sense of belonging in services to local community ensures information sharing and partnership is fundamental in urban, regional and remote settings. However, services must recognise that some Aboriginal knowledge might not be public. Culturally appropriate programs have occurred in regard to climate change, such as a positive initiative in the Torres Strait where partners are working with local communities to prepare for the increases in tidal disruptions (www.tsra.gov.au/_data/assets/pdf_file/0016/2491/Inundation-Managment-on-Saibai-Boigu-and-Iama-Islands.pdf).

There is a great deal that can be learned from local and Indigenous approaches when tackling climate change events, especially in relation to disaster risk management. A final example is provided in the handbook of good practice from Pacific Integrated Island Management (Jupiter *et al.* 2013), which highlighted several principles in tackling climate-related disasters such as flooding. These principles include:

1. adopting long-term integrated social and ecological approaches
2. engaging with stakeholders through participatory governance that ensures gender, social and cultural equity
3. acknowledging uncertainty and applying flexible management approaches that are regularly monitored
4. ensuring decision-making is based on both evidence and evaluations
5. observing this situation across the whole ecosystem, which is multi-layered/diverse and involves a transdisciplinary approach.

Final comments

Climate change is one of the greatest health and community issues of our time. As Davis (2013, p. 499) put it so succinctly, health and community services:

> … will be increasingly challenged to embrace a 'more ecologically focused' approach to health, which would 'enable the 'health of country' and its inextricable links with human health to be considered in climate impact assessments …

In this chapter, the challenge we pose to the health and community sector is to incorporate, promote and build partnerships with earth-centred approaches to climate change. This will require a shift in thinking, leadership and an incorporation of Aboriginal and Torres Strait Islander peoples' perspectives into any solution.

References

Albrecht G (2005) Solastalgia: a new concept in health and identity. *Philosophy, Activism and Nature* 3, 41–55.

Albrecht G, Sartore GM, Connor L, Higginbotham N, Freeman S, Kelly B, Stain H, Tonna A, Pollard G (2007) Solastalgia: the distress caused by environmental change. *Australasian Psychiatry* **15**(Suppl. 1), S95–S98. doi:10.1080/10398560701701288

Albrecht G, Higginbotham N, Connor L, Freeman S (2008) Social and cultural perspectives on eco-health. In *International Encyclopedia of Public Health*. (Eds K Heggenhougen and S Quah) pp. 57–63. Academic Press, San Diego.

Alexander C, Bynum N, Johnson E, King U, Mustonen T, Neofotis P, Oettlé N, Rosenzweig C, Sakakibara C, Shadrin V, Vicarelli M, Watehouse J, Weeks B (2011) Linking Indigenous and scientific knowledge of climate change. *Bioscience* **61**(6), 477–484. doi:10.1525/bio.2011.61.6.10

Arabena K (2006) The universal citizen: an Indigenous citizenship framework for the twenty-first century. *Australian Aboriginal Studies* **2**, 36–46.

Arabena K (2009) Inquiry into community stores in remote Aboriginal and Torres Strait Islander communities. *House Standing Committee on Aboriginal and Torres Strait Islander Affairs.* Submission no. 108. <http://www.aph.gov.au/binaries/house/committee/atsia/communitystores/subs/sub0102.pdf

AHRC (Australian Human Rights Commission) (2013) *Questions and Answers about Aboriginal & Torres Strait Islander Peoples.* Australian Human Rights Commission, Sydney, <http://www.humanrights.gov.au/publications/questions-and-answers-about-aboriginal-torres-strait-islander-peoples>

Berkes F, Jolly D (2001) Adapting to climate change: social-ecological resilience in a Canadian Western Artic community. *Conservation Ecology* **5**(2), 18.

Berry H, Butler JRA, Burgess CP, King UG, Tsey K, Cadet-James YL, Rigby CW, Raphael B (2010) Mind, body, spirit: co-benefits for mental health from climate change adaptation and caring for country in remote Aboriginal Australian communities. *NSW Public Health Bulletin* **21**, 139–145. doi:10.1071/NB10030

Bardsley DK, Wiseman ND (2012) Climate change vulnerability and social development for remote indigenous communities of South Australia. *Global Environmental Change* **22**, 713–723. doi:10.1016/j.gloenvcha.2012.04.003

Burgess CP, Berry HL, Gunthorpe W, Bailie RS (2008) Development and preliminary validation of the 'Caring for Country' questionnaire: measurement of an Indigenous Australian health determinant. *International Journal for Equity in Health* **7**(26), 1–14.

Campbell D, Stafford Smith M, Davies J, Kuipers P, Wakerman J, McGregor MJ (2008) Responding to health impacts of climate change in the Australian desert. *Rural and Remote Health* **8**(3), online, <http://www.rrh.org.au/articles/subviewnew.asp?ArticleID=1008>.

Charlesworth M (2005) Introduction. In *Aboriginal Religions in Australia*. (Eds M Charlesworth, F Dussart and H Morphy) pp. 1–28. Ashgate Publishing Limited, Surrey.

Charron DF (Ed.) (2012) *Ecohealth Research in Practice: Innovative Applications of an Ecosystem Approach to Health*. Springer, New York.

Choy DL, Clarke P, Jones D, Serrau-Neumann S, Hales R, Koschade O (2013) 'Aboriginal reconnections: understanding coastal urban and peri-urban Indigenous people's vulnerability and adaptive capacity to climate change'. National Climate Change Adaptation Research Facility, Gold Coast.

Cochran P, Huntington OH, Pungowiyi C, Tom S, Chapin FS, III, Huntington HP, Maynard NG, Trainer SF (2013) Indigenous frameworks for observing and responding to climate change in Alaska. *Climatic Change* **120**, 557–567. doi:10.1007/s10584-013-0735-2

Collins LM, Murtha M (2009) Indigenous environmental rights in Canada: the right to conservation implicit in treaty and Aboriginal rights to hunt, fish, and trap. *Indigenous Environmental Rights in Canada* **47**(4), 959–991.

Connor L, Albrecht G, Higginbotham N, Smith W, Freeman S (2004) Environmental change and human health in Upper Hunter communities of New South Wales, Australia. *EcoHealth* **1**(Supp. 2), 47–58. doi:10.1007/s10393-004-0053-2

CSIRO (2011) *Indigenous Seasonal Indicators and Climate Change*, CSIRO, Clayton South, Victoria, http://www.csiro.au/Organisation-Structure/Divisions/Ecosystem-Sciences/Ngadju.aspx.

Davis M (2013) Climate change impacts to Aboriginal and Torres Strait Islander communities in Australia. In *Climate Change and Indigenous Peoples: The Search for Legal Remedies.* (Eds RS Abate and EA Kronk) pp. 493–507. Edward Elgar, Cheltenham, Victoria.

Devkota R, Marasini T, Cockfield G, Devkota LP (2013) Indigenous knowledge for climate change induced flood adaptation in Nepal. *The International Journal of Climate Change: Impacts and Responses* **5**, 35–46.

Dockery AM (2010) Culture and wellbeing: the case of Indigenous Australians. *Social Indicators Research* **99**, 315–332. doi:10.1007/s11205-010-9582-y

Ford JD (2012) Indigenous health and climate change. *American Journal of Public Health* **102**(7), 1260–1266. doi:10.2105/AJPH.2012.300752

Frumkin H (2002) Urban sprawl and public health. *Public Health Reports (Washington, D.C.)* **117**(3), 201–217. doi:10.1016/S0033-3549(04)50155-3

Gerrard E (2008) Climate change and human rights: issues and opportunity for Indigenous peoples. *The University of New South Wales Law Journal* **31**(3), 941–952.

Gislason MK (2013) Expanding the social: Moving towards the ecological in social studies of health. In *Ecological Health: Society, Ecology and Health*. (Ed. MK Gislason) pp. 3–22. Emerald Group Publishing Limited, UK.

Green D, King U, Morrison J (2009a) Disproportionate burdens: the multidimensional impacts of climate change on the health of Indigenous Australians. *The Medical Journal of Australia* **190**(1), 4–5.

Green D, Jackson S, Morrison J (2009b) 'Risks from climate change to Indigenous communities in the tropical north of Australia'. Australian Government, Canberra.

Green D, Billy J, Tapim J (2010) Indigenous Australians' knowledge of weather and climate. *Climatic Change* **100**, 337–354. doi:10.1007/s10584-010-9803-z

Green D, Niall S, Morrison J (2012) Bridging the gap between theory and practice in climate change vulnerability assessment for remote Indigenous communities in Northern Australia. *Local Environment* **17**(3), 295–315. doi:10.1080/13549839.2012.665857

Greenpeace International (2013) 'A review of factors that put pollinators and agriculture in Europe at risk'. Greenpeace Research Laboratories, Amsterdam.

Heinamaki L (2009) Rethinking the status of Indigenous peoples in international environmental decision-making: pondering the role of Arctic Indigenous peoples and the challenge of climate change. *Environment and Policy* **50**, 207–262. doi:10.1007/978-1-4020-9542-9_9

Holmes M, Jampijinpa W (2013) Law for country: the structure of Warlpiri ecological knowledge and its application to natural resource management and ecosystem stewardship. *Ecology and Society* **18**(3), 19. doi:10.5751/ES-05537-180319

Hughes L, Steffen W (2013) 'Be prepared: climate change and the Australian bushfire threat'. Climate Council, Canberra.

Jampijinpa WP-K, Holmes M, Box AL (2008) 'Ngurra-Kurlu: a way of working with Warlpiri people'. Desert Knowledge CRC, Alice Springs, Northern Territory.

Johnston F, Jacups SP, Vickery AJ, Bowman DMJS (2007) Ecohealth and Aboriginal testimony of the nexus between human health and place. *EcoHealth* **4**(4), 489–499. doi:10.1007/s10393-007-0142-0

Jupiter S, Jenkins A, Lee Long W, Maxwell S, Watson J, Hodge K, Govan H, Carruthers T (2013) 'Pacific Integrated Island Management: Principles, case studies and lessons learned'. Secretariat of the Pacific Regional Environment Program and United Nations Environment Program, Samoa.

Kingsley J, Townsend M, Henderson-Wilson C, Bolam B (2013a) Developing an exploratory framework linking Australian Aboriginal peoples' connection to country and concepts of wellbeing. *International Journal of Environmental Research and Public Health* **10**(2), 678–698. doi:10.3390/ijerph10020678

Kingsley J, Townsend M, Henderson-Wilson C (2013b) Exploring Aboriginal people's connection to country to strengthen human–nature theoretical perspectives. In *Ecological Health: Society, Ecology and Health.* (Ed. MK Gislason) pp. 45–64. Emerald Group Publishing Limited, UK.

Lanyi G (2012) Climate change and human rights: An unlikely relationship? *Alternative Law Journal* **37**(4), 269–271.

Layton RH (2001) Hunter–gatherers, their neighbours and the nation state. In *Hunter–Gatherers: An Interdisciplinary Perspective.* (Eds C Panter-Brick, RH Layton and P Rowley-Conwy) pp. 292–321. Cambridge University Press, Cambridge, UK.

Löf A, Carriere N (2011) 'Learning from our elders: Aboriginal perspectives on climate change and reindeer/caribou habitat in the circumboreal forest'. Umeå University (Faculty of Social Science), Sweden, <http://www.diva-portal.org/smash/get/diva2:564600/FULLTEXT01.pdf>.

Lutschini M (2005) Engaging with holism in Australian Aboriginal health policy: a review. *Australia and New Zealand Health Policy* **2**(15), 1–10.

McIntyre-Tamwoy S, Fuary M, Buhrich A (2013) Understanding climate, adapting to change: indigenous cultural values and climate change impacts in North Queensland. *Local Environment* **18**(1), 91–109. doi:10.1080/13549839.2012.716415

McNamara KE, Westoby R (2011) Solastalgia and the gendered nature of climate change: an example from Erub Island, Torres Strait. *EcoHealth* **8**(2), 233–236. doi:10.1007/s10393-011-0698-6

Milton K (2006) Cultural theory and environmentalism. In *The Environment in Anthropology.* (Eds N Haenn and R Wilk) pp. 351–354. New York University Press, London.

NAHSWP (National Aboriginal Health Strategy Working Party) (1989) *A National Aboriginal Health Strategy.* Australian Government Publishing Service, Canberra.

NITV (National Indigenous Television) (2014) *The Tipping Points – The Last Frontier* <http://www.sbs.com.au/ondemand/video/225538627513>.

Parkes MW (2010) *Ecohealth and Aboriginal Health: A Review of Common Ground.* National Collaborating Centre for Aboriginal Health, Prince George, Canada.

Parkes MW, Horwitz P (2009) Water, ecology and health: ecosystems as settings for promoting health and sustainability. *Health Promotion International* **24**(1), 94–102. doi:10.1093/heapro/dan044

Pearson N (2006) Walking in two worlds, *The Australian*, 28 October 2008, <http://www.theaustralian.com.au/opinion/noel-pearson-walking-in-two-worlds/story-e6frg6zo-1111112429555>.

Petheram L, Zander KK, Campbell BM, High C, Stacey N (2010) 'Strange changes': Indigenous perspectives of climate change and adaptation in NE Arnhem Land (Australia). *Global Environmental Change* **20**, 681–692. doi:10.1016/j.gloenvcha.2010.05.002

Pilgrim SE, Cullen LC, Smith DJ, Pretty J (2008) Ecological knowledge is lost in wealthier communities and country. *Environmental Science & Technology* **42**(4), 1004–1009. doi:10.1021/es070837v

Prout S (2012) Indigenous wellbeing frameworks in Australia and the quest for quantification. *Social Indicators Research* **109**(2), 317–336. doi:10.1007/s11205-011-9905-7

Read P (1996) *Returning to Nothing: The Meaning of Lost Places.* Cambridge University Press, Melbourne.

Rose D (2005) Rhythms, pattern, connectivities: Indigenous concepts of seasons and change. In *A Change in the Weather: Climate and Culture in Australia.* (Eds T Sherratt, T Griffiths and L Robin) pp. 32–41. National Museum of Australia Press, Canberra.

Rose DB (2013) Fitting into country. In *The Routledge Companion to Landscape Studies.* (Eds P Howard, I Thompson and E Waterton) pp. 8–11. Routledge Taylor & Francis Group, London.

Sakakibara C (2008) 'Our home is drowning': Iñupiat storytelling and climate change in Point Hope, Alaska. *Geographical Review* **98**(4), 456–475. doi:10.1111/j.1931-0846.2008.tb00312.x

Shiva V (1995) *Monocultures of the Mind: Perspectives on Biodiversity and Biotechnology.* Third World Network, Malaysia.

Spartz JT, Shaw BR (2011) Place meanings surrounding an urban natural area: a qualitative inquiry. *Journal of Environmental Psychology* **31**, 344–352. doi:10.1016/j.jenvp.2011.04.002

Steffen W, Hughes L, Perkins S (2014) Climate Council: heatwaves are getting hotter and more frequent. *The Conversation*, 18 February 2014, <http://theconversation.com/climate-council-heatwaves-are-getting-hotter-and-more-frequent-23253>.

Strand LB, Tong S, Aird R, McRae D (2010) Vulnerability of eco-environmental health to climate change: the views of government stakeholders and other specialists in Queensland, Australia. *BMC Public Health* **10**, 441–450. doi:10.1186/1471-2458-10-441

Strang V (2008) Cosmopolitan natures: paradigms and politics in Australian environmental management. *Nature and Culture* **3**(1), 41–62. doi:10.3167/nc.2008.030104

Strang V (2009) *Gardening the World: Agency, Identity, and the Ownership of Water.* Berghahn Books, New York.

Swan P, Raphael B (1995) 'Ways forward: national Aboriginal and Torres Strait Islander mental health policy'. National Consultancy Report, Canberra.

Tam BY, Gough WA, Edwards V, Tsuji LJS (2013) The impact of climate change on the well-being and lifestyle of a First Nation community in the western James Bay region. *The Canadian Geographer* **57**(4), 441–456. doi:10.1111/j.1541-0064.2013.12033.x

Taylor J (2008) Indigenous peoples and indicators of well-being: Australian perspectives on United Nations global frameworks. *Social Indicators Research* **87**, 111–126. doi:10.1007/s11205-007-9161-z

Tran T, Strelein L, Weir J, Stacey C, Dwyer A (2013) 'Changes to country and culture, changes to climate: strengthening institutions for Indigenous resilience and adaptation'. Australian Institute of Aboriginal and Torres Strait Islander Studies and National Climate Change Adaptation Research Facility, Canberra.

Tse S, Lloyd C, Petchkovsky L, Manaia W (2005) Exploration of Australian and New Zealand indigenous people's spirituality and mental health. *Australian Occupational Therapy Journal* **52**(3), 181–187. doi:10.1111/j.1440-1630.2005.00507.x

Turner NJ, Clifton H (2009) 'It's so different today': climate change and Indigenous lifeways in British Columbia, Canada. *Global Environmental Change* **19**, 180–190. doi:10.1016/j.gloenvcha.2009.01.005

United Nations (2007) 'Declaration on the rights of indigenous peoples'. 22 September 2009 <http://www.un.org/esa/socdev/unpfii/documents/DRIPS_en.pdf>

Vitebsky P (2005) *Reindeer People: Living with Animals and Spirits in Siberia.* Houghton Mifflin, Boston.

Wahbe TR, Jovel EM, Silva Garcia DRS, Pilco Llagcha VE, Rose Point NR (2007) Building international Indigenous people's partnerships for community-driven health initiatives. *EcoHealth* **4**(4), 472–488. doi:10.1007/s10393-007-0137-x

Wilcox B, Kueffer C (2008) Transdisciplinarity in ecohealth: status and future prospects. *EcoHealth* **5**, 1–3. doi:10.1007/s10393-008-0161-5

Wiseman ND, Bardsley DK (2013) Climate change and indigenous natural resource management: a review of socio-ecological interactions in the Alinytjara Wilurara NRM region. *Local Environment: The International Journal of Justice and Sustainability* **18**(9), 1024–1045. doi:10.1080/13549839.2012.752799

Zander KK, Petheram L, Garnett ST (2013) Stay or leave? Potential climate change adaptation strategies among Aboriginal people in coastal communities in north Australia. *Natural Hazards* **67**, 591–609. doi:10.1007/s11069-013-0591-4

9

Support for adaptation in culturally and linguistically diverse communities

Alana Hansen, Scott Hanson-Easey and Peng Bi

Key points

- Climate change adaptation strategies need to consider the diversity of cultural groups within Australian society.
- While diversity does not imply vulnerability, some sub-populations can be disadvantaged by socio-economic and linguistic barriers to adaptation.
- Community-based health and social service organisations can play an important role in meeting the information needs of migrants and refugees.
- A culture-centred communication approach to engagement can be used to augment adaptation to heatwaves and climate change within culturally and linguistically diverse communities.

Introduction

Australia is a culturally diverse country – indeed, it is difficult to overemphasise how different waves of immigrants and refugees have shaped and enriched modern Australia. Nearly one-third of the population was born in another country, and of that sub-population, 8.5% were born in non-English speaking countries. Underpinning this society is a multicultural tenet that provides for ethnic and cultural groups to maintain their unique systems of beliefs, knowledge, values and practices, while simultaneously being expected to integrate into the Australian polity at large. Undoubtedly, the linguistic, cultural and ethnic diversity of Australia's population has fostered a rich and vibrant multicultural society. This diversity is also manifest in the communication needs of different groups; culture shapes perceptions and moulds social constructs such as risk perception. It is well documented that if communicators do not meet these needs, inequitable access to crucial risk and emergency information will ensue, endangering the health and wellbeing of Australia's culturally diverse citizens.

In this chapter, we address the question of how community-based health and social service organisations can engage, support and advocate for clients whose communication and support needs may not be adequately met by traditional approaches. Drawing on findings from our research program on climate change adaptation and culturally and linguistically diverse (CALD) communities, we propose that a 'culture-centred' communication approach (Dutta-Bergman 2004; Dutta 2007; Sellnow *et al.* 2009) can be harnessed to augment adaptation capacity within CALD communities. Community-based agencies play a critical role in providing CALD communities with a platform from which to shape how adaptation information is constructed and disseminated to their own cultural group.

History of Australian immigration

To understand the dimensionality of intercultural risk communication, a brief overview of Australia's immigration history is necessary. The cultural diversity apparent in Australia's urban centres would be unrecognisable to most white Australians living in the years before World War II. Up until 1945, most of immigrants to Australia emigrated from the United Kingdom (Jupp 1991); the White Australia Policy ensured that potential immigrants, not endowed with British descent, were systematically barred from gaining citizenship. In 1945, the beginning of the end for the exclusionist White Australia Policy was foreshadowed by Arthur Caldwell, Minister for Immigration in the Chifley government, when he announced a mass immigration program to bolster the Australian population. Conceived in a climate of fear of invasion from the north (under the 'populate or perish' slogan), and with a view of creating a larger domestic market and manufacturing sector, the program sought to attract 70 000 migrants a year. It soon became apparent that British 'stock' alone would not singularly meet the required annual intake goal, and the immigration net was thrown more widely to include emigrants leaving central and eastern Europe (Lack & Templeton 1995). Although most of the immigrants to Australia until the 1970s were from Britain and northern Europe, the composition of the immigration program changed markedly in the 1950s. Wide-scale unemployment and poverty in the wake of World War II drove a diaspora from Italy, Greece, Yugoslavia and other southern European countries. Over two decades, from 1951 to 1971, Australia accepted over 478 000 settlers from southern European countries. This represented a cultural transformation of the Australian population, from one of the most monocultural societies in the world to one that was well on its way to becoming truly multicultural.

Australia also has a long history of accepting humanitarian refugees escaping the cataclysm of war and political persecution. Australia first offered refugee assistance in 1938, when the Lyons government offered a safe haven to 15 000 Jews fleeing Nazi Germany. From 1947 till 1954, 170 000 displaced persons were accepted by Australia. Yet, Australia's post-war refugee policy was, ostensibly, discriminatory and hierarchical in terms of 'race', gender and age. Priority was given in the first instance to 'blonde and blue-eyed' Baltic people; in particular, favouring fit men of working age (to feed Australia's burgeoning labouring needs) over families and the elderly (Lack & Templeton 1995). It was not until 1981, in accordance with the 1951 United Nations Convention relating to the Status of Refugees that Australia formalised an ongoing humanitarian resettlement program (Hugo 2002). Previous refugee intakes had been in direct response to geo-political upheavals and, before 1975, these had focused on the resettling of refugees from central Europe. The conclusion of the Vietnam (American) War in 1975 and the refugee crisis resulting from the war in Lebanon (1982) instigated a new wave of refugee migration.

More recently, the Australian Government's humanitarian program for refugees has witnessed the arrival of groups from the Middle East, South-East Asia and sub-Saharan African countries. Taken together, the 'offshore' and 'onshore' components of the program resettled over 20 000 refugees in the last reporting period (2012–13), including 7500 protection and humanitarian visas being granted to asylum seekers (DIAC 2013a). As we will go on to argue, this population may be particularly vulnerable to the risks posed by climate change. New and emerging refugee communities are in the process of reconstructing their lives in a new country, sometimes after long periods of time in refugee camps. They can experience linguistic and socio-economic problems, mitigating their ability to take adaptive action in times of extreme weather and emergencies. Moreover, refugees and asylum seekers have often suffered experiences that have an impact on their physical health (Burnett & Peel 2001), making them particularly susceptible to health risks associated with climate change.

Australia has a steadily expanding general migration program, with a total outcome for the 2012–13 reporting period of 190 000 places (DIAC 2013b). Of this total, 67% were accepted under the Skilled Migration Program, which seeks to fill critical skill needs, especially within regional areas. The 'Professional' category comprised the largest component of the skills stream, followed by 'Tradespeople & Related Workers' and 'Managers'. A second major component of the immigration program, the family stream, accounted for 31% of the total migration outcome. Family stream migrants are selected on the basis of their close family relationship with a 'sponsor' in Australia (DIBP 2014).

The top three source countries of migrants to Australia in the period 2012–13 were India, China and the United Kingdom, testifying to the changing nature of Australia's immigration program (DIAC 2013b).

What this brief reading of Australia's immigration history underwrites is the diverse and dynamic formulation of the CALD population. Clearly, this population is far from static or homogenous; it represents a heterogeneous collective not adequately encompassed by the CALD acronym. Clearly, long-settled emigrants from Italy, Greece, Vietnam and Lebanon differ in their support needs from more recent humanitarian refugees from Sudan, Iran and Afghanistan. Not only are CALD sub-populations differentiated by culture, but importantly, by socio-economic factors that parallel various immigration pathways. This group differentiation has important implications for service provision. As we will go on to elucidate, the sociocultural and economic backgrounds of distinct CALD groups necessitate the development of a culture-centred approach (Dutta 2007) that not only considers the structure of climate change risk and adaptation messages, but the context and cultural conditioning of their intended audiences (Perelman & Olbrechts-Tyteca 1971).

Climate change in the local context

The Intergovernmental Panel on Climate Change Fifth Assessment Report states that, consistent with global trends, the Australasian region is demonstrating long-term changes in temperature and rainfall patterns (Reisinger *et al.* 2014). Australia's mean surface air temperature has risen by 0.9°C since 1910 (with most of the warming occurring since 1950) and minimum temperatures have increased by 1.1°C. More extreme heat and heatwaves have occurred and extreme fire weather has increased. This trend is expected to continue with temperatures projected to rise by 0.6 to 1.5°C by 2030 compared with 1980–1999, with these changes manifesting in more hot days and fewer cool days (CSIRO & Bureau of Meteorology 2014). The effects of the changing climate form different patterns in different parts of Australia. The following discussion shows the pattern forming in South Australia.

South Australia is home to 1.6 million people and, like many regions of Australia, will face considerable risks from climate change including rising sea levels, extreme heat events and reduced rainfall (CSIRO & Bureau of Meteorology 2014; Steffen & Hughes 2013). As

South Australia is one of the driest inhabited places on earth, extended droughts are expected to place further pressure on the state's finite water resources (DCCEE 2011). The summer of 2013 recorded a total area-average rainfall of only 25.9 mm (a 58% departure from the mean, and the lowest since 1985) (Bureau of Meteorology 2013). Hot dry summers are common in South Australia. Adelaide, the capital city, experienced 38 separate 'heat events' (three or more days greater or equal to 35°C) between 1994 and 2009 (Williams *et al.* 2012). With a predicted increase in the frequency and magnitude of extreme high temperatures (Reisinger *et al.* 2014) the average number of days above 35°C in South Australia could increase from 17 days currently to 21–26 days by 2030, and to 24–47 days by 2070 (CSIRO & Bureau of Meteorology 2007). This will coincide with an increase in the number of days when bushfires may threaten lives and property (Lucas *et al.* 2007).

Extreme heatwaves are, and will continue to be, associated with increased morbidity and mortality. In the absence of an effective adaptation response, it is projected that heat-related deaths in South Australia will double over the next decade (DCCEE 2011) with vulnerability dependent on several factors, including the distribution of wealth across society, workforce participation and access to information (Reisinger *et al.* 2014).

The impacts of climate change on CALD communities

Recent history has shown that people in cultural minority groups and migrants can be disproportionately affected by extreme weather events. For example, in the Cayman Islands where tropical cyclones are common, new migrants are vulnerable because they lack past exposure to cyclones and can underestimate the hazard severity (Tompkins *et al.* 2009). In 2005 Hurricane Katrina wreaked havoc across New Orleans and provided a graphic depiction of how impoverished, minority communities can be most severely affected by extreme weather events. Inadequate emergency communication, lack of trust in authorities, socio-economic disadvantage, existing social disparities in health, and cultural and language barriers contributed to minorities experiencing far worse outcomes during and after the disaster in comparison to white, middle-class citizens (Andrulis *et al.* 2007; Samovar *et al.* 2011; Vaughan & Tinker 2009). Also in the United States, the mortality rate of African American residents during a Los Angeles heatwave was nearly double that of the city's average (Shonkoff *et al.* 2009), while in the Chicago heatwave of 1995, African Americans had the highest proportional death rates of any group (Klinenberg 2003). Although similar evidence is scarce in Australia, ethnicity has been found to be one of the risk factors for heat-related adverse health outcomes in the cities of Perth, Melbourne and Adelaide (Loughnan *et al.* 2013) with socio-economic and linguistic factors being among the major contributing factors to vulnerability (Hansen *et al.* 2013a).

Visions of ample sunshine and leisurely days at the beach in Australia can be appealing to people in countries abroad. However, Australia's weather brings with it potentially life-threatening health impacts that may be multiplied with climate change. Of all natural weather hazards including floods, cyclones and bushfires, it is heatwaves that account for the most deaths (Coates 1996) and hence are often termed the 'silent' killer. However, the concept of dying from heat is unknown to many from cooler climates. Hence, a climate change driven increase in heatwaves will pose considerable health risks for vulnerable subgroups in society in the absence of adaptation.

Our research has shown that incoming migrants are resourceful and generally adapt well upon resettlement in Australia, with a high level of resilience to stressors, which may include environmental stressors. However, some people within CALD communities can be particularly vulnerable during heatwaves due to several interrelated factors that can pose barriers to adaptation (Hansen *et al.* 2013b). Socio-economic disadvantage and lack of employment prospects can result in low-income migrants living in poor quality rental housing with no access to cooling. If air-conditioning is available, high energy costs can make usage prohibitive. Other studies have also shown that housing characteristics can contribute to vulnerability (Ebi 2012; Maller & Strengers 2011) as temperatures inside thermally inefficient homes can get hotter than outdoor air temperatures (Scovronick & Armstrong 2012; Strengers 2008). As practised in their country of origin, some migrants open doors and windows during the day in an attempt to cool the house, an action that is counterproductive when temperatures outdoors are higher than indoors. It can also be a health hazard as a study in France showed that opening windows during the afternoon during an extreme heatwave increased the risk of heat-related death (Vandentorren *et al.* 2006).

With people being physiologically and behaviourally acclimatised to the weather in their homeland (Knowlton *et al.* 2009; Kovats & Hajat 2008) foreign climates can pose a risk to health for new arrivals and tourists. Indeed, for deaths attributed to high temperatures in Australia, a lack of familiarity with local environmental conditions can be a significant risk factor for tourists (Green *et al.* 2001). Our findings showed that migrants often report that the dry, unrelenting heat is different from that experienced abroad. To counter the high temperatures and low humidity, there is a need for people to dress lightly, which may not be the cultural norm, and stay well hydrated by increasing their fluid intake. However, this is not well understood and dehydration is a risk for some, particularly those cautious of drinking tap water due to experiences of poor water quality in other countries and refugee camps.

Linguistic and social isolation is linked with vulnerability to extreme heat (Hansen *et al.* 2013a; Uejio *et al.* 2011) and can be common for migrants not proficient in English. Limited transport options can increase isolation and be a barrier for people seeking cooler places. Furthermore, language barriers, together with low literacy rates, can result in inequities in access to extreme heat warnings, weather information and healthcare services. These factors can place some migrants and refugees at risk during extreme heat and compound vulnerability for those with poor health and pre-existing risk factors such as older age and/or chronic mental or physical conditions, particularly in humanitarian entrants. Additionally, the stresses associated with the settlement process can impact on the health and wellbeing of new arrivals (Renzaho 2008). The people most vulnerable can include humanitarian entrants in new and emerging communities, recent arrivals and older migrants. It is these groups in particular that can benefit from assistance offered by community-based health and social service agencies.

Climate change poses climate risks that threaten the wellbeing of populations and while adaptation will be required, socio-economic and cultural factors will affect adaptive capacity such that minority groups and vulnerable populations will be disproportionately affected (Patrick *et al.* 2012). Essentially, vulnerability to climate change is a function of hazard exposure, sensitivity to change and adaptive capacity (Spickett *et al.* 2011). However, central to the concept of vulnerability in one group (e.g. CALD communities) and not another is categorisation, which can incite the notion of 'other' (Dutta 2007). This inference of vulnerability can therefore be divisive and cause offence. Rather, climate change will affect whole populations collectively and cultural connectedness together with coordinated community action will be necessary to facilitate adaptation.

Some immigrant populations can be among the poorest and lowest educated in society (Kreps & Sparks 2008). Vulnerability to climate change can be increased due to inequality, which can place people of CALD backgrounds at risk despite their high adaptive capacity. Inter-cultural communication barriers can limit adaptive capacity, and stakeholders in our research (Hansen *et al.* 2013b) asserted that with equal access to resources, including appropriate risk information and emergency warnings, CALD community members may be no more vulnerable than others in the population. As we argue, however, to meaningfully engage cultural groups, risk and adaptation messages need to be accessible, formulated, and delivered using a culture-centred framework that addresses the individual needs of communities (Dutta 2007; Sellnow *et al.* 2009).

Given the considerable heterogeneity in cultural and educational backgrounds among and within migrant populations, it is not surprising there are different levels of knowledge and understanding about climate change. While some people are well versed, to others it is a new concept that was not discussed in their home countries. Indeed, in some languages 'climate change' is not even easily translatable and the term 'heatwave' is unknown. Information and risk communication about climate change and its impacts should therefore be available to people of all cultural and linguistic backgrounds, while acknowledging diversity in approaches to risk perception.

Risk communication and the culture-centred approach

For over 60 years, researchers in the fields of psychology and sociology have sought to understand how people evaluate hazardous activities and on what basis these judgements are made. The field of risk perception (e.g. Slovic 1999) grew out of an observation that public risk opinions diverged significantly from those of experts. In general, risk perception psychology (Lichtenstein *et al.* 1978; Slovic 2000) aims to understand how cognitive (thoughts that may be or may not be considered 'biased') and affective (emotional) factors determine laypeople's beliefs about what is (or is not) 'risky'. For example, studies have shown that people commonly judge nuclear power as presenting a large degree of risk. However, when compared to the risks inherent in everyday activities, such as driving a car, the actual danger this technology presents to one's health (qualified in terms of mortality rates) is relatively small. Risk psychology attempts to unpack why some hazards mean different things to different people, and what factors determine these perceptions. This research aims to assist policymakers and hazard communicators to improve their interventions with the public.

The role of sociocultural factors in risk perception and communication has also been addressed. According to Dessai *et al.* (2004), risk perception is determined by interactions between *external*, technical definitions of risk, such as those emanating from government research facilities (e.g. CSIRO), and *internal* variables, such as trust in the communicator, pre-existing personal 'world views', personal experience, values and culture (Dessai *et al.* 2004). Similarly, an individual's culture is conceived to act as a powerful lens for perceiving and interpreting risk signals and the field of inter-cultural communication (Samovar *et al.* 2011) has highlighted the fundamental role of culture in framing people's perception of risk. Recognising the role that culture plays in risk perception, a sizable body of international disaster research (chiefly from the United States) has shown how preparatory, response and recovery efforts can be improved when risk communication addresses cultural understanding and socio-economic disadvantage (Andrulis *et al.* 2007; Fothergill *et al.* 1999; Holmes 2008). On the whole, this literature proposes that culturally specific communication has the potential to promote self-protective behaviours, effectively building resilience and adaptation capacity in vulnerable communities.

Expanding on the idea that sociocultural components of risk communication need to be addressed if messages are to resonate with different CALD groups, inter-cultural communication scholars (e.g. Dutta 2007; Dutta-Bergman 2004; Lupton 1994; Sellnow *et al.* 2009) have advocated for a community-based participatory research (henceforth, CBPR) model (Israel *et al.* 2012) that engages communities, researchers and stakeholders as equal partners in developing risk and health messages. This approach constitutes a departure from traditional 'message oriented' theories, such as the health belief model (Rosenstock *et al.* 1988), that depend on persuasion to alter attitudes, knowledge and behaviours of target populations (Murray-Johnson *et al.* 2003).

Lupton (1994) challenges the mainstream health communication paradigm because it excludes social groups outside of dominant social, political and cultural structures. An implication of this exclusion is that minority groups are prevented from equitably voicing their views, concerns and needs (Dutta-Bergman 2004), perpetuating social and political inequality. To correct this exclusion, a CBPR-informed, culture-centred approach, aims to re-engage minority groups in a dialogue on what risk means and what possible solutions community members themselves could provide. This approach champions cultural values, context and community strengths in developing research partnerships with CALD groups. In the health promotion arena, Dutta-Bergman (2004, p. 243) explains that the culture-centred approach 'begins by asking how the community defines health and what factors prevent community members achieving health as they conceptualize it'. In other words, multiple cultural publics have the opportunity to participate in the message construction and the means by which the information is presented to their constituencies (Sellnow *et al.* 2009).

Implications of the culture-centred model for communicating about climate change adaptation

The culture-centred paradigm has important implications for the conceptualisation and design of adaptation messages for CALD groups. First, this approach avoids the notion that a one-size-fits-all message will appeal to all cultural groups, as if they constituted a homogenous audience that construes information in fundamentally similar ways. Second, it goes beyond the 'cultural-sensitivity' model (Resnicow *et al.* 2002) that seeks to identify and incorporate 'relevant' cultural characteristics – determined by the expert – into the development and delivery of persuasive communication interventions. In contrast, a culture-centred approach places under-represented groups at the centre of message development and engagement (Sellnow *et al.* 2009). Accordingly, researchers and experts take a less formative, directive role in defining what culture *is* and what cultural variables might be used to gain adherence of the audience to persuasive messages. Community members are more than merely consulted, rather they are integral to defining what the problem under discussion means for the community within the context of existing social conditions.

The culture-centred approach focuses on the particular needs of the communities by centrally integrating them in the process of defining the problem, the construction of messages and the delivery of interventions to their social groups (Sellnow *et al.* 2009). Preparatory and crisis response messages are designed in collaboration with cultural brokers (leaders or trusted members) and government or non-government stakeholders. Communication is conceived as a two-way and interactive process, harnessing many of the principles of grassroots community development theory. In this way, CALD communities are not presumed to be lacking in expertise and knowledge of the risk phenomenon; instead, they are acknowledged as holders of valued local knowledge and resources vital to the development of effective interventions and policy.

In essence, we propose that the culture-centred approach can fruitfully inform climate change adaptation communication interventions, building stronger social capital, resilience and participation in CALD communities. Specifically, climate change adaptation interventions would benefit though a process that:

- considers the needs of particular cultural audiences, instead of assuming that one universal message will be sufficient
- uses cultural brokers who are affiliated with, and trusted by, the community
- designs messages from 'the ground up', minimising the risk of misinterpretation or irrelevance
- challenges pre-existing power inequalities by treating CALD communities as key producers of knowledge and problems solvers for effecting change.

As we will go on to elaborate, CALD groups may be differentially affected by the threats posed by climate change, yet with ongoing culture-centred engagement, vulnerable groups can participate in providing adaptive solutions that are aligned with their particular needs and cultural values.

Evidence-supported interventions to support adaptation

There are several ways community-based non-government organisations and health and social service organisations can pragmatically assist climate change adaptation by Australia's migrants and refugees. Underpinned by the culture-centred approach, we first discuss building social capital, then community information programs, heatwave strategies, formulating and disseminating risk messages, and finally, programs aimed to aid adaptation in specific subgroups.

It is acknowledged that service providers and agencies have limited funding and are constrained in terms of human resources. This is one of the reasons health and social service organisations often use a generalist approach and may not be able to provide services which adequately meet the specific linguistic needs of diverse communities and individuals (Renzaho 2008). Ideally, bi-cultural staff could be employed who not only speak the language of the community, but have a relationship with, and a deep understanding of, the group's culture and historical background, while also being acculturated into Australian society. Notwithstanding, some languages have multiple dialects and interpreters may still need to be called upon. This can also be problematic if there is a language–interpreter mismatch, and sensitivity issues arise in relation to communication between clients and interpreters (Renzaho 2008). However, facilitating and delivering programs in multiple languages can incur considerable expenses and to avoid the costs associated with professional interpreters and translators, the assistance of volunteers and community members may be considered.

Moreover, it is useful for agencies to be aware of services provided by similar organisations and refugee and migrant support services, so that service delivery options can be maximised and useful partnerships developed (Renzaho 2008). Service providers can also benefit by liaising with local government, which is becoming increasingly involved in strategies to address climate change adaptation. Council community engagement officers can be a valuable source of knowledge and resources regarding local community needs, and council venues such as libraries may be used for meetings.

Building social capital

Isolation can be common in newly arrived refugee families, although some readily form social relationships and connections with community and religious groups (Sheikh-Mohammed *et al.* 2006). For young refugees, social inclusion, a sense of belonging and being able to contribute to society, and building bonding relationships with others in their community and the broader community are essential to wellbeing (Correa-Velez *et al.* 2010). Social connectedness is also particularly important for women who may be largely confined to the home and lack support networks, and those in small marginal communities. Strengthening social capital (i.e. social relationships that enable coordinated community action) can help minimise adverse health impacts of climate change through adaptation (Ebi & Semenza 2008). Bonding social capital outlines interpersonal relationships between individuals within cultural, religious or ethnic groups, where bridging social capital is networked relationships between groups (Ebi & Semenza 2008; Pelling & High 2005), for example, like-minded agencies or service providers.

Agencies can assist in increasing local adaptive capacity and social capital by initiating community-based groups (Ebi & Semenza 2008) such as social groups or clubs where people can meet others in their community and socialise in a supportive environment. Another is the support of community gardens, as many migrants have experience in cultivating their own food but lack the means to do so when there is instability in housing circumstances. These strategies can help build social capital and increase a sense of belonging to a community, while also reducing the likelihood of isolation, a known risk factor for heat-related mortality (Reid *et al.* 2009; Vaneckova *et al.* 2008).

Resonating with the culture-centred approach, such contexts also provide a setting for CALD groups to voice their ideas, concerns and potential solutions to climate change adaptation. Further, social clubs and activities are valuable contexts in which relationships between agencies and CALD groups can be developed with a view to fostering participation. As we have noted, ongoing and meaningful social engagement is partially based on providing community members opportunities to voice their concerns in reference to their day-to-day experiences and local environment. It is within this everyday context where pragmatic climate change adaptation responses are negotiated and potential solutions found.

Community information programs

Adaptation involving community engagement has been identified previously as a means of bolstering a community's resilience to climate stressors (Ebi & Semenza 2008). There is an unmet need for information regarding climate change and heatwaves amongst CALD communities (Hansen *et al.* 2013b) and sharing information via community engagement programs is one way organisations can help in this regard. Again, this requires due consideration of the community's circumstances, needs and cultural values (Boughtwood *et al.* 2012) and should incorporate a trust-centred community-based participatory approach as mentioned previously. Adaptation options will be most effective if the community is at the centre of the interventions and can assist in their design and implementation (Ebi & Semenza 2008).

To meet this objective, it is suggested that organisations formulate a culture-centred strategy that prioritises relationships with CALD leaders and the sharing of knowledge through peer communication (Kreps & Sparks 2008). If rapport has already been established with certain CALD communities, invitations to attend the sessions can be extended to key bi-cultural community leaders who speak the community's language. Cultural agents can then articulate messages to members of their respective communities; thereby ensuring dialogue is delivered in a culturally appropriate and salient manner. Audiences may be the communities in general or specific groups based on age, gender, nationality or religion. Alternatively, meetings on a one-to-one basis may be more acceptable depending on the level of individualism or collectivism within the cultural group (Dutta 2007).

In the first instance, for example, sessions could focus on understanding how group members experience extreme heat, enact low-cost heat-adaptive behaviours and consider the impacts of climate change in general. Setting this scene is fundamental to a participatory approach that prioritises the expertise of CALD communities and the critical importance of equitable dialogue.

As well as extreme heat, other aspects of climate change can be discussed including underlying key concepts and causes, and impacts of other extreme weather events including bushfires, floods and droughts. While most of the overseas-born population reside in cities, with the recent introduction of regional migration incentive schemes (Massey & Parr 2012) many move to regional (and remote) areas where there can be less access to health services and greater climate risks such as bushfires, with which overseas-born people may not be familiar. Risk communication about the dangers of bushfires would therefore take on greater importance in this instance and discussions may involve the need for bushfire emergency plans. Relating climate change to conditions in homelands can improve salience as issues such as rising sea levels, warmer temperatures, flooding rains and longer droughts may well resonate with pre-migration climate experiences. The focus may then shift to the Australian context and the need for adaptive measures. Storylines or scenarios of possible climate change impacts such as extended or unprecedented heatwaves can be compelling narratives that highlight potential risks to health and the need for adaptive measures (Ebi & Semenza 2008).

Notwithstanding, it is recognised that agency personnel may themselves not be well versed in the environmental complexities of climate change and/or cross-cultural competency. In order for this model to be successful, there may be a need for staff and healthcare providers to undertake professional development to improve cross-cultural communication skills (Renzaho 2008; Kreps & Sparks 2008). Information on climate change is available via numerous online sources including the websites of the Intergovernmental Panel on Climate Change (www.ipcc.ch), the CSIRO (www.csiro.au) and the Bureau of Meteorology (www.bom.gov.au). A barrier to successful information programs on these topics is that attendance numbers may be low due to the fact that in the context of the many stressors facing new arrivals upon resettlement (e.g. housing instability, unemployment, financial difficulties and learning English), climate change and extreme heat can take a low priority.

Heatwave strategies

New arrivals may never have experienced the likes of an Australian heatwave in their country of origin. Consequently, when high temperatures persist for several days and nights, those unfamiliar with the adaptive behaviours necessary to maintain thermal comfort can find the heat oppressive. It can also be a threat to health and wellbeing. Harm minimisation strategies are not necessarily intuitive and new arrivals living in rental homes that lack cooling mechanisms can be unaware that keeping the home closed up with blinds drawn can help keep the heat out; and that drinking adequate quantities of water, avoiding physical activity and dressing lightly are recommended by health authorities. If people have the means, some may choose to seek cooler environs such as shopping centres. However, this practice is not universally favoured, as research has shown that some people with dark skin in mainly Caucasian societies can feel uncomfortably conspicuous spending extended periods of time in public spaces (Hansen *et al.* 2013b; Colic-Peisker 2009). Instead, organisations and agencies may be able to offer or arrange, cooled spaces where people in new and emerging communities can relax and socialise during the heat in a culturally appropriate and welcoming environment. This may be in the form of an air-conditioned community centre, hall, council library or other central meeting area.

Agencies can assist community members adapt and build resilience by preparing for heatwaves. This forward planning may involve encouraging families to have a heat plan, part of which may involve checking on other community members, particularly the elderly, to ensure their wellbeing during heatwaves. Agencies can help initiate and facilitate strategies such as a 'telephone tree' (i.e. a person calls a friend who then calls another and so forth) to avert the need for individuals to make multiple calls.

Extreme heat over consecutive days can be considered an emergency situation due to the health risks. An added burden on the health sector results from increased ambulance callouts, hospital admissions, emergency department visits and cause-specific deaths during heatwaves (Bi *et al.* 2011; Nitschke *et al.* 2007). For this reason, authorities in South Australia (SA SES 2010) and elsewhere (Department of Health 2009; WA Department of Health 2010) issue extreme heat warnings, which are broadcast, mainly in English, via the media and the internet. Our studies have recognised the importance of equitable access to emergency and preparatory heatwave messages for people in CALD communities and that these messages need to be formulated in a culturally competent manner.

Risk communication messages

The conceptualisation, design and implementation of evidence-based, culturally appropriate risk communication interventions are crucial to building resilience to extreme weather events and have been discussed to some degree. Suffice to say that as there is no homogeneity within and between cultural groups, it is advisable to assume diversity (Renzaho 2008) and recognise there is no one-size-fits-all approach that provides an appropriate means of risk communication to people of all cultures and languages. A stakeholder workshop held as part of our research (Hansen *et al.* 2013b) offered insights into how heat-health messages to a multicultural audience could be improved. Topics discussed included the content of messages, how and when they should be delivered and who should be involved. The outcomes of the workshop are discussed below.

First, when communicating risk, the message content should be clear, basic, expressed in simple terms in migrants' own language (Hansen *et al.* 2013b) while avoiding catchy or unfamiliar phrases, and pre-tested with community representatives (Kreps & Sparks 2008). Key messages to minimise the risk of harm during extreme heat should include practical advice such as 'find a cool space'; 'pull down blinds'; 'take baths or showers to cool down'; 'watch out for the elderly' and 'seek medical help if feeling unwell'. The importance of drinking fluids (mainly water) needs to be relayed, especially to migrants from areas with poor water quality. The messages must be in migrants' own language and cater for unique linguistic nuances between CALD groups. Culturally appropriate images can be used to demonstrate key points. Often emergency heatwave messages and warnings will be distributed by government authorities and emergency services. This offers the opportunity for a coordinated approach between government and the non-government organisations, community groups and health and social service agencies that play an important role in the dissemination of the messages to people in CALD communities. This may involve having printed resources in several languages available on notice boards, in libraries and community centres or in waiting rooms in, for example, migrant resource centres, migrant health services and doctors' surgeries. If pamphlets need translation, a credible source should be used and the final product checked for accuracy and acceptability by native speakers of the language. An English version of pamphlets also needs to be on hand for bilingual speakers. It is worth noting, however, that in some cultures it is not acceptable to take something that is not given and that pamphlets would need to be handed to individuals.

There is diversity in how different cultures engage with information. Some prefer oral-aural communication whereas others relate better to graphic or literal methods, and these differences need to be acknowledged. For instance, humanitarian arrivals and others may have low educational attainment or disrupted schooling and hence literacy in an individual's own language should not be assumed. As mentioned previously, visual images depicting key messages can be useful when literacy levels are poor. Messages therefore need to be structured to the understanding and cultural norms of the target audience (Renzaho 2008) with printed resources for those who are literate, and information on a one-to-one basis for people with low literacy levels.

When relaying messages or information one-on-one or to a group, it is important that cultural competence is taken into consideration. Patronising tones should be avoided and the method of delivery should not engender xenophobia or use prejudicial discourses. An information sharing approach where key points are explained appropriately and related to personal situations is recommended, rather than a didactic and paternalistic top down approach. Messages should relate to key beliefs, values and attitudes of the cultural group, using familiar language, images and examples to highlight key points. Participatory communication strategies (Kreps & Sparks 2008) should involve community members or a trusted key community service provider who can take a leadership role in message dissemination, as mentioned previously. Many migrant communities have a strong religious base and key faith leaders respected among the community may be able to assist in risk communication, particularly in emergency situations. Bi-cultural community health workers are ideally placed to inform clients/patients individually about the health impacts of extreme heat. In general, these strategies may benefit from a multi-sectoral approach incorporating liaison with interpreters, volunteers, support services, local government and other like-minded agencies and service providers.

Other communication methods and channels can also be considered. In an emergency situation messages can be sent to community members via mobile and landline telephones or via multilingual messages on ethnic radio. Preparedness messages could be disseminated via language-specific newspapers or flyers in letter boxes. The Australian Red Cross website lists a useful range of documents that are 'designed to help people better prepare, respond to and recover from emergencies', including information on how to cope with hot weather (Australian Red Cross 2014). The resources in 'easy English' would be helpful to people for whom English is a second language. The website also contains a list of resources specifically designed for agencies, educators and people working in emergencies.

Innovative means of communication should also be considered. For example, if funding is available, DVDs and YouTube clips could be developed as a visual means of relaying risk communication and prevention messages. Fridge magnets in multiple languages could be produced with key hints on keeping safe in the heat. Additionally, community events such as cultural festivals and markets can be used as a vehicle to provide information about extreme heat.

Younger generations can assist in informing other family members. For example, heat awareness and adaptation strategies could be incorporated into curricula at schools, English language classes and TAFE colleges. While the internet is not accessed universally among CALD communities, messages via social media can be disseminated further by word of mouth. Other groups that could be targeted include people in new and emerging communities, refugees, asylum seekers, women (including those on 'women at risk' visas), skilled migrants, hard-to-reach sub-groups, CALD groups with few support networks and perhaps even non-English speaking tourists. Two particular groups at risk in terms of climate change and extreme heat are new arrivals and older migrants. Specific strategies to address these groups are discussed in the following sections.

Importantly, underpinning these communication recommendations is the premise that CALD groups themselves should be integral to any new communication intervention. Hence, message design and dissemination channel considerations should be made with respect to the existing social structures that cultural groups face in their everyday lives. In this way, CALD groups can articulate human agency when developing solutions to coping with climate change impacts.

New arrivals

Often new migrants are either not aware of the potential dangers of extreme heat or have been given piecemeal advice about Australia's sun intensity before arrival. Hence, stakeholders in our research study (Hansen *et al.* 2013b) believed it would be beneficial to provide information to new arrivals about local climate information, climate risks including extreme heat, adaptive behavioural strategies, and tips on beach safety.

The information could be available pre-migration in the country of origin, or as part of an orientation package upon arrival in Australia (or at least provided before summer). This may involve organisations/agencies with responsibility for new arrivals and be relayed via settlement workers who could assist in clarifying or interpreting key points. Alternatively, health service providers may consider speciality clinics for new arrivals or information sessions for families, where heat information can be distributed. It should be noted, however, that newly arrived refugees may be reluctant to seek health services for reasons which include language barriers, lack of information, financial handicap and poor experiences of health care overseas (Sheikh-Mohammed *et al.* 2006).

By promoting simple changes in behaviour and the use of adaptive strategies, agencies can help to reduce the risk to wellbeing in new arrival families during extreme heat events. Building relationships with newly arrived individuals and new and emerging communities, and developing an understanding of needs, will help engender trust in agency staff. Formulating new, or fine-tuning existing, communication materials requires cultural sensitivity as some messages and concepts may challenge individuals' underlying maladaptive health beliefs, misconceptions, practices and assumed knowledge about heat. Preferably, a culture-centred communication intervention can be developed that draws on pre-existing experiences of extreme weather conditions. Ideally via participatory communication methods incorporating communicators from the same CALD background, extreme heat and climate change adaptation messages will recognise and acknowledge the social and financial limitations experienced by the community in the local context together with the stressors associated with resettlement. Theoretically, these strategies embody aspects of both the cultural sensitivity and the culture-centred approach (Dutta 2007) in the communication of heat-health messages.

Older CALD community members

Older people are a group well recognised as being amongst the most vulnerable during heatwaves (Worfolk 2000) although they often do not accept they are at risk (Abrahamson *et al.* 2009). While an earlier chapter discusses elderly people and climate change impacts, the focus herein is specifically on the aged in CALD communities. This group comprises elders in new and emerging communities as well as older, long standing migrants.

Ageing has been associated with increased stability (i.e. inflexibility) of attitudes over time (Krosnick & Alwin 1989). This phenomenon can negatively affect the acculturation process, as recently resettled older people find it comparatively more difficult to adapt to their new cultural setting than younger people. Importantly, the process of aging may also hamper second language acquisition, compounding social isolation and vulnerability. Additionally, many long-standing migrants, for example, post-World War II arrivals from Europe who may have been once fluent in English, often revert to their first language with age (Hansen *et al.* 2011; Schmid & Keijzer 2009). These linguistic barriers can lead to isolation if the person is unable to communicate with their social contacts or family members, thereby increasing their vulnerability during periods of extreme heat.

Additionally, many older humanitarian entrants and longstanding migrants have low educational attainment and poor literacy levels. Therefore the most effective interventions for older people in CALD communities are those that involve oral-aural communication in their first language. For service providers this can be a problem as bilingual staff or volunteers, or interpreters may need to be recruited.

Identifying the vulnerable within communities can be difficult. Some organisations use a vulnerability audit or checklist with questions such as:

- Does the client speak English?
- Do they live alone?
- Do they have mobility restrictions?
- Do they need assistance with household chores?
- Do they drive a car?

This is then used to set up registers of vulnerable older clients who could be checked on during heatwaves by volunteers, either in person, or by telephone. A drawback, however, is that some older migrants do not like using telephones. The South Australian Red Cross has a Telecross REDi service which involves regular telephone calls to registered vulnerable people during heatwaves (Australian Red Cross 2012). The service is offered in several languages, but the concept can be seen as unusual to those from cultures where the elders normally live with the extended family. Nevertheless, it is older people living alone, isolated, without family connections and unnoticed by service providers that can be most at risk during a heatwave, and hence the importance of community spirit and checking on elderly family members and neighbours.

Once vulnerable persons have been identified, offering assistance during extreme heat can be challenging for organisations, particularly as some older people can be wary of service providers, thinking there will be costs involved. It is important therefore that agencies place emphasis on gaining the trust and respect of older people in CALD communities. Checks on the client's wellbeing should include that an air-conditioner, if present, is in use and set to 'cool', and that they are drinking adequate amounts of water to avoid dehydration, a common condition in the elderly (Olde Rikkert *et al.* 2009). Some agencies and organisations receive funding to provide clients with community care services; however, in some cases services may cease during the festive season. This can be problematic if a heatwave falls during this time and client vulnerability is increased. It is best practice therefore that each service that deals with the elderly has a heatwave strategy in place.

Conclusion

In this chapter we have proposed that a one-sized-fits-all approach to engaging various cultural publics on the topic of climate change adaptation will not comprise an efficient nor equitable strategy. For many sociocultural reasons, some within CALD communities face barriers when adapting to extreme heat in Australia, yet most current interventions and emergency messages are aimed at locally born, English speaking residents. Community-based health and social service agencies, including local governments, can play a key role in building adaptive capacity of migrants to increasing frequency and intensity of heatwaves and other impacts of climate change. By using a culture-centred approach to engagement, agencies are well positioned to build upon existing client–service provider relationships. We argue that it is within the context of these relationships that the design and dissemination of climate change risk communication and adaptation messages have their best chance of meeting the needs of CALD publics. Indeed, this is because many of the limitations inherent in traditional top-down communications models are side-stepped when communities are asked to co-participate in defining the problem as it affects them. CALD groups can be empowered and mobilised into action when their voices are provided equal space in the social realm. Organisations will require commitment and additional resources to address the unmet information needs of CALD communities through a culture-centred lens – in a changing climate, however, this is less a choice than a social justice imperative.

References

Abrahamson V, Wolf J, Lorenzoni I, Fenn B, Kovats S, Wilkinson P, Adger WN, Raine R (2009) Perceptions of heatwave risks to health: interview-based study of older people in London and Norwich, UK. *Journal of Public Health (Oxford, England)* **31**, 119–126. doi:10.1093/pubmed/fdn102

Andrulis DP, Siddiqui NJ, Gantner JL (2007) Preparing racially and ethnically diverse communities for public health emergencies. *Health Affairs* **26**, 1269–1279. doi:10.1377/hlthaff.26.5.1269

Australian Red Cross (2012) *Telecross REDi*. Australian Red Cross, Carlton, Victoria, <http://www.redcross.org.au/telecross-redi.aspx>.

Australian Red Cross (2014) *Resources*. Australian Red Cross, Carlton, Victoria, <http://www.redcross.org.au/emergency-resources.aspx>.

Bi P, Williams S, Loughnan M, Lloyd G, Hansen A, Kjellstrom T, Dear K, Saniotis A (2011) The effects of extreme heat on human mortality and morbidity in Australia: implications for public health. *Asia-Pacific Journal of Public Health* **23**(Supp 2), 27S–36S. doi:10.1177/1010539510391644

Boughtwood D, Shanley C, Adams J, Santalucia Y, Kyriazopoulos H, Pond D, Rowland J (2012) Dementia information for culturally and linguistically diverse communities: sources, access and considerations for effective practice. *Australian Journal of Primary Health* **18**, 190–196. doi:10.1071/PY11014

Bureau of Meteorology (2013) *Australia in summer 2012–13*. Bureau of Meteorology, Melbourne, <http://www.bom.gov.au/climate/current/season/aus/archive/201302.summary.shtml>.

Burnett A, Peel M (2001) Asylum seekers and refugees in Britain: health needs of asylum seekers and refugees. *British Medical Journal* **322**, 544–547. doi:10.1136/bmj.322.7285.544

Coates L (1996) An overview of fatalities from some natural hazards in Australia. In *Conference on Natural Disaster Reduction (NDR96): Conference Proceedings*. (Eds RL Heathcote, C Cuttler and J Koetz) pp. 49–54. Institution of Engineers Australia, Barton, Australian Capital Territory.

Colic-Peisker V (2009) Visibility, settlement success and life satisfaction in three refugee communities in Australia. *Ethnicities* **9**, 175–199. doi:10.1177/1468796809103459

Correa-Velez I, Gifford SM, Barnett AG (2010) Longing to belong: social inclusion and wellbeing among youth with refugee backgrounds in the first three years in Melbourne, Australia. *Social Science & Medicine* **71**, 1399–1408. doi:10.1016/j.socscimed.2010.07.018

CSIRO & Bureau of Meteorology (2007) 'Climate change in Australia. Technical report 2007'. CSIRO and the Bureau of Meteorology, Melbourne.

CSIRO & Bureau of Meteorology (2009) 'Climate change in Australia science update 2009 – Issue two', CSIRO and the Bureau of Meteorology, Melbourne, <http://ccia2007.climate-changeinaustralia.gov.au/documents/resources/ClimateScienceUpdate2009_2.pdf>.

CSIRO & Bureau of Meteorology (2014) *The State of the Climate 2014*. Bureau of Meteorology, Melbourne, <http://www.bom.gov.au/state-of-the-climate >.

DCCEE (Department of Climate Change and Energy Efficiency) (2011) 'Climate change risks to coastal buildings and infrastructure. A supplement to the first pass national assessment'. DCCEE, Canberra.

Department of Health (2009) 'Heatwave plan for Victoria 2009–2010. Protecting health and reducing harm from heatwaves'. Victorian Government, Melbourne, Victoria.

Dessai S, Adger WN, Hulme M, Turnpenny J, Köhler J, Warren R (2004) Defining and experiencing dangerous climate change. *Climatic Change* **64**, 11–25. doi:10.1023/B:CLIM.0000024781.48904.45

DIAC (Department of Immigration and Citizenship) (2013a) 'Australia's offshore humanitarian program: 2012–13'. Department of Immigration and Citizenship, Canberra.

DIAC (Department of Immigration and Citizenship) (2013b) 'Migration program report: program year to 30 June 2013'. Department of Immigration and Citizenship, Canberra.

DIBP (Department of Immigration and Border Protection) (2014) *Fact Sheet 29 – Overview of Family Stream Migration*. Department of Immigration and Border Protection, Canberra, <http://www.immi.gov.au/media/fact-sheets/29overview_family.htm>.

Dutta MJ (2007) Communicating about culture and health: theorizing culture-centered and cultural sensitivity approaches. *Communication Theory* **17**, 304–328. doi:10.1111/j.1468-2885.2007.00297.x

Dutta-Bergman MJ (2004) The unheard voices of Santalis: communicating about health from the margins of India. *Communication Theory* **14**, 237–263. doi:10.1111/j.1468-2885.2004. tb00313.x

Ebi KL (2012) High temperatures and cause-specific mortality. *Occupational and Environmental Medicine* **69**, 3–4. doi:10.1136/oemed-2011-100259

Ebi KL, Semenza JC (2008) Community-based adaptation to the health impacts of climate change *American Journal of Preventive Medicine* **35**, 501–507. doi:10.1016/j.amepre.2008. 08.018

Fothergill A, Maestas EGM, Darlington JD (1999) Race, ethnicity and disasters in the United States: a review of the literature. *Disasters* **23**, 156–173. doi:10.1111/1467-7717.00111

Green H, Gilbert J, James R, Byard RW (2001) An analysis of factors contributing to a series of deaths caused by exposure to high environmental temperatures. *The American Journal of Forensic Medicine and Pathology* **22**, 196–199. doi:10.1097/00000433-200106000-00018

Hansen A, Bi P, Nitschke M, Pisaniello D, Newbury J, Kitson A (2011) Perceptions of heat-susceptibility in older persons: barriers to adaptation. *International Journal of Environmental Research and Public Health* **8**, 4714–4728. doi:10.3390/ijerph8124714

Hansen A, Bi L, Saniotis A, Nitschke M (2013a) Vulnerability to extreme heat and climate change: is ethnicity a factor? *Global Health Action* **6**, 21364. doi:10.3402/gha.v6i0.21364

Hansen A, Bi P, Saniotis A, Nitschke M, Benson J, Tan Y, Smyth V, Wilson L, Han G-S (2013b) 'Extreme heat and climate change: adaptation in culturally and linguistically diverse (CALD) communities'. National Climate Change Adaptation Research Facility, Gold Coast.

Holmes BJ (2008) Communicating about emerging infectious disease: the importance of research. *Health Risk & Society* **10**, 349–360. doi:10.1080/13698570802166431

Hugo G (2002) From compassion to compliance? Trends in refugee and humanitarian migration in Australia. *GeoJournal* **56**, 27–37. doi:10.1023/A:1021752802043

Israel BA, Eng E, Schulz AJ, Parker EA (2012) Introduction to methods in community-based participatory research for health. In *Methods in Community-Based Participatory Research for Health*. (Eds BA Israel, E Eng and AJ Schultz) pp 3–37. John Wiley & Sons, Somerset, New Jersey.

Jupp J (1991) *From White Australia to Woomera: The Story of Australian Immigration*. Cambridge University Press, Cambridge, UK.

Klinenberg E (2003) *Heat Wave: A Social Autopsy of Disaster in Chicago*. University of Chicago Press, Chicago.

Knowlton K, Rotkin-Ellman M, King G, Margolis HG, Smith D, Solomon G, Trent R, English P (2009) The 2006 Californian heat wave: impacts on hospitalizations and emergency department visits. *Environmental Health Perspectives* **117**, 61–67. doi:10.1289/ehp.11594

Kovats RS, Hajat S (2008) Heat stress and public health: a critical review. *Annual Review of Public Health* **29**, 41–55. doi:10.1146/annurev.publhealth.29.020907.090843

Kreps GL, Sparks L (2008) Meeting the health literacy needs of immigrant populations. *Patient Education and Counseling* **71**, 328–332. doi:10.1016/j.pec.2008.03.001

Krosnick JA, Alwin DF (1989) Aging and susceptibility to attitude change. *Journal of Personality and Social Psychology* **57**, 416–425. doi:10.1037/0022-3514.57.3.416

Lack J, Templeton J (1995) *Bold Experiment: A Documentary History of Australian Immigration Since 1945*. Oxford University Press, Melbourne.

Lichtenstein S, Slovic P, Fischhoff B, Layman M, Combs B (1978) Judged frequency of lethal events. *Journal of Experimental Psychology. Human Learning and Memory* **4**, 551. doi:10.1037/0278-7393.4.6.551

Loughnan ME, Tapper NJ, Phan T, Lynch K, McInnes JA (2013) 'A spatial vulnerability analysis of urban populations during extreme heat events in Australian capital cities'. National Climate Change Adaptation Research Facility, Gold Coast.

Lucas C, Hennessy K, Mills G, Bathols J (2007) 'Bushfire weather in southeast Australia: recent trends and projected climate change impacts. Consultancy report prepared for The Climate Institute of Australia'. Bushfire CRC and Australian Bureau of Meteorology, CSIRO Marine and Atmospheric Research, Melbourne.

Lupton D (1994) Toward the development of critical health communication praxis. *Health Communication* **6**, 55–67. doi:10.1207/s15327027hc0601_4

Maller CJ, Strengers Y (2011) Housing, heat stress and health in a changing climate: promoting the adaptive capacity of vulnerable households, a suggested way forward. *Health Promotion International* **26**, 492–498. doi:10.1093/heapro/dar003

Massey SJL, Parr N (2012) The socio-economic status of migrant populations in regional and rural Australia and its implications for future population policy. *Journal of Population Research* **29**, 1–21. doi:10.1007/s12546-011-9079-9

Murray-Johnson L, Witte K (2003) Looking toward the future: health message design strategies. In *Handbook of Health Communication*. (Eds TL Thompson, AM Dorsey, KI Miller and R Parrot) pp. 473–495. Lawrence Erbaum Associates, Mahwah, New Jersey.

Nitschke M, Tucker G, Bi P (2007) Morbidity and mortality during heatwaves in metropolitan Adelaide. *The Medical Journal of Australia* **187**, 662–665.

Olde Rikkert MG, Melis RJ, Claassen JA (2009) Heat waves and dehydration in the elderly. *British Medical Journal* **339**, b2663. doi:10.1136/bmj.b2663

Patrick R, Capetola T, Townsend M, Nuttman S (2012) Health promotion and climate change: exploring the core competencies required for action. *Health Promotion International* **27**, 475–485. doi:10.1093/heapro/dar055

Pelling M, High C (2005) Understanding adaptation: what can social capital offer assessments of adaptive capacity? *Global Environmental Change* **15**, 308–319. doi:10.1016/j.gloenvcha.2005.02.001

Perelman C, Olbrechts-Tyteca L (1971) *The New Rhetoric: A Treatise on Argumentation*. University of Notre Dame Press, Notre Dame, Indiana.

Reid C, O'Neill MS, Gronlund C, Brines SJ, Brown DG, Diez-Roux AV, Schwartz J (2009) Mapping community determinants of heat vulnerability. *Environmental Health Perspectives* **117**, 1730–1736.

Reisinger A, Kitching RL, Chiew F, Hughes L, Newton PCD, Schuster SS, Tait A, Whetton P (2014) Australasia. In *Climate Change 2014: Impacts, Adaptation, and Vulnerability. Part B: Regional Aspects. Contribution of Working Group II to the Fifth Assessment Report of the Intergovernmental Panel on Climate Change*. (Eds VR Barros, CB Field, DJ Dokken, MD Mastrandrea, KJ Mach, TE Bilir, M Chatterjee, KL Ebi, YO Estrada, RC Genova, B Girma, ES Kissel, AN Levy, S MacCracken, PR Mastrandrea and LL White) pp. 1371–1438. Cambridge University Press, Cambridge UK and New York, NY, USA.

Renzaho A (2008) Re-visioning cultural competence in community health services in Victoria. *Australian Health Review* **32**, 223–235. doi:10.1071/AH080223

Resnicow K, Braithwaite RL, Dilorio C, Glanz K (2002) Applying theory to culturally diverse and unique populations. In *Health Behavior and Health Education: Theory, Research and Practice*. 3rd edn. (Eds K Glanz, BK Rimer and FM Lewis) pp. 485–509. Jossey-Bass, San Francisco.

Rosenstock IM, Strecher VJ, Becker MH (1988) Social learning theory and the health belief model. *Health Education & Behavior* **15**, 175–183. doi:10.1177/109019818801500203

Samovar L, Porter R, McDaniel E (2011) *Intercultural Communication: A Reader*. Cengage Learning, Boston.

SA SES (South Australian State Emergency Service) (2010) *Extreme Heat Plan*. Government of South Australia, Adelaide, <http://www.ses.sa.gov.au/site/community_safety/heatwave_information/extreme_heat_plan.jsp>.

Schmid MS, Keijzer M (2009) First language attrition and reversion among older migrants. *International Journal of the Sociology of Language* **200**, 83–101.

Scovronick N, Armstrong B (2012) The impact of housing type on temperature-related mortality in South Africa, 1996–2015. *Environmental Research* **113**, 46–51. doi:10.1016/j.envres.2012.01.004

Sellnow TL, Ulmer RR, Seeger MW, Littlefield R (2009) *Effective Risk Communication: A Message-Centered Approach*. Springer, New York.

Sheikh-Mohammed M, Macintyre CR, Wood NJ, Leask J, Isaacs D (2006) Barriers to access to health care for newly resettled sub-Saharan refugees in Australia. *The Medical Journal of Australia* **185**, 594–597.

Shonkoff S, Morello-Frosch R, Pastor M, Sadd J (2009) 'Environmental health and equity impacts from climate change and mitigation policies in California: a review of the literature'. California Environmental Protection Agency, Sacramento.

Slovic P (1999) Trust, emotion, sex, politics, and science: surveying the risk-assessment battlefield. *Risk Analysis* **19**(4), 689–701. doi:10.1111/j.1539-6924.1999.tb00439.x

Slovic P (2000) *Trust, Emotion, Sex, Politics, and Science: Surveying the Risk-Assessment Battlefield. The Perception of Risk*. Earthscan Publications, London.

Spickett JT, Brown HL, Katscherian D (2011) Adaptation strategies for health impacts of climate change in Western Australia: application of a health impact assessment framework. *Environmental Impact Assessment Review* **31**, 297–300. doi:10.1016/j.eiar.2010.07.001

Steffen W, Hughes L (2013) 'The critical decade 2013: climate change science, risks and responses'. Climate Commission Secretariat (Department of Industry, Innovation, Climate Change, Science, Research and Tertiary Education), Canberra.

Strengers Y (2008) Comfort expectations: the impact of demand-management strategies in Australia. *Building Research and Information* **36**, 381–391. doi:10.1080/09613210802087648

Tompkins EL, Hurlston L-A, Poortinga W (2009) Foreignness as a constraint on learning: the impact of migrants on disaster resilience in small islands. *Environmental Hazards* **8**, 263–277. doi:10.3763/ehaz.2009.0018

Uejio CK, Wilhelmi OV, Golden JS, Mills DM, Gulino SP, Samenow JP (2011) Intra-urban societal vulnerability to extreme heat: the role of heat exposure and the built environment, socioeconomics, and neighborhood stability. *Health & Place* **17**, 498–507. doi:10.1016/j.healthplace.2010.12.005

Vandentorren S, Bretin P, Zeghnoun A, Mandereau-Bruno L, Croisier A, Cochet C, Riberon J, Siberan I, Declercq B, Ledrans M (2006) August 2003 heat wave in France: risk factors for death of elderly people living at home. *European Journal of Public Health* **16**, 583–591. doi:10.1093/eurpub/ckl063

Vaneckova P, Beggs PJ, de Dear RJ, McCracken KW (2008) Effect of temperature on mortality during the six warmer months in Sydney, Australia, between 1993 and 2004. *Environmental Research* **108**, 361–369. doi:10.1016/j.envres.2008.07.015

Vaughan E, Tinker T (2009) Effective health risk communication about pandemic influenza for vulnerable populations. *American Journal of Public Health* **99**, S324–S332. doi:10.2105/AJPH.2009.162537

WA Department of Health (Western Australia Department of Health) (2010) *Circular details: Heatwave Policy.* Government of Western Australia, Perth, <http://www.health.wa.gov.au/circularsnew/circular.cfm?Circ_ID=12612>.

Williams S, Nitschke M, Sullivan T, Tucker GR, Weinstein P, Pisaniello DL, Parton KA, Bi P (2012) Heat and health in Adelaide, South Australia: assessment of heat thresholds and temperature relationships. *The Science of the Total Environment* **414**, 126–133. doi:10.1016/j.scitotenv.2011.11.038

Worfolk JB (2000) Heat waves: their impact on the health of elders. *Geriatric Nursing* **21**, 70–77. doi:10.1067/mgn.2000.107131

10

Rural communities experiencing climate change: a systems approach to adaptation

Glenda Verrinder and Lyn Talbot

Key points

- There is an inextricable link between rural people and the effects of climate on their daily lives and their health.
- Rural Australians are currently experiencing inequitable access to health care, and there is reluctance to deal with this challenge.
- Only interdisciplinary, collaborative approaches that draw on the wisdom of rural people are likely to be effective in health enhancement.
- Some creative programs have been designed that give optimism about meeting the health needs of rural people, despite their low numbers and geographic isolation.

Introduction

Rural Australia is diverse: climatically, geographically, economically, demographically and culturally. Consequently, the impact of climate change on the wellbeing of humans in rural Australia is varied. Adaptation to climate change is necessarily different in different locations.

This chapter is presented in three sections. The first section presents an overview of the rural context; the second section outlines the impacts of climate change on rural and remote communities in Australia; and the third section presents recommendations for future practice with the aim of alleviating or minimising the impacts of climate change and building community resilience.

Background

Extreme weather events, including droughts, floods, severe storms and bushfires are experienced in arid, temperate and tropical regions of Australia. In the north, cyclones and tornadoes occur. Australia is the world's second driest continent but there are large seasonal variations in rainfall and temperature across the country. Fire and drought are affecting the southern areas of Australia more often and more severely, whereas it is becoming wetter in the north (CSIRO & Bureau of Meteorology 2014).

Communities in rural areas are particularly vulnerable to climate change because of their direct dependence on natural resources, weather-dependent activities such as farming and lack of access to services. Communities are affected by the climate in two main ways: first, the impact on infrastructure and loss of life caused by extreme weather events; and second, the impact on ecosystems and agriculture of prolonged drought, as demonstrated in the recent droughts in south-eastern Australia. Protracted drought affects livelihoods, infrastructure and patterns of settlement. Adaptation needs to address these vulnerabilities. Conversely, connectedness to nature and the land can have positive health effects (Hegney *et al.* 2007; Pereira 2008). There are psychological and spiritual benefits (Townsend & Mahoney 2004), which rural Australians have greater exposure to.

The health of rural Australians is dependent upon a range of factors: the environment, population size and demographics, dispersal and density, cultures, economic activity and viability, employment and labour patterns, industry (including agriculture), infrastructure, access to services (Dade Smith 2007), and now climate change. Rurality exacerbates the health status inequalities that increase with remoteness (AIHW 2008).

Direct and indirect health impacts of climate change are associated with climatic conditions that amplify the health challenges already faced by rural communities (Blashki 2008). However, research on the current impacts of climate change in inland Australian settlements is at an early stage, and somewhat fragmented, and we are only beginning to understand the issues. The Intergovernmental Panel on Climate Change (Dasgupta *et al.* 2014) reports that some, but not all, indigenous peoples and farmers have higher than average exposure to climate change impacts due to their reliance on primary industries and strong connections to the natural environment. They face specific constraints to adaptation. Greater vulnerability to climate change in rural areas is associated with the degree of remoteness; in particular, in cropping areas of south-western and south-eastern Australia, and also for Indigenous Australians in remote communities whose health is already affected by poor housing and water quality, and deficits in other basic infrastructure such as access to appropriate health services. (See Chapter 8 for detailed discussion about climate change and Indigenous communities.)

More extreme weather events have exacerbated the environmental, economic and social decline experienced in many parts of rural Australia since the 1970s (Talbot & Verrinder 2014). The impact of structural adjustment programs in the agricultural sector and the knock-on effects of lower incomes, fewer jobs and reduced educational opportunities and services, are cumulative, leading to chronic rural disadvantage and reduced capacity to cope with changes (Hall & Scheltens 2005; Whittaker *et al.* 2012). Predicted further substantial pressure on agriculture in parts of Australia 'requires high levels of adaptive responses' (Gray *et al.* 2009, p. 140).

Rural communities have always been adaptive to climate variability; however, building community resilience requires health and social services in rural areas to be cognisant of global, national and regional policies and planning frameworks, and to develop knowledge of, and attachment to, rural culture. Health and social service workers need to have capacity to relate to rural communities and have a specific rural focus (Hall & Scheltens 2005).

Rural populations

Definitions of 'rural' vary depending on the issues under examination and the criteria used (Wakerman & Humphreys 2008). However, knowing what is meant by 'rural' in the context of climate change is important for several reasons. Knowing where particular groups of people are geographically, how they are economically, and who they are demographically and culturally enables health and social services to identifying specific characteristics and needs, which assists in developing appropriate adaptation responses to climate change. Rural Australia is diverse. Farmer *et al.* (2012, p. 187) report from the OECD (2006) that 'if you have seen one rural place, you have seen one rural place' and suggest that we should be acknowledging place more in rural health research and practice.

Broad descriptions of populations (ABS 2013) do not describe the disparities in health between rural and urban populations, nor the unique characteristics of rurality. For example, the distribution of the population in rural areas is sparse and numbers are declining. There are more Indigenous Australians and older people and there is more unemployment. Climate change adds to the burden of long-term trends due to environmental change, and social and economic policies that have contributed to rural poverty, the health status of Indigenous Australians and inequalities in services to rural areas. Climate change exacerbates chronic rural disadvantage and appropriate responses to rurality must now encompass adaptation to climate change (Costello *et al.* 2009).

The risks and adaptations to climate change that are relevant to the population sub-groups discussed in the other chapters in this book are relevant to rural populations. In different locations people's health is at greater or lesser risk, but over and above these vulnerabilities, all sub-groups are influenced by the characteristics of rural settings. Adaptation potential will vary between locations.

A conceptual model for understanding rural and remote health

Bourke *et al.* (2012a,b) developed a conceptual framework (after Giddons 1979, 1986) to increase understanding of the issues challenging rural and remote health and wellbeing, and the potential drivers for change. Bourke *et al.* (2012a) argue that using the conceptual

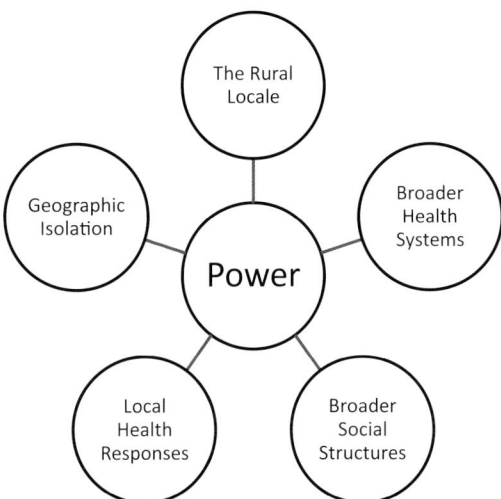

Figure 10.1.   Conceptual framework for understanding issues challenging rural and remote health and wellbeing (Bourke *et al.* 2012a).

framework to understand the complexities of rural health and to make changes to improve wellbeing requires recognition and renegotiation of power and appreciation of the dynamic interaction of five further key concepts: geographical isolation; rural locale; local health actions; broader health systems; and broader social structures (Fig. 10.1). We propose that this systems approach be used in developing and implementing climate change adaptation in rural areas. We now outline each of the key concepts in this approach and, later in this chapter, use them to provide an overview of 'ways forward' through adaptation.

Geographic isolation

This concept encompasses the geographic isolation and distance from other centres that characterise the life experience for rural and remote residents (Bourke *et al.* 2012a,b).

Climate change amplifies the chronic disadvantage experienced by rural dwellers. Opportunities for education have always been limited in rural areas but this is further exacerbated by the loss of population from these areas brought about by the economic transition over the past 40 years. Declining incomes, property amalgamations and various cost-cutting measures taken by businesses and rural institutions have resulted in depopulation of rural areas and subsequent contraction of financial and human resources (Alston & Kent 2004). Demographic changes may mean the loss of viability of small rural primary schools, bringing consequent discontinuation of junior sports and other local activities. Students have to travel significant distances to attend secondary schooling or take some of their chosen subjects by distance education or video link in order to ensure viable class sizes. Other subject choices may not be available at all, which further limits tertiary education options. Tertiary completion rates for rural students are already significantly lower than for their urban counterparts. With the impact of climate change on the financial viability of farms and other rural businesses, many young people are reluctant to enrol in tertiary courses because of the impact that the financial commitment the family has to make to tuition, accommodation and travel costs (YACV 2008). This may change with online courses becoming readily available from universities, but currently these examples are the reality of geographical isolation's impact on the rural locale. These pressures on individuals, families and communities diminish their capacity to adapt (Talbot 2004).

The rural locale

This concept relates to the particular social and geographic setting where rural social relations occur (Bourke *et al.* 2012a).

There are direct and indirect, short-term and long-term health consequences for rural Australians in a changing climate. Social disruption due to extreme weather has increased. Recent droughts, floods, heatwaves and bushfires in Australia have caused premature deaths, severe damage to infrastructure and settlements, and resulted in substantial environmental, economic and social losses. These events are predicted to continue with increased frequency and intensity (Reisinger *et al.* 2014).

Extreme weather

Fires. The devastating fires in recent times came on top of 40 years of deregulation in agriculture and, in some areas, several years of drought. All together this resulted in a vastly changed economic and social landscape. Short- and long-term post-traumatic stress, depression and anxiety have been reported by health services following fires (Whittaker *et al.* 2012). The horrific experiences of some people and the devastating effects on landscapes, assets and animals that have been graphically reported in the media have meant that most people have heightened awareness of fire risks.

Better understanding of fire behaviour in extreme weather conditions changed the focus of fire awareness messages towards early preparedness and taking individual action to avoid risk, recognising that conditions can change rapidly and there are insufficient resources to guarantee safety. While the immediate impact of fires on social and community health may create a short-term environment of social cohesion, Whittaker *et al.* (2012) discovered this new-found cohesion was not necessarily all encompassing or long lasting.

Heat. Very hot days and lack of overnight cooling can result in heat exhaustion for vulnerable community members in any location. The public health impacts of heatwaves particularly affect old and frail people as well as people with fewer resources (such as air-conditioners) and rural communities and farmers who work outdoors and are more exposed to extreme events (McMichael 2009). Indirect effects can include excess demand on electricity supply, with blackouts expected to be more frequent.

Flood. In line with expectations, weather patterns in temperate areas are changing, bringing more storms rather than steady rain. Floods have affected many locations where flooding was previously rare. Roads and rail for freight are lifelines for rural communities for goods such as fresh food, and cash-strapped governments struggle with the cost of flood prevention works, particularly in sparsely populated municipalities.

Drought. There are clear associations between social distress, mental illness and extreme weather events such as droughts and heatwaves (Nicholls *et al.* 2006, Nitschke *et al.* 2011 and Hope 2010 in Hughes & McMichael 2011). Because of the additional demands created by day-to-day coping with drought, usual social structures and networks are disrupted, exacerbating the other adversity created by economic changes (Hart *et al.* 2011). A survey of rural Australians in 2007 found that drought-affected areas had twice the rate of self-reported mental illness and slightly higher rates of self-reported poor health compared with areas not in drought (Edwards *et al.* 2008 in Baxter *et al.* 2011). Older farmers find adaptation to changed climate particularly hard to cope with (McMichael 2011) and studies of the social impacts of drought in Australia provide evidence of the impact of environmental decline on the social fabric and wellbeing of communities (Alston & Kent 2004, 2006; Alston 2012). The importance of maintaining opportunities for children's social connections is emphasised. School forums recommended a more coordinated support system in rural schools, and accessible information about how to access support services provided by professionals who understand the rural and drought contexts (Carnie *et al.* 2011).

Community-based social service organisations and health services are affected during, and in the aftermath of, extreme weather events. There are direct and indirect repercussions, such as damage to service properties and the failure of critical infrastructure such as roads, transport networks, water supply and telecommunications. Power outages seriously disrupt operations (Mallon *et al.* 2013). In an emergency, making alternative arrangements for service delivery takes time. Mallon *et al.* (2013) report that it took 50% of organisations at least a week to make these arrangements, and 25% between two weeks and a month, while 25% said they would never be able to provide services again.

Local health responses

This concept describes the local formal and informal responses and programs established to meet the health needs of rural and remote residents (Bourke *et al.* 2012a). These will change as the impacts of climate change evolve.

Vulnerability to the health impacts of climate change increases in populations that:

- have poorer health status
- are already vulnerable due to culture or life conditions, such as indigenous communities and ethnic minorities, displaced populations and farmers
- have limited access to resources and services, including health care
- are subject to flood risk zones, drought risk zones, water stressed zones and fire.

All vulnerabilities are more prevalent in rural communities (WHO n.d.).

Rural people of various ages tend to avoid following through on health problems, or seeking appropriate services, because of the cost and other reasons such a lack of appropriate services and stoicism (Rich *et al*. 2012). As described, mental illness is widely reported in drought conditions. This is within the context that most farmers are older and male, a group traditionally reluctant to seek help for any health condition, but especially mental illness. The rate of suicide in areas affected by extreme weather conditions is high. Green *et al*. (2009) report that mental and emotional stress increases with loss of a sense of place. Attachment to place is often disrupted after environmental disasters such as fire and flood (Norris *et al*. 2008). Likewise, drought has 'a cumulative impact on the ability of adolescents to cope with the stress', resulting 'significantly in higher levels of distress and behavioural difficulties than adolescents in the general population' (Dean & Stain 2010, p. 35). These findings highlight the vulnerability of young people to the impacts of climate change.

Evidence of the health consequences of a changing climate for rural Australians is growing and health systems need to respond. However, it is challenging for broader health systems responses given the dispersal and density of populations, cultures, economic activity and level of infrastructure characteristic of rural areas.

Broader health systems

This concept describes the 'organisations, institutions, resources and people aiming to improve health through shaping the ways in which health care services are designed, delivered, managed, monitored and funded' (Bourke *et al*. 2012a, p. 319).

Rural and remote residents experience limited access to health promotion and screening services which increases their risks and reduces their survival rates (Corboy *et al*. 2011; Sabesan *et al*. 2012). People who suffer with chronic conditions that require regular monitoring and treatments, such as cancer, incur significant out-of-pocket expenses for travel and accommodation, and must arrange for longer disruption to their daily lives in order to access services located in distant places. These barriers to equitable health services access in part explain the disparity in cancer survival and other health status measures, including mental health outcomes, between urban and rural residents (Corboy *et al*. 2011; Sabesan *et al*. 2012).

Compounding the limited mental and physical health services availability in rural areas is the low uptake of some services specifically designed for cancer patients, including counselling services (Corboy *et al*. 2011). Further research is needed to better understand the supportive care needs and service use by specific subgroups, but Corboy *et al*. (2011) found that rural men held specific values about the usefulness of support services, and this influenced their likelihood to take part.

Health and social services have been aware of the access discrepancies, and under-recognition of mental and physical health impacts of drought and climate change. Financial strategies are insufficient to overcome these 'human' problems (McMichael 2011).

Well-documented, carefully evaluated evidence of the models of care that provide the best physical and mental health outcomes and financial options for rural and remote residents is needed. Reluctance to document a suitable model is perhaps based on fear of the financial implications of doing so (Wakerman 2009).

Broader social structures

These are existing rules, resources, relations and social organisations that are maintained to reproduce particular, social, cultural and economic behaviours and understandings (Bourke *et al.* 2012a).

Australia has always had a climate that is subject to drought. Long-term weather records document drought affecting some areas more than a century ago. These records lead many in rural areas to consider recent drought conditions to be part of the normal cycle of climate variation that is characteristic of Australia. Historically, broadacre farmers have recognised there will be good and bad years and have adapted to these cycles, buying equipment and saving money in the good years to tide them over difficult times (Talbot 2004).

In some cases the gradual change in climate has increased productivity, such as increased wheat yields in some regions since the 1950s (Bi & Parton 2008). In others, the increased frequency and severity of droughts is having serious negative impacts on farm productivity (Hennessy *et al.* 2008). The overall temperature increase, including warmer winters, is reducing the fertilisation and maturation of some blossom crops. Research indicates that the slower onset and chronicity of prolonged droughts means they have a greater social impact than other natural disasters.

More frequent and more severe droughts means many farmers are unable to plan for extreme weather cycles, many farms have become unviable and some farmers have lost hope and a sense of control and just walked away from their enterprise (Polain *et al.* 2011). Similarly, many farms on what was considered 'marginal' land have become unrealistic to farm, and some previously viable land has now become marginal. 'Get big or get out' seems to be the way many farmers cope. Larger farms are needed to ensure economic viability, and with this the demands on the personal lives of famers increases, with consequent effects on family and personal lives of farmers and members of rural communities (Polain *et al.* 2011). In contrast to previous generations, it is increasingly common for farmers to advise their children to find alternative career paths, a consequence of which is the increasing average age of farmers and rural residents in Australia (Polain *et al.* 2011; DPRESP 2008). Another of the impacts is the greater inequality this change in agricultural planning has created between owners of large agglomerated farm enterprises, and the lowly paid (often casual) employees. The casual staff are less likely to become full-time residents of the rural communities meaning viability of local schools, clubs and businesses is further weakened (Talbot 2004).

Economic, social and environmental systems are interdependent. The last prolonged drought in Australia resulted in farm closures, increased poverty, and increased off-farm work, and hence, involuntary separation of families, increased social isolation, rising stress and associated health impacts, including suicide (especially of male farmers). It accelerated rural depopulation and closure of key services (Alston & Kent 2004; Edwards *et al.* 2008; Edwards & Gray 2009; Alston 2012; Hanigan *et al.* 2012 in Reisinger *et al.* 2014). These problems for rural communities are predicted to continue. As outlined, farmers are getting older, and rural communities more generally are experiencing the effects of a loss of young people from their communities (Polain *et al.* 2011; DPRESP 2008). Many young people leave home after secondary school to seek tertiary education and employment in cities and regional centres. They are even less likely to return to small towns than previously. Several regional centres are seeing the benefits of a peak in young adults (ABS 2011).

Loss of young people from small rural communities has significant effects on local economic activity. Young people generally have more disposable income and spend the highest proportion on social activities. They are also the most physically able to assist in local activities, most likely to join sporting clubs and build and furnish a home. Loss of their financial exchange is significant and undermines the viability of rural businesses and industries. When people need to leave their small rural community to attend health services or make business transactions, they also tend to make other purchases in the rural centres (the 'sponge centre effect'). Climate change magnifies these effects. The hollowed out age distribution in rural areas has been illustrated in the preceding discussion. Many commentators are concerned for the future viability of family farms and small business in rural areas in Australia.

Farmers and rural businesses tend to stay on in the role longer than the usual retirement age now, perhaps with the expectation that the service, or the farm, will discontinue when they go. With family succession many farmers find themselves with insufficient funds for independent retirement because their assets are tied up in farm land and equipment. There is a domino effect on the viability of businesses in small rural towns. The large multi-site businesses, such as banks, move to the regional larger towns and centres, but the small local businesses really struggle.

The economic impact of drought can be substantial. During the last major drought in Australia there was a reduction in the national GDP. The average income on farms fell, unemployment and farmer debt rose and there was a negative impact on services (RBA 2006 and Wittwer & Griffith 2011 in Reisinger *et al.* 2014; Kiem *et al.* 2010).

Health and social service organisations need to be prepared for possible transformation in the face of climate change, as they are the 'shock absorbers for every-day adversity as well as crises' (Mallon *et al.* 2013, p. 4).

Power

This concept relates to the capacity to make a difference. It is a particularly important concept in the framework (Fig. 10.1) because it explains how change is achieved or restricted (Bourke *et al.* 2012a).

National agricultural policies have not been kind to rural Australia as the evidence already discussed suggests, and this lack of support has depleted regions of sociocultural, economic and environmental capital. This is a social justice issue.

With regards to equitable access to healthcare services, it can be argued that current Australian funding arrangements for delivery of health and community services that share power and costs between federal and state governments actually reinforce the disadvantaged access to services being experienced by rural and remote residents. This is because each is able to blame the other for health system failures or funding insufficiencies (Wakerman 2009). Brett (2011 cited in Farmer *et al.* 2012, p. 188) suggests that the country has been pushed aside and perceptions of what it means to be a rural Australian have changed:

> *It wasn't always thus. Once the country believed itself to be the true face of Australia; sunburnt men and capable women raising crops and children, enduring isolation, hardship and fickle environment, carrying the nation on their sturdy backs.*

In recent decades economic and trade policy expressed in neoliberalism that has been the philosophy guiding most international economies, has progressively undermined the ability of small industries and primary producers to control trade of their commodities. The stress of reduced commodity prices, along with climate challenges, means many rural people and other producers are 'powerless in the face of government and global forces (that) intensify the emotional outcomes when the rural way of life is under threat' (Dean & Stain 2010, p. 32). And yet, a well-developed economic system is critically important in building community resilience (Bajayo 2012).

The diversity of rural Australia makes nation-wide adaptation planning challenging. Climate projections (Reisinger *et al.* 2014) outline vast differences between geographical regions in the impacts of climate change, with a range from very little change to a 40% change in rainfall runoff in south-eastern Australia with 2°C rise in temperature. A 40% change signals severe implications for rural communities and would necessitate transformational adaptation for some communities but it cannot be 'one size fits all'. Places are affected by 'forces and flows and different scales' (Farmer *et al.* 2012, p. 188). National, state and local conditions and responses will be experienced differently in different places.

Past successes in coping with climate variability may be a strength in the face of climate change; however, it cannot be assumed that past response strategies will be sufficient to deal with the range of projected changes (Jones *et al.* 2014).

Climate change and drought conditions can undermine the personal psychological resilience of a full range of people, including, but not limited to farmers. People who are dependent on the weather may be forced to move out of their industry thus depriving them of an occupation that may be their preference and love. Similarly, backyard gardeners may be deprived of the emotional rewards and productivity that gardening brings, and neither group has power to change the situation (Pereira 2008). Higher suicide rates result from a combination of factors including farmers' limited ability to acknowledge or seek help (Judd *et al.* 2006), poorer access to health care and a culture of independence (Alston & Kent 2006; Alston 2012; Whittaker *et al.* 2012).

Adaptation

The process of adaptation is not straightforward. As Eckermann *et al.* (2006, p. 3) point out 'it is based on problem solving, steeped in creativity and characterises the process of coping when environments … change'.

The model (Bourke *et al.* 2012a; Fig. 10.1) used in the previous section is used here to provide an overview of 'ways forward'. Although stakeholders within rural communities differ in their vulnerabilities and adaptive capacities, they are bound by similar dependence upon critical infrastructure and resources, economic conditions, government policy direction and societal expectations (Loechel *et al.* 2013).

We encourage use of a range of theories that describe mechanisms for successful change in communities. In essence we need to:

- assist health and social service agencies to adapt their services to meet the needs generated by the impacts of climate change
- build on rural communities' local knowledge, resources and networks
- assist communities to plan and implement adaptation strategies for major vulnerabilities such as agriculture, service management and communication.

Geographic isolation

As populations in rural and remote areas continue to decline, the problems of geographic isolation will be more pronounced. Adaptation strategies will be most successful when they can overcome the negative effects of distance and isolation. A range of strategies has been used including transport assistance, fuel subsidies, mobile health and screening services and, especially, telemedicine.

Diffusion of knowledge to and from rural areas has improved with technology although great inequalities remain with access to high-speed internet. Access to high-speed internet connection has the potential to significantly reduce the negative impacts of geographic isolation and enable adaptation on a range of fronts, including farming and health education, counselling services and improved business opportunities.

The rural locale

Adaptation will require input from all levels and sectors of society (Costello *et al.* 2009; Rissik & Reis 2013). Some communities are preparing for the future; however, inland settlements lag behind their coastal and urban counterparts (Beer *et al.* 2013). Investment in infrastructure and increased capacity for firefighting, water supply, transport and flood preparedness is underway in preparation for emergencies, and changed farming practices are entirely consistent with climate change adaptation; still, adjustments are ad hoc rather than well-developed strategic plans for adaptation. The reason for this inconsistency is thought to be that people in rural communities may not accept the evidence of anthropogenic climate change (Beer *et al.* 2013) although Leviston *et al.* (2014) found that people who lived in capital cities were only slightly more likely to think that climate change was happening than rural residents.

Rural Australians have long-established credibility in adaptation to climate variability. Furthermore, participation in volunteer work and feelings of safety are higher in rural communities, which builds higher social cohesiveness (Edwards *et al.* 2008). The Human Rights and Equal Opportunity Commission 2000 National Inquiry, 'Bush Talks', suggested that despite great disadvantage, there is 'enormous energy, creativity, vibrancy and commitment to a future for rural Australia' (Dade Smith 2007, p. 17).

Perceptions of risks and loss associated with climate change are dependent upon the value placed on specific places and activities. Reisinger *et al.* (2014) suggest that the deeper the connection with place, the stronger the perception of loss and conversely, a greater resolve to address the environmental threat, which affects people's ability to make healthy adaptations, particularly in disadvantaged communities (e.g. Rogan *et al.* 2005; Raymond & Brown 2011).

Transformational change in agricultural practice and land use is reported to be occurring in response to recent and/or expectations of future climate or policy change (Kenny 2011; Gaydon *et al.* 2010; Park *et al.* 2012) and these changes will need to continue, with significant implications for rural communities (Kiem & Austin 2013). Transformative adaptation in water management is needed most in the Murray–Darling Basin and far south-eastern and far south-western regions of Australia. Reform is currently underway in the Murray–Darling Basin to return water to riverine ecosystems and develop flexible and adaptive water sharing mechanisms to cope with current and future climates. Conflicts between water users will need resolution requiring multifaceted approaches. In times of hardship, such as drought, communities often pull together and share resources to ensure mutual benefit. Yet, when the extreme conditions subside traditional conflicts and rivalries re-emerge (Benjaminsen *et al.* 2012). Fair distribution of resources needs strong, trusted leadership with capacity to facilitate constructive engagement between groups (Meier *et al.* 2007). Resilience to predicted changes will emerge if policies that ensure more efficient water use and participatory approaches in decision-making are in place.

Adaptation can occur if efforts are made to engage communities and build on remaining strengths by fostering a strong sense of community. Adaptation will require an approach where all stakeholders are at the decision-making table. A range of information is needed to support people to make well-informed decisions about matters such as financial security, farming and business diversification, and drought/hardship relief.

The federal government has moved away from providing drought assistance, with the view that many farms are already on the borderline of being unviable in 'normal' times. Providing emergency assistance in times of drought only artificially prolongs their demise. Drought assistance is now allocated for specifically declared geographic areas with limited conditional loans rather than grants. Initiatives are required to support the development of community-based capacity-building strategies that maintain and/or re-build social capital.

Social capital protects mental health because it builds the ties and mutual rewards that are a part of informal social networks (Talbot & Walker 2007). Building community resilience and ownership of the local challenges, with the active involvement of local community members, increases the chances that the capacities will remain after professional support has finished (Hart *et al.* 2011). Even in times of drought, adolescents living in rural areas maintain a strong sense of connection to place. Strategies that help them to sustain a rural lifestyle and inter-family connections are effective in building their resilience despite increasing challenges brought by climate change (Dean & Stain 2010).

Local community members and non-health consultants such as agricultural and financial advisers are valuable in bringing together assets, sharing skills and building relationships; they understand the local context, they have privileged access to local communities and they can reduce the stigma associated with mental illness for example, and reduce the likelihood of a drastic outcome (Hart *et al.* 2011). Local networks also provide valuable up-to-date information about local community needs back to relevant agencies.

Local health responses

The challenge of maintaining local healthcare and other services in rural areas often depends on 'a few committed champions ... and a very supportive local community' (Bourke *et al.* 2012a, p. 496).

Health and social service organisations have been overlooked so far in climate change adaptation policy development. Mallon *et al.* (2013) found that these organisations are vulnerable to, and not well prepared for, responding to extreme weather events despite their skills, assets and capabilities that would contribute to rural community resilience. What is needed is:

> *a comprehensive, sector-specific adaptation and preparedness program, which includes mechanisms to institutionalise knowledge and skills, streamlined tools appropriate to the needs and capacity of a diverse range of organisations and a benchmarking system to allow progress towards resilience and preparedness to be monitored. (Mallon* et al. *2013, p. 2)*

The impact of climate change can be minimised with public sector and community planning aimed at decreasing susceptibility to the impacts and increasing resilience within communities. Beer *et al.* (2013) argue that rural Australia lags behind other settlements in 'formal place-based plans', possibly because most funding for local governments in Australia to identify and manage climate change risks has gone to urban and coastal settlements.

Climate change adaptation frameworks need to follow sound program planning rules. These are, broadly, research and assessment, decision-making and planning, implementation and evaluation (Talbot & Verrinder 2014). These are not new rules but it is important to contextualise them.

The 'Adaptation Good Practice' project report (Rissik & Reis 2013, p. 12) states that there are five critical success factors for any climate change adaptation planning and implementation:

- leadership
- good engagement with stakeholders
- strong connectivity with other relevant activities
- sustainability
- consideration of short- and long-term costs.

Leadership that facilitates the development of tools and frameworks that could be used by others to prevent 'reinventing the wheel' is important. Sector specific or locally specific information and examples are preferred. This enables direct applicability to operations for short-term emergencies and longer-term management of agriculture, biodiversity and water. In this process, climate champions are important especially within organisation. Work with 'green teams' in local health and social service organisations suggests that 'the biggest cog in this chain of events was the CEO – genuine unconditional support from the top opens doors and cascades throughout an organisation' (McBurney 2013).

Engagement refers to the extent and mechanisms of communication, collaboration and partnerships at all levels of adaptation planning. Stakeholder engagement is critically important so that everyone knows about, contributes to, and agrees upon, the local issues and planning process. Different communication strategies are needed to cater for different stakeholders; for example, long trips to attend public meetings in rural areas may exclude some people.

An example of successful community engagement is the 'Climate Change Community Conversations' program on the Mornington Peninsula (an urban coastal community) where a long-running community engagement program fosters engagement through social media, newsletters, an Eco Living Display Centre and Green Business network (Rissik & Reis 2013). Rissik and Reis (2013) report that an 'open and transparent' process in development and management is important. Collaboration enables wider use of resources and expertise, which is an important consideration for resource-poor rural councils and sustainability, which also refers to long-term benefits. Cost refers to both short- and long-term factors. Seed funding from government sources is an important stimulus to planning, as is allocation of a dedicated project manager. Rissik and Reis (2013) also refer to the need for realistic time frames and flexibility in planning given the complexity of the task. In the longer term, strong connectivity with other relevant activities enables action to be embedded across sectors and within organisational strategic plans.

Some programs within the health sector have been successful in local areas. Help-seeking, especially by men, is facilitated when the problem and the management behaviour is normalised, and this is best created in local settings (Corboy *et al.* 2011) by local people and health professionals who understand the rural context (Hart *et al.* 2011; Sartore *et al.* 2005; Hall & Scheltens 2005). The Rural Adversity Mental Health Program has demonstrated that a combination of increased understanding of mental health literacy through 'mental health first aid' training, good coordination and collaboration between health and social service providers and agencies (including with the non-health sector), and community education about coping with drought and climate change can bring communities together and build capacity and resilience (Hart *et al.* 2011).

Broader health systems

'In Australia and elsewhere, macro level health systems tend to disadvantage rural residents' (Bourke *et al.* 2012a, p. 500). However, three general strategies are likely to minimise the adverse mental health outcomes in times of drought (Sartore *et al.* 2005). These correspond with strategies that have been proposed by other researchers, and are appropriate as a systems approach to providing the essential support to rural and remote communities generally and for dealing with the impacts of climate change.

The three strategies are:

1. *Community building and education about the physical, mental, financial (and social) health effects of drought.* As previously outlined, community development principles are used to engage local community members. Farm family gatherings allow a range of relevant information, including mental health information, to be disseminated widely, not just to those perceived to be at risk. Such gatherings promote communication and collaboration between agencies and community groups. These and other opportunities can be used to provide information packs. For best effectiveness skilled professionals familiar with the rural context are essential.
2. *Cooperation between, and coordination of, agencies* with a role in supporting rural and remote communities to deal with drought and climate change. Permanent employment of rural health specialists with expertise in mental health is a way of coordinating services and increasing use and uptake by vulnerable community members. Various forms of media can be useful. Drought support workers can be employed by local councils to undertake this coordination role.
3. *Continuity of planning of improved, coordinated services*, especially through expansion of services with demonstrated effectiveness. Collating evaluation evidence and collaborating with university programs with direct links with rural communities ensures that best practice guidelines are developed and adapted as circumstances change and/or greater understanding is developed.

Important progress has also been made in the development and refinement of new models of service provision (Wakerman 2009). Some have been designed to overcome barriers to equitable access currently experienced by rural and remote residents requiring specialist medical care. A range of telephone and internet-based support and treatment options are being used more widely with success. Call centre counselling services have been a useful alternative that fosters a sense of control over the counselling process (Hall & Scheltens 2005).

Telemedicine using videoconferencing technology has proven effective in delivering a range of inpatient and outpatient care services, including emergency management for cancer patients. It can be adapted to meet the cultural needs and preferences of people and thus has proven widely acceptable to many, including Indigenous Australians.

Telemedicine has potential for much wider use with reduced cost for service providers and clients, resulting in less disruption and high satisfaction for health professionals, clients and their families (Sabesan *et al.* 2012).

Corboy *et al.* (2011) reported that younger men and those with limited social support in other aspects of their lives are more likely participate in support services, and participants reported less emotional distress in dealing with cancer than non-participants. In their study 'highest rates of participation in a support service were for cancer organisations that offer support via telephone or the internet' (Corboy *et al.* 2011, p. 188). Users valued the ease of access and anonymity, which helped overcome the stigma associated with seeking help for emotional and mental health problems that is higher with rural residents. Telephone support services also need to be free to further reduce barriers to accessing the service (Hart *et al.* 2011). Health professionals and health planners need to be aware of the possibly greater reluctance of rural residents, particularly older people, to use formal mental health support services (Corboy *et al.* 2011). Working with rural callers requires a rural specialist focus (Hall & Scheltens 2005).

Rural health and social service professionals see, and attend to, many unique health and community issues, with scarce resources and greater personal awareness of the impact of their actions and the outcomes on community (Strasser et al. 2000 in Veitch 2009). Specific preparation of, and support for, rural professionals is needed to implement these strategies.

Broader social structures

Economic dependence on the climate and natural resources in agriculture, along with the need for infrastructure and services beyond the capacity of governments and communities to deliver, exacerbates the problem of climate change and yet, a well-developed economic system is critically important to build community resilience.

Nascent climate change adaptation policies and practices are underway in developed and developing countries. Some are reported to be government-led and others by non-government organisations, communities, companies and individuals. However, evaluation of adaptation plans in Australia, the United Kingdom and the United States suggest the plans are ad hoc and under-developed (Preston *et al.* 2011; Bierbaum *et al.* 2013; Berrang Ford *et al.* 2011 in Jones *et al.* 2014) and there is a gap between planning and implementation. Barriers to adaptation include perceptions, knowledge and lack of human and financial resources. Uncertainty about risks and limited integration of different levels of government tends to affect rural areas more (Jones *et al.* 2014).

There is a suggestion that Australian farming has been under 'adaptive management' for 40 000 years, first with fire, and then through the development of techniques resulting in industrial farming, and that this 'adaptation, has led to the present, significant, environmental problems that are continent-wide' (Gray *et al.* 2009, p. 148). Arguably, these behaviours and understandings in farming now are tied to broader social structures, including agricultural policies, which, since the latter part of the last century, have led to systematic inequalities in the health and wellbeing of rural Australians.

Farm diversification may not increase resilience or reduce vulnerability and can increase inequity within communities (Osbahr *et al.* 2008 in Dasgupta *et al.* 2014). Many attempts at diversification have failed; some, such as biofuels have been controversial (German *et al.* 2011 in Dasgupta *et al.* 2014). This implies analysing each case in its context, including production for both local and global markets, and factoring in concerns for social, cultural and economic costs (Dasgupta *et al.* 2014).

Gray *et al.* (2009) argue that rural communities will cope with climate change the way they have coped in the past. The solution for some will be technology. Others will reduce spending and hope for better seasons. Others will survive on 'off-farm' income and some will leave. Research shows that 'selling up is the last dramatic step' (Gray *et al.* 2009, p. 149). Barriers to adaptation are reported to come from macro influences, such as trade policies, and local barriers, such as knowledge and information (Jones *et al.* 2014). Finance is a key factor. 'When in financial stress – some producers will make a conscious and 'rational' decision to exploit natural resources as a 'tide over' mechanism' (Gray *et al.* 2009, p. 143). This choice will not be any easier in the face of climate change.

Changes in farming practices are occurring and innovation in rural communities could be supported by national and local government policies and practical support from other government and non-government agencies. There is a growing body of literature on adaptation practices led by individuals, companies and non-government agencies. The choice of whether to stay farming or in a rural community has already been made by many.

Organisations such as Soils for Life are ready to support changes in farming practices. Local governments, Catchment Management Authorities, health and social service organisations and other government bodies can facilitate connection to organisations such as these. Soils for Life is a non-profit, non-government registered environment organisation which aims to 'facilitate positive and sustained change in how the Australian landscape is managed to ensure a thriving natural environment for the benefit of all Australians' (Soils for Life n.d.). The organisation provides leadership in regenerative landscape management by identifying and communicating leading practice and providing education and support to farmers and policymakers. The case studies in Soils for Life report positive economic, environmental and social outcomes from adopting regenerative landscape management practices.

By lifting local barriers, including lack of information and knowledge, and facilitating connection with organisations such as Soils for Life, broader social structures and local authorities can support farming families and small communities in climate change adaptation to suit local conditions. The cautionary advice of Gray *et al.* (2009, p. 151) however, is that climate science may reinforce an 'expansionist productivist framework' of farming. They argue that scientific and local knowledge may collide, 'creating a struggle in which power relations may determine which type of knowledge can claim legitimate rationality'.

Power

The final but central concept relates to the capacity to make a difference. Power needs to be thought about in terms of the dynamic interaction of the other five key concepts discussed above.

Primary health care policy and programs that are based on equity, social justice and empowerment (Talbot & Verrinder 2014) and designed to meet the needs of rural and remote residents require the flexibility to fit well in the local context and existing structures and networks. Inter-professional teams need to be supported to work in collaboration with communities to develop specific programs to meet local needs. As previously set out, broad changes are having the effect of undermining social capital in rural communities, including farm amalgamation, diversification, and loss of young people (Talbot & Walker 2007). However, one of the most important strategies to build sustainability of their communities is efforts to build and strengthen social capital. In times of personal and collective hardship social capital can provide protection for wellbeing. Bajayo (2012) argues that social capital is one of four primary sets of networked resources needed to build community resilience to climate change. Economic development, information and communication, and community competence form a community resilience framework. If well prepared, health and social service organisations can help build community resilience. Information technology is key to this approach because it reduces the impact of distance and allows individuals to take greater control over their own health decisions (Wakerman 2009). This power sharing relates to research and assessment, decision-making and planning, implementation and evaluation in improving social and environmental capital.

In her analysis of climate change research since 1995, Bell (2013, p. 2) argues that published research about the impacts of climate changes needs to be more integrated with 'wider socio-economic and human system factors' and that a potential way to bridge the current gap is to make more widely available the research and wisdom from community-based researchers and practitioners.

Furthermore, environmental activism has already had significant success in bringing matters of concern to public and political notice. In the changing political climate and with governments' reluctance to move away from an economic growth imperative, such activism and support for activism will continue to be a necessity to protect environmental sustainability at policy level locally, nationally and internationally (Pereira 2008).

Conclusion

It is the case that rural Australians experience the climate differently to city dwellers. Dorothea Mackellar's poem, 'My Country'[1] not only describes the geographic and climatic extremes of Australia but also the human experience of the relationship with the land.

Adaptation for health in rural communities in the face of climate change cannot be too prescriptive because rural Australia is diverse: climatically, geographically, economically, demographically and culturally. However, adaptation potential may be realised if we look through the lens of frameworks that encompass the complexity of rural life. In this chapter we have examined climate change and rural communities in terms of the interrelationships between six key concepts: rural locale; local health responses; broader health systems; geographical isolation; and broader social structures, in the context of power relations. We propose that this understanding may be pre-requisite to building the environmental and social capital and resilience required for adaptation to climate change.

The cascading effect of environmental degradation and social change has contributed to transformed settlements, which have compounded inequalities in health and social services to rural areas. Further transformation is underway and we need to use robust theories and examples that describe mechanisms for successful change in communities if we are to make a difference.

Endnote

1. Dorothea Mackellar (1885–1968) http://www.dorotheamackellar.com.au/archive/mycountry.htm

References

AIHW (Australian Institute of Health and Welfare) (2008) *Rural, Regional and Remote Health: Indicators of Health Status and Determinants of health.* Cat. no. PHE 97). Australian Institute of Health and Welfare, Canberra, <http://www.aihw.gov.au/publication-detail/?id=6442468076>

Alston M (2012) Rural male suicide in Australia. *Social Science & Medicine* **74**(4), 515–522. doi:10.1016/j.socscimed.2010.04.036

Alston M, Kent J (2004) 'The social impacts of drought'. Centre for Rural Social Research, Charles Sturt University, Wagga Wagga, NSW.

Alston M, Kent J (2006) 'The impact of drought on secondary education access in Australia's rural and remote areas: a report to DEST and the rural education program of FRRR'. Centre for Rural and Social Research, Charles Sturt University, Wagga Wagga, NSW.

ABS (Australian Bureau of Statistics) (2011) *Australian Statistical Geography Standard (ASGS) Volume 5 – Remoteness Areas, July 2011*. Cat. no. 1270.0.55.005. Australian Bureau of Statistics, Canberra.

ABS (Australian Bureau of Statistics) (2013) *Regional Population Growth, Australia, 2011–2012*. Cat. no. 3218.0. Australian Bureau of Statistics, Canberra, <http://www.abs.gov.au/ausstats/abs@.nsf/Products/3218.0~2011-12~Main+Features~Main+Features?OpenDocument>.

Bajayo R (2012) Building community resilience to climate change through public health planning. *Health Promotion Journal of Australia* **23**(1), 30–36.

Baxter J, Hayes A, Gray M (2011) *Families in Regional, Rural and Remote Australia*. Fact sheet 2011 (March). Australian Institute of Family Studies, Melbourne, <http://www.aifs.gov.au/institute/pubs/factssheets/2011/fs201103.html>.

Beer A, Tually S, Kroehn M, Martin J, Gerritsen R, Taylor M, Graymore M, Law J (2013) 'Australia's country towns 2050: What will a climate adapted settlement pattern look like?' National Climate Change Adaptation Research Facility, Gold Coast.

Bell EJ (2013) Climate change and health research: has it served rural communities? *Rural and Remote Health* **13**, 2343.

Benjaminsen TA, Alinon K, Buhaug H, Useth JT (2012) Does climate change drive land-use conflicts in the Sahel? *Journal of Peace Research* **49**(1), 97–111. doi:10.1177/0022343311427343

Bi P, Parton K (2008) Effect of climate change on Australian rural and remote regions: what do we know and what do we need to know? *The Australian Journal of Rural Health* **16**, 2–4 doi:10.1111/j.1440-1584.2007.00945.x

Bierbaum R, Smith JB, Lee A, Blair M, Carter L, Chapin FS III, Fleming P, Ruffo S, Stults M, McNeeley S (2013) A comprehensive review of climate adaptation in the United States: more than before, but less than needed. *Mitigation and Adaptation Strategies for Global Change* **18**, 361–406. doi:10.1007/s11027-012-9423-1

Blashki G (2008) Climate change and rural health workforce. *The Australian Journal of Rural Health* **16**, 1. doi:10.1111/j.1440-1584.2007.00944.x

Bourke L, Humphreys J, Wakerman J, Taylor J (2012a) Understanding rural and remote health: a framework for analysis Australia. *Health & Place* **18**, 496–503. doi:10.1016/j.healthplace.2012.02.009

Bourke L, Humphreys J, Wakerman J, Taylor J (2012b) Understanding drivers of rural and remote health outcomes: a conceptual framework in action. *The Australian Journal of Rural Health* **20**, 318–323. doi:10.1111/j.1440-1584.2012.01312.x

CSIRO & Bureau of Meteorology (2014) *The State of the Climate 2014*. Bureau of Meteorology, Melbourne, <http://www.bom.gov.au/state-of-the-climate/>.

Carnie TL, Berry HL, Blinkhorn SA, Hart CR (2011) In their own words: young people's mental health in drought-affected rural and remote New South Wales. *The Australian Journal of Rural Health* **19**, 244–248. doi:10.1111/j.1440-1584.2011.01224.x

Corboy D, McLaren S, McDonald J (2011) Predictors of support service use by rural and regional men with cancer. *The Australian Journal of Rural Health* **19**, 185–190. doi:10.1111/j.1440-1584.2011.01210.x

Costello A, Abbas M, Allen A, Ball S, Bell S, Bellamy R, Friel S, Groce N, Johnson A, Kett M, Lee M, Levy C, Maslin M, McCoy D, McGuire B, Montgomery H, Napier D, Pagel C, Patel J, de Oliveira JAP, Redclift N, Rees H, Rogger D, Scott J, Stephenson J, Twigg J, Wolff J, Patterson C (2009) Managing the health effects of climate change. *Lancet* **373**, 1693–1733. doi:10.1016/S0140-6736(09)60935-1

Dade Smith J (2007) *Australia's Rural and Remote Health: A Social Justice Perspective.* Tertiary Press, Croydon, Victoria.

Dasgupta P, Morton JF, Dodman D, Karapinar B, Meza F, Rivera-Ferre MG, Toure Sarr A, Vincent KE (2014) Rural areas. In *Climate Change 2014: Impacts, Adaptation, and Vulnerability. Part A: Global and Sectoral Aspects. Contribution of Working Group II to the Fifth Assessment Report of the Intergovernmental Panel on Climate Change.* (Eds CB Field, VR Barros, DJ Dokken, KJ Mach, MD Mastrandrea, TE Bilir, M Chatterjee, KL Ebi, YO Estrada, RC Genova, B Girma, ES Kissel, AN Levy, S MacCracken, PR Mastrandrea and LL White). Final draft. Intergovernmental Panel on Climate Change, Geneva, <http://ipcc-wg2.gov/AR5/report/final-drafts/>.

Dean J, Stain H (2010) Mental health impact for adolescents living with prolonged drought. *The Australian Journal of Rural Health* **18**, 32–37. doi:10.1111/j.1440-1584.2009.01107.x

DPRESP (Drought Policy Review Expert Social Panel) (2008) 'It's about people: changing perspective. A report to government by an expert social panel on dryness'. Department of Agriculture, Fisheries and Forestry, Canberra.

Eckermann A-K, Dowd T, Chong E, Nixon L, Gray R, Johnson S (2006) *Binan Goonj. Bridging Cultures in Aboriginal Health.* 2nd edn. Elsevier, Marrickville, New South Wales.

Edwards B, Gray M (2009) A sunburnt country: the economic and financial impact of drought on rural and regional families in Australia in an era of climate change. *Journal of Labor Economics* **12**(1), 109–131.

Edwards B, Gray M, Hunter B (2008) 'Social and economic impacts of drought on farm families and rural communities'. Australian Institute of Family Studies, Melbourne.

Farmer J, Munoz S-A, Threlkeld G (2012) Theory in rural health. *The Australian Journal of Rural Health* **20**, 185–189. doi:10.1111/j.1440-1584.2012.01286.x

Gaydon RS, Beecher HG, Reinke R, Crimp S, Howden SM (2010) Rice. In *Adapting Agriculture to Climate Change. Preparing Australian Agriculture, Forestry and Fisheries for the Future.* (Eds C Stokes and S Howden) pp. 67–83. CSIRO Publishing, Collingwood.

Giddons A (1979) *Central Problems in Social Theory Action. Structure and Contradiction in Social Analysis.* The McMillan Ltd Press, London.

Giddons A (1986) *The Constitution of Society.* University of California Press, California, USA.

Gray L, Lawrence G, Sinclair P (2009) The sociology of climate change for regional Australia. In *Climate Change in Regional Australia: Social Learning and Adaptation.* (Eds J Martin, M Rogers and C Winter) pp. 136–154. Victorian University Regional Research Network Press, Ballarat.

Green D, Jackson S, Morrison J (2009) 'Risks from climate change to Indigenous communities in the tropical north of Australia'. Department of Climate Change, Canberra.

Hall G, Scheltens M (2005) Beyond the drought: towards a broader understanding of rural disadvantage. *Rural Society* **15**(3), 348–358. doi:10.5172/rsj.351.15.3.348

Hart CR, Berry HL, Tonna AM (2011) Improving the mental health of rural New South Wales communities facing drought and other adversities. *The Australian Journal of Rural Health* **19**, 231–238. doi:10.1111/j.1440-1584.2011.01225.x

Hegney DG, Buikstra E, Baker P, Rogers-Clark C, Pearce S, Ross H, King C, Watson-Luke A (2007) Individual resilience in rural people: a Queensland study. *Rural and Remote Health* 7, 620[online].

Hennessy KJ, Fawcett R, Kirono DGC, Mpelasoka F, Jones D, Bathols JM, Whetton PH, Stafford Smith M, Howden M, Mitchell CD, Plummer N (2008) 'An assessment of the impact of climate change on the nature and severity of exceptional climate events'. CSIRO Marine and Atmospheric Research, Aspendale.

Hughes L, McMichael AJ (2011) 'The critical decade: climate change and health'. Climate Commission Secretariat (Department of Climate Change and Energy Efficiency), Canberra.

Jones RN, Patwardhan A, Cohen SJ, Dessai S, Lammel A, Lempert RJ, Mirza MMQ, von Storch H (2014) Foundations for decision making. In *Climate Change 2014: Impacts, Adaptation, and Vulnerability. Part A: Global and Sectoral Aspects. Contribution of Working Group II to the Fifth Assessment Report of the Intergovernmental Panel on Climate Change.* (Eds CB Field, VR Barros, DJ Dokken, KJ Mach, MD Mastrandrea, TE Bilir, M Chatterjee, KL Ebi, YO Estrada, RC Genova, B Girma, ES Kissel, AN Levy, S MacCracken, PR Mastrandrea and LL White). Final draft. Intergovernmental Panel on Climate Change, Geneva, <http://ipcc-wg2.gov/AR5/report/final-drafts/>.

Judd F, Jackson H, Fraser C, Murray G, Robins G, Komiti A (2006) Understanding suicide in Australian farmers. *Social Psychiatry and Psychiatric Epidemiology* **41**, 1e10. doi:10.1007/s00127-005-0007-1

Kenny G (2011) Adaptation in agriculture: lessons for resilience from eastern regions of New Zealand. *Climatic Change* **106**(3), 441–462. doi:10.1007/s10584-010-9948-9

Kiem A, Austin E (2013) Drought and the future of rural communities: opportunities and challenges for climate change adaptation in regional Victoria, Australia. *Global Environmental Change* **23**(5), 1307–1316. doi:10.1016/j.gloenvcha.2013.06.003

Kiem AS, Askew LE, Sherval M, Verdon-Kidd DC, Clifton C, Austin E, McGuirk PM, Berry H (2010) 'Drought and the future of rural communities: drought impacts and adaptation in regional Victoria, Australia'. National Climate Change Adaptation Research Facility, Gold Coast.

Leviston Z, Price J, Malkin S, McCrea R (2014) 'Fourth annual survey of Australian attitudes to climate change: interim report'. CSIRO, Perth, Australia.

Loechel B, Hodgkinson J, Moffat K (2013) Climate change adaptation in Australian mining communities: comparing mining company and local government views and activities. *Climatic Change* **119**(2), 465–477. doi:10.1007/s10584-013-0721-8

Mallon K, Hamilton E, Black M, Beem B, Abs J (2013) 'Adapting the community sector for climate extremes: Extreme weather, climate change & the community sector – risks and adaptations'. National Climate Change Adaptation Research Facility, Gold Coast.

McBurney I (2013) *When a Green Team Stumbles … and then Flies.* Ian McBurney blog, <http://ianmcburney.com/blog/2013/11/8/when-a-green-team-stumbles-and-then-flies>.

McMichael A (2009) 'Climate change in Australia: risks to human wellbeing and health'. Austral Special Report 09–03S, 23 March 2009. Nautilus Institute Australia.

McMichael AJ (2011) Drought, drying and mental health: lessons from recent experiences for future risk-lessening policies. *The Australian Journal of Rural Health* **19**, 227–228[Editorial]. doi:10.1111/j.1440-1584.2011.01217.x

Meier P, Bond D, Bond J (2007) Environmental influences on pastoral conflict in the Horn of Africa. *Political Geography* **26**(6), 716–735. doi:10.1016/j.polgeo.2007.06.001

Norris FH, Stevens SP, Pfefferbaum BJ, Wyche KF, Pfefferbaum RL (2008) Community resilience as a metaphor, theory, set of capacities and strategy for disaster readiness. *American Journal of Community Psychology* **41**, 127–150. doi:10.1007/s10464-007-9156-6

OECD (Organisation for Economic Co-operation & Development) (2006) *The New Rural Paradigm: Policies and Governance*. Organisation for Economic Co-operation & Development, Paris.

Park SE, Marshall NA, Jakku E, Dowd AM, Howden SM, Mendham E, Fleming A (2012) Informing adaptation responses to climate change through theories of transformation. *Global Environmental Change* **22**(1), 115–126. doi:10.1016/j.gloenvcha.2011.10.003

Pereira RB (2008) Population health needs beyond ratifying the Kyoto Protocol: a look at occupational deprivation. *Rural and Remote Health* **8**, 927[Online].

Polain J, Berry H, Hoskin J (2011) Rapid change, climate adversity and the next 'big dry': older farmers' mental health. *The Australian Journal of Rural Health* **19**, 239–243. doi:10.1111/j.1440-1584.2011.01219.x

Preston B, Yuen E, Westaway R (2011) Putting vulnerability to climate change on the map: a review of approaches, benefits, and risks. *Sustainability Science* **6**, 177–202. doi:10.1007/s11625-011-0129-1

Raymond CM, Brown G (2011) Assessing spatial associations between perceptions of landscape value and climate change risk for use in climate change planning. *Climatic Change* **104**(3–4), 653–678. doi:10.1007/s10584-010-9806-9

Reisinger A, Kitching RL, Chiew F, Hughes L, Newton PCD, Schuster SS, Tait A, Whetton P (2014) Australasia. In *Climate Change 2014: Impacts, Adaptation, and Vulnerability. Part B: Regional Aspects. Contribution of Working Group II to the Fifth Assessment Report of the Intergovernmental Panel on Climate Change*. (Eds CB Field, VR Barros, DJ Dokken, KJ Mach, MD Mastrandrea, TE Bilir, M Chatterjee, KL Ebi, YO Estrada, RC Genova, B Girma, ES Kissel, AN Levy, S MacCracken, PR Mastrandrea and LL White). Final draft. Intergovernmental Panel on Climate Change, Geneva, <http://ipcc-wg2.gov/AR5/report/final-drafts/>.

Rich J, Wright S, Loxton D (2012) Patience, hormone replacement therapy and rain! Women, ageing and drought in Australia. Narratives from the mid-age cohort of the Australian Longitudinal Study on Women's Health. *The Australian Journal of Rural Health* **20**, 324–328. doi:10.1111/j.1440-1584.2012.01294.x

Rissik D, Reis N (2013) 'Climate change adaptation good practice – synthesis report. Key lessons from practitioners' experiences'. National Climate Change Adaptation Research Facility, Gold Coast.

Rogan R, O'Connor M, Horwitz P (2005) Nowhere to hide: awareness and perceptions of environmental change, and their influence on relationships with place. *Journal of Environmental Psychology* **25**(2), 147–158. doi:10.1016/j.jenvp.2005.03.001

Sabesan S, Larkins S, Evans R, Varma S, Andrews A, Beuttner P, Brennan S, Young M (2012) Telemedicine for rural cancer care in North Queensland: bringing cancer care home. *The Australian Journal of Rural Health* **20**, 259–264. doi:10.1111/j.1440-1584.2012.01299.x

Sartore G, Hoolahan B, Tonna A, Kelly B, Stain H (2005) Wisdom from the drought: recommendations from a consultative conference. *The Australian Journal of Rural Health* **13**, 315–320. doi:10.1111/j.1440-1584.2005.00723.x

Soils for Life (n.d.) *About Soils for Life*. Soils for Life, Fairbairn, Australian Capital Territory, <http://www.soilsforlife.org.au/about.html>.

Talbot L (2004) Well, you can't eat a tractor! A study of changing patterns of social capital in a Victorian rural community. Doctor of Public Health thesis. La Trobe University, Melbourne.

Talbot L, Verrinder G (2014) *Promoting Health. The Primary Health Care Approach*, 5th edn. Elsevier, Chatswood, New South Wales.

Talbot L, Walker R (2007) Community perspectives on the impact of policy change on linking social capital in a rural community. *Health and Place* **13**(2), 482–492.

Townsend M, Mahoney M (2004) Ecology, people, place and health. In *Understanding Health: A determinants approach*. (Eds H Keleher and B Murphy) pp. 269–275. Oxford University Press, Melbourne.

Veitch C (2009) Impact of rurality on environmental determinants and hazards. *The Australian Journal of Rural Health* **17**(1), 16–20. doi:10.1111/j.1440-1584.2008.01031.x

Wakerman J (2009) Innovative rural and remote primary health care models: what do we know and what are the research priorities? *The Australian Journal of Rural Health* **17**, 21–26. doi:10.1111/j.1440-1584.2008.01032.x

Wakerman J, Humphreys J (2008) Rural and remote health – definitions, policy and priorities. In *Textbook of Australian Rural Health*. (Eds ST Liaw and S Kilpatrick) pp. 13–30. Australian Rural Health Network, Canberra.

Whittaker J, Handmer J, Mercer D (2012) Vulnerability to bushfires in rural Australia: a case study from East Gippsland, Victoria. *Journal of Rural Studies* **28**, 161–173. doi:10.1016/j.jrurstud.2011.11.002

WHO (World Health Organization) (n.d.) 'Protecting health from climate change: vulnerability and adaptation assessment'. World Health Organization, Geneva, <http://www.who.int/globalchange/publications/Final_Climate_Change.pdf>

YACV (Youth Affairs Council of Victoria) (2008) 'Talking about the big dry: young people and the impact of drought. Forum report'. North Central Local Learning and Employment Network, Youth Affairs Council of Victoria Inc.

PART 3:

ORGANISATIONAL ADAPTATION

11

Community-based health and social services: managing risks from climate change

Karl Mallon and Emily Hamilton

Key points

- People experiencing poverty and social disadvantage are hardest hit by extreme weather events, which are projected to become more intense and frequent with climate change.
- The impacts are severely compounded by the lack of resilience across the community services organisations who provide critical support for highly vulnerable people.
- Community services organisations have distributed and specialised resources to support community resilience to climate extremes; however, the majority lack the skills and resources to adapt effectively.
- If community services organisations are resourced and supported to build their adaptive capacity, it increases the resilience of high-risk groups and the community as a whole.

Introduction

In the past decade, multiple extreme weather events in North America, Europe and Australia have indicated that people affected by poverty and social disadvantage are particularly hard hit. This chapter presents the finding from world-first Australian research, which demonstrates that a systemic lack of resilience across the community services sector has the potential to worsen the impacts from extreme weather events and therefore climate change experienced by the individuals and groups experiencing poverty and social disadvantage within developed countries.

In Australia, the community services sector is made up of tens of thousands of organisations that provide frontline social, health and welfare services, which form a critical component of the social safety net for people at risk of poverty and other forms of disadvantage. It makes up a large part of the broader not-for-profit sector, a growing and economically significant sector that contributes 5% of GDP and 8% of employment to the Australian economy annually (Knight & Gilchrist 2014).

The research presented here shows clearly that a failure to ensure that this critical sector is well adapted to climate change would likely worsen the impacts experienced by its clients with the projected increases in the severity and frequency of extreme weather events. Specifically, the results of a national survey of approximately 600 organisations suggests that with few exceptions, community services organisations (CSOs) are not sufficiently prepared to be able to function through extreme weather events or the consequential disruption to critical infrastructure. Worse, many may be forced to permanently cease operation, thus impacting the community recovery process.

The silver lining to this research is that many CSOs have human, physical and network resources that are able to support community resilience to climate extremes as well as to the longer-term impacts of climate change. However, while there is growing recognition across the community services sector that climate change is not just an environmental issue but a threat to core business, a majority of CSOs lack the specific skills and resources to adapt effectively and efficiently.

To rectify this skills and knowledge gap, CSOs must be resourced and supported to build adaptive capacity and to implement the internal and cross sector adaptation necessary to assure resilience.

Community sector consequences of extreme weather events

Extreme weather has dramatic impacts on health and social wellbeing around the globe, but time and again the evidence shows that even in modern developed countries, vulnerable groups are disproportionately affected. In the United States Hurricane Katrina and the failure of critical defensive infrastructure in its aftermath was one of the most deadly extreme weather events to hit the United States in recent memory. While no analysis has been done of the socio-economic status of the 971 people who lost their lives, a black person was 2.5 times more likely to lose their life than a white person; 70 people died in hospitals, many of which were suffering power failures; over 70 died in nursing homes; and the median age of those who died was 69 years.

A few years later, in October and November 2012, Hurricane Sandy left thousands of public housing residents in New York stranded in freezing apartments without power, heating, hot water, food or access to medical care for almost two weeks (Lipton & Moss 2012).

In Australia, the Victorian heatwave and bushfires in January and February 2009 had similar dramatic consequences. During the heatwave that preceded the bushfires in the Melbourne metropolitan area, the increase in severe health impacts included:

- a 2.8-fold increase in cardiac arrest cases
- a 25% increase in the number of emergency cases responded to by Ambulance Victoria (and a 46% increase in cases over the three hottest days)
- a 62% increase in all-cause mortality or 374 more deaths than would be expected (Department of Human Services 2009).

On 9 February 2009, 173 people died in the bushfires. Of these:

- 24% had a chronic disability
- 5% had an acute disability
- 16% were aged 70 or over.

Chronic conditions that increased the vulnerability of people experiencing emergency conditions were found to include severe mobility restrictions caused by degenerative conditions, injuries and obesity. Mental conditions found to contribute to the increased vulnerability of those that died included depression, anxiety, brain injury and post-traumatic stress disorder (Centre for Risk and Community Safety & Bushfire CRC 2010, p. 15).

There is also the problem that at-risk people can be plunged into poverty by extreme events. Analysis conducted by the Queensland Council of Social Service (QCOSS) found that the Queensland floods in 2010–11 also had a disproportionate effect on people experiencing poverty. Prior to the floods, 10% of Queenslanders lived in poverty, with a further 20% at risk of poverty. According to QCOSS, the broader impacts of the floods included loss of employment, lack of insurance payouts and difficulty finding affordable housing, all of which have the potential to trigger significant increases in poverty. In light of the disproportionate impacts of disasters on people affected by or at risk of poverty, QCOSS concluded that unless community resilience to extreme weather events is enhanced, up to 30% of Queenslanders, or 1.2 million people, could be forced into poverty due to the state's high exposure to climate and weather extremes (QCOSS 2011, p. 3).

Stories and statistics like these provide a clear warning that even in modern developed countries, we are not properly equipped to protect people who may be physically or economically vulnerable, marginalised or dependent on social, health and community services for their wellbeing.

A myth-busting objective

People experiencing poverty and disadvantage in both developing and developed countries are at the greatest risk of harm from climate change, including increasingly frequent and intense extreme weather events. They have the least capacity to cope, to adapt, to move and to recover.

The danger is that this gives rise to the myth that such people will inevitably suffer the greatest actual harm as climate impacts escalate. In a developing country this view might be valid. But in a developed country there are numerous mechanisms in place to compensate for financial and social adversity: a person whose house burns down is not left destitute and without support by virtue of state assistance and access to appropriate insurance cover; a child born into a poor family is not automatically denied access to a decent education; a person with poor eyesight does not have to spend her life unable to see. So why should we accept that simply because a person experiences poverty or disadvantage, he should suffer disproportionately the negative impacts of climate change and be denied access to the opportunities that both risk mitigation and climate change adaptation present?

Well-functioning, modern societies provide 'insurance' against poverty and disadvantage for their citizens via the social safety net: a mosaic of community supports, ranging in size and scope from a single volunteer coordinating help for her neighbours through to government-funded public health services or transnational aid organisations. This diverse social safety net should also be adequate to ensure that people experiencing poverty and disadvantage have adequate capacity to cope with the impacts of extreme weather and have access to the opportunities presented by successful climate change adaptation. That this system fails during extreme weather events indicates that something is going wrong.

Something that needs to be fixed if we are to ensure that all climate change adaptation is distributed equitably among all in the community.

In attempting to bust the myth that vulnerability to climate change is the result only of a person's underlying economic or social vulnerability, the hypothesis (for which the research presented provides the evidence) is that the lack of embedded resilience in the community service sector (community-based health and social service organisations, referred to in this chapter as community services organisations or CSOs) gives rise to a systemic risk to climate impacts among the people and communities they service, particularly those most at risk of poverty and disadvantage. Put simply, though a modern developed society has systems and organisations to protect people who are experiencing disadvantage, these are failing during extreme events. The objective for this study was to provide an evidence base for the processes of strain or failure and their consequences, and to identify both the policy and organisational responses required to address them.

The study was undertaken by a partnership between the Australian Council of Social Service (ACOSS) and Climate Risk Pty Ltd and was funded by the National Climate Change Adaptation Research Facility. All of the data and figures presented in this chapter are drawn from the final research report, entitled 'Adapting the community sector for climate extremes' (Mallon *et al.* 2013).

Methods
Literature review

The project commenced with an integrative review of Australian and international literature about the community services sector's capacity to cope with climate change (including incremental and extreme impacts), which sought to identify any important gaps in the research. The following research questions were explored:

- Are people experiencing poverty and social disadvantage in developed countries more susceptible than the general community to climate change impacts, particularly extreme weather events?
- Is there evidence that CSOs increase the resilience of people experiencing poverty and social disadvantage?
- Has the potential role and importance of the community sector in climate change adaptation been addressed in the literature?
- Are CSOs at risk of failure or strain from climate change, particularly impacts to infrastructure?
- Do specific adaptation strategies exist for CSOs to allow them to continue carrying out their role in supporting people experiencing poverty and social disadvantage under climate change?

The research questions were addressed through the review of more than 350 academic and 'grey' (non-peer reviewed) articles, tools, guides and reports from credible sources. The review deliberately excluded much of the international development literature because, although an extensive and important body of research, adaptation issues faced in developing countries are highly compounded by other pre-existing conditions such as extreme poverty, low levels of education and inadequate health and community service systems which rendered the research pertaining to the adaptation needs of these countries too different to be reasonably compared.

A large survey of community sector climate resilience

The literature review revealed that globally almost no surveys have been conducted to assess the climate change resilience of CSOs. To rectify this gap in the research, this project harnessed the extensive sector networks and collaborative approach of the combined national, state and territory Councils of Social Service to attempt the largest survey possible. Through these networks, thousands of organisations were invited to participate, from large national organisations delivering multi-million dollar government-funded services, to small, voluntary groups delivering specialist programs in regional and remote communities.

The response was very strong. Nearly 600 organisations or groups attempted the survey, many seeking to address climate change and resilience questions for the first time. Approximately 500 successfully completed the survey to the standard required by the researchers. Though one of the first of its kind, this is a large dataset for the community sector by any measure and in our view this level of participation was only achieved because it was run by the community sector for the community sector. Figures 11.1 to 11.4 show the characteristics and distribution of participants. The percentage figures refer to the percentage of respondents. Some questions allowed for multiple responses.

Figure 11.1 shows the diversity of services the community services sector provides and reflects the breadth of participation by service types in the national survey. A majority of respondents that selected 'information, advice and referral services' and 'advocacy other than legal services' also provided at least one other type of service. The prevalence of these two service types highlights the sector's inherent capability to deliver community education and emergency preparedness information to its clients.

Most of survey respondents delivered services from a centre or office (Fig. 11.2). As will become apparent later in the chapter, this reliance on the built environment for service delivery is a significant source of the sector's vulnerability to extreme weather impacts. The findings suggest that strategies to diversify methods of service delivery across the sector and to develop service continuity plans that cover circumstances in which organisations' service centres are rendered inaccessible by extreme weather events are particularly important to building the sector's adaptive capacity and resilience.

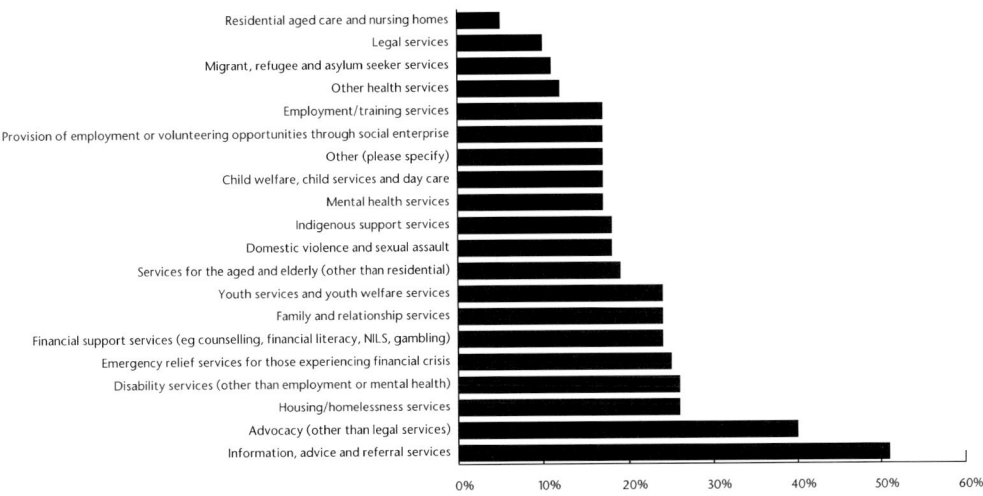

Figure 11.1.   Main types of services provided.

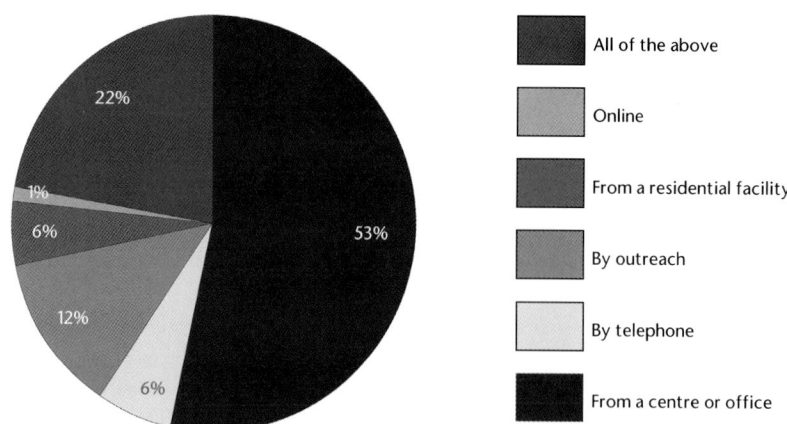

Figure 11.2.   Main method of service delivery.

The survey respondents worked for organisations that delivered services across every Australian state and territory, demonstrating the truly national scope the survey achieved. A small number of organisations represented in the sample delivered services across all states and territories (Fig. 11.3). Survey participants were also distributed across urban, regional and remote areas, with ~45% of respondents delivering services only in regional and remote areas, and ~30% of respondents delivering services only in urban areas. The high representation of organisations operating in regional and remote areas may reflect higher levels of experience and awareness about extreme weather events in across regional and remote Australia.

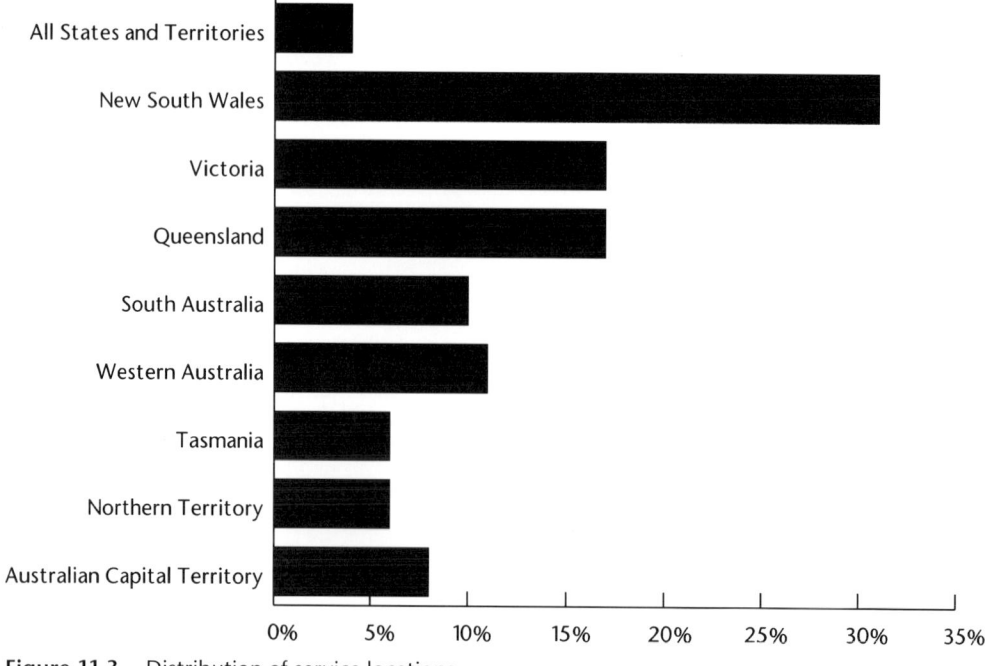

Figure 11.3.   Distribution of service locations.

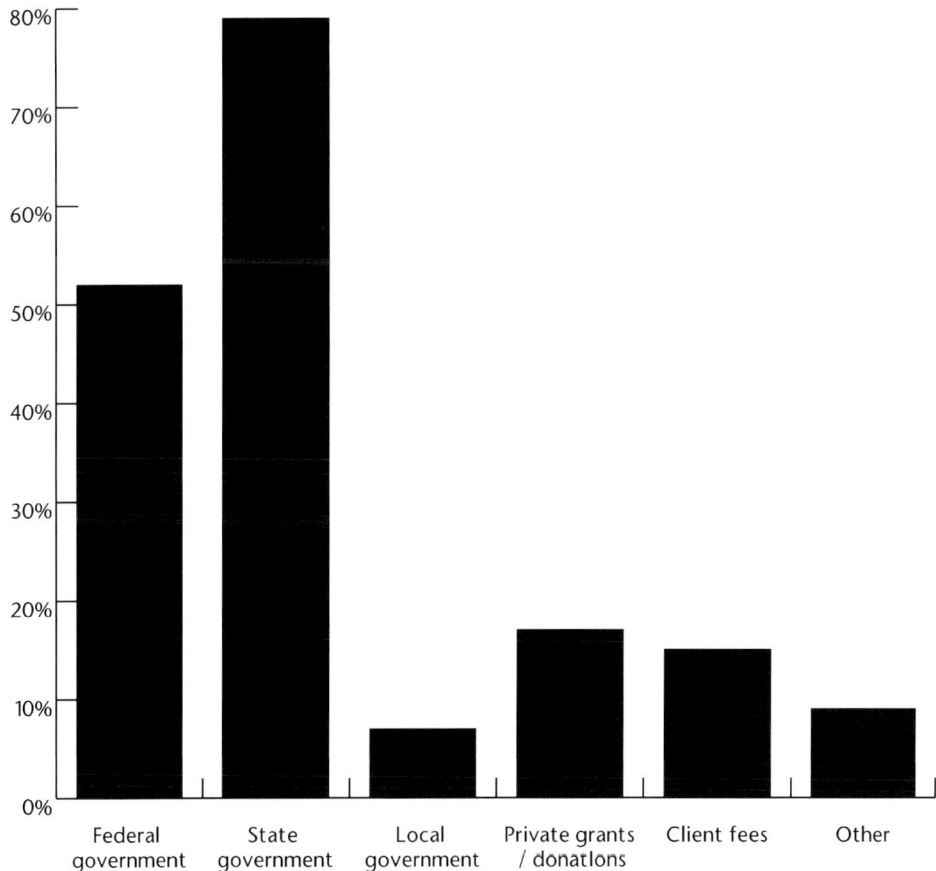

Figure 11.4.   Main source of organisational income.

As shown in Figure 11.4, respondent organisations were heavily reliant on the federal and state governments for funding. This reflects the increasing practice by governments of contracting out health and social service delivery to CSOs. It also suggests that governments have a critical role to play in ensuring CSOs are well prepared to respond to increasingly frequent and severe climate extremes.

National program of workshops

The workshop approach was derived from training and investigation workshops developed for insurance companies to engage small and medium-sized enterprises on climate change risks. In many ways, small and medium-sized enterprises are the private sector equivalent for CSOs: staff numbers range from a handful to a few hundred; business types are too diverse to enable purpose-designed solutions; and many lack the capacity for sophisticated risk management and continuity planning in comparison to larger entities.

To facilitate this project, the ACOSS was given access to Zurich Financial Services workshop materials, which were adapted for use with CSOs. Using these materials, a total of 10 workshops were held with over 150 participants from CSOs in every Australian state and territory. Most of these workshops were structured as 'Welfare Professional Climate Workshops', which supported researchers and participants to work together to:

1. identify the particular risks from climate change and extreme weather events to CSOs
2. develop example adaptation strategies
3. consider the opportunities that might arise from successful adaptation and evolution.

In this way the workshops were structured as an exchange in which information relevant to the research task was gained, while participants learned about emergent risks using tools adapted from the private and local government sectors.

The two final workshops were held to test the preliminary results emerging from the national survey of CSOs with sector experts and to identify the adaptation priorities for the sector and the resources required to achieve them.

Literature results: a bad place for a blind spot

The third sector [community sector] holds the key to success in tackling climate change. It can provide individuals with the collective opportunities to act that are so vital to securing individual action. Its responsibility – and opportunity – is to provide the leadership that can secure the action needed from governments and others to put us on track to securing a sustainable future. (HM Government 2010, p. 26)

In response to the research questions outlined earlier, the literature review found:

1. People experiencing poverty and inequality in developed countries are more susceptible than the general community to climate change impacts and particularly extreme weather events. While the evidence, in the form of academic literature in peer-reviewed journals, was strong, the review also found more literature focused on the needs of particular groups, such as elderly and frail people, than on others, such people experiencing homelessness.
2. CSOs increase the resilience of people experiencing poverty and social disadvantage. This finding was based on evidence in the form of analysis conducted by the community services sector and in grey literature. There is a gap in the academic literature.
3. Evidence from sector analysis and grey literature suggests the potential roles and importance of CSOs in climate change adaptation. However, the evidence base is limited and there is a significant gap in the academic literature.
4. CSOs are at risk of failure or strain from climate change. There is a significant gap in the academic literature addressing this issue and this finding was reached using proxy-based evidence from the disaster management, health and small and medium-sized enterprise sectors.
5. There is a major gap in policy and research into how CSOs need to adapt to climate change to continue carrying out their role in providing support to people experiencing poverty and inequality. The evidence is limited to research being undertaken by the sector in the United Kingdom and Australia.

The key finding was that around the developed world there appears to be an almost universal blind spot in connecting the vulnerability of people affected by poverty and disadvantage, and the climate resilience of the CSOs that provide support in response to everyday adversity as well as in times of crisis. This key finding leads to the conclusion that developed countries, including Australia, have been largely unaware of the criticality of the resilience of CSOs to the resilience of communities, particularly those most disadvantaged within them. And because CSO resilience is being neither measured nor managed, we have very little idea of how the sector will perform in response to both immediate and future events, which are predicted to become increasingly frequent and severe as climate change progresses.

Survey results: a sector unprepared

To avoid any obscure suppositions on how climate change and extreme weather events might affect CSOs, the survey focused on the effects of disruption to enabling essential services such as power, telecommunications, roads and water.

Such 'critical infrastructure' is often disrupted or damaged during extreme weather events and there is an emerging trend of pre-emptive closure. Power is turned off before possible flooding (to prevent electrocution) and during extreme heat events (to prevent a pole fire triggering wide-scale bushfire). Roads are closed to prevent people entering areas of high risk and becoming trapped.

But how are CSOs able to operate when critical infrastructure services are unavailable? What do they know about extreme weather events driven by climate change? Have they experienced such events in the past? How prepared are they for future events? As shown in Figures 11.5 to 11.7, a majority of survey respondents self-assessed their knowledge of climate change risks as 'moderate', while over 50% had experienced at least one extreme weather event in the past. Organisations that reported moderate and high levels of understanding about climate change and those that reported past experience of an extreme event also reported higher levels of action to respond to risks.

A common scenario of business continuity planning is loss of access to premises – perhaps the access road has been closed, or the roof has been damaged in a storm. The survey asked CSOs of many different types how long they could manage to operate without access to premises (Fig. 11.8).

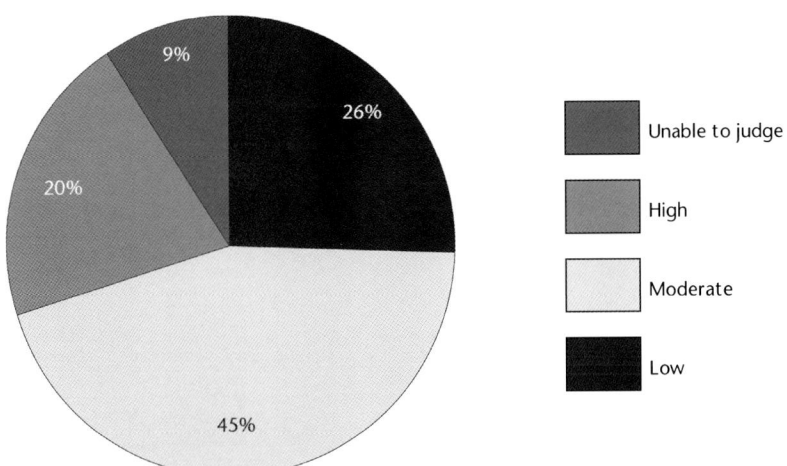

Figure 11.5.   Knowledge of local climate change risks.

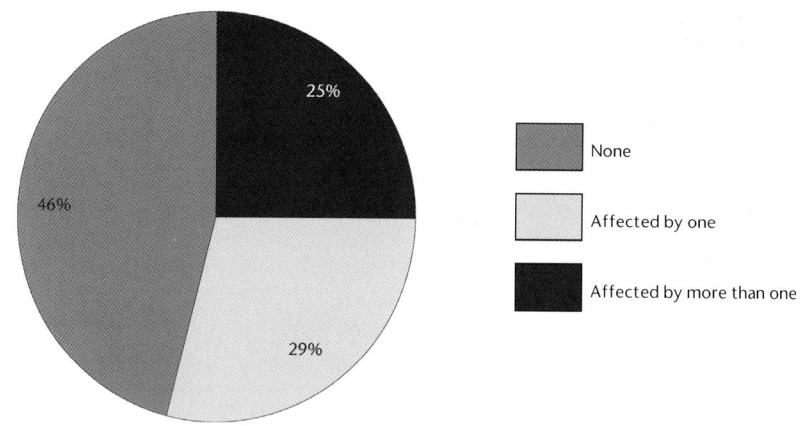

Figure 11.6. Past experience of extreme weather events.

Clearly the impact of a disruption to critical infrastructure depends on the type of CSO and services provided. A little over a quarter were confident they could have alternative arrangements in place within a week. However, even the loss of operation for a single day

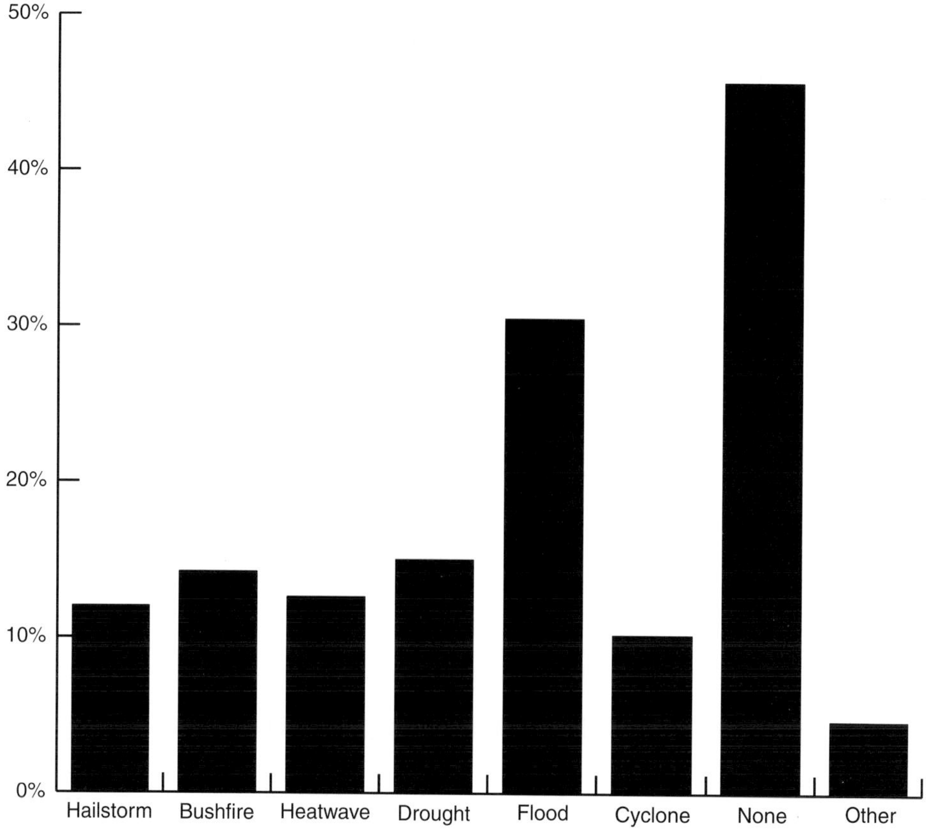

Figure 11.7. Types of extreme events experienced.

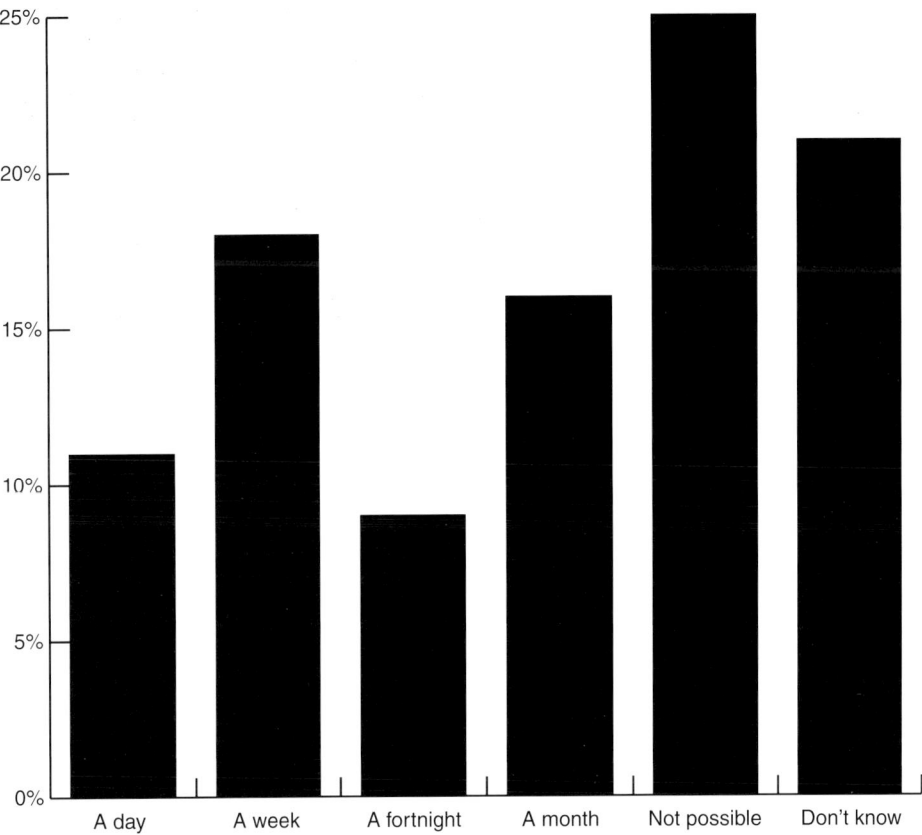

Figure 11.8.   Length of time to resume service delivery if premises become inaccessible.

could have serious consequences for clients who receive food or vital medicine. More worrying is the 60% of organisations that would need at least a month to recover operations if their premises were damaged, or the 25% that might never open their doors again – a figure which is very similar to statistics for small and medium business enterprises in the private sector.

An equivalent pattern is seen with other critical infrastructure (Fig. 11.9). Imagine operating an organisation within a persistent power blackout or without phones or the ability of staff to access workplaces and clients due to road closures.

When we look at that estimated levels of organisational disruption, the scale of abandonment of the vulnerable seen following Hurricanes Sandy and Katrina starts to make sense. When a service provider organisation becomes dysfunctional there is very little its staff can do to systematically fill the service gap, particularly in circumstances in which they and their families may have been directly impacted by an event as well.

In the business sector many organisations might be equally unprepared for unexpected disruptive events – that's where insurance comes in. A typical 'wild-card' cover for the private sector is business interruption cover, designed to allow a business to re-open after unforeseen closure. Other insurance coverage can extend to staff, contracts and income protection. Unfortunately, as shown in Figure 11.10, insurance cover in the community services sector is characterised by a 'house and contents' approach rather than 'business continuity'.

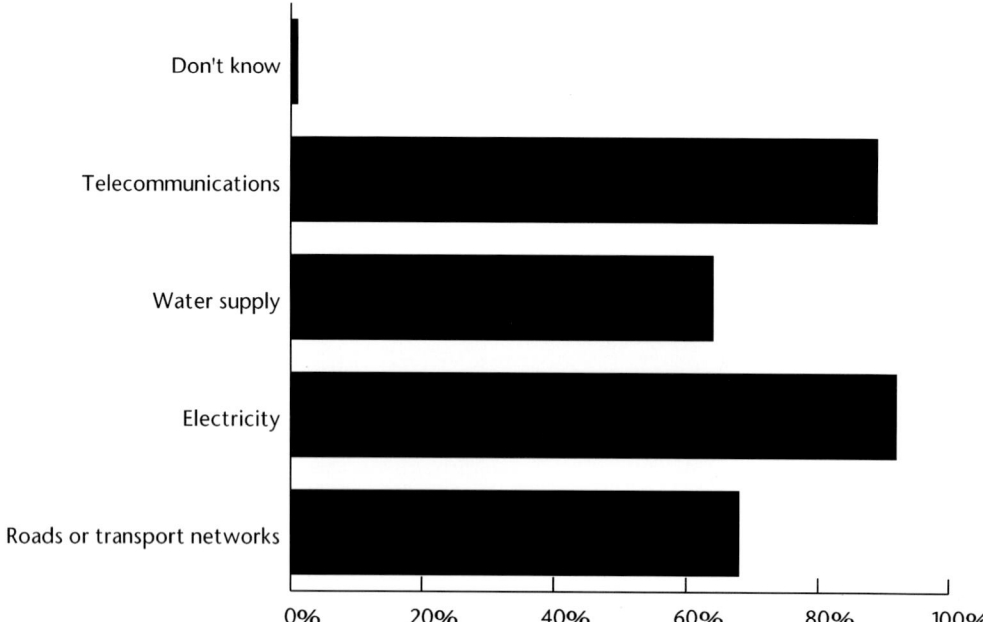

Figure 11.9. Percentage of organisations experiencing serious disruptions to service delivery as a result of critical infrastructure failure.

Figure 11.10 shows that the community sector is missing the opportunity to cover its preparedness weakness through insurance, which indicates that it is generally not appreciating the materiality of the risks to its ability operate or recover.

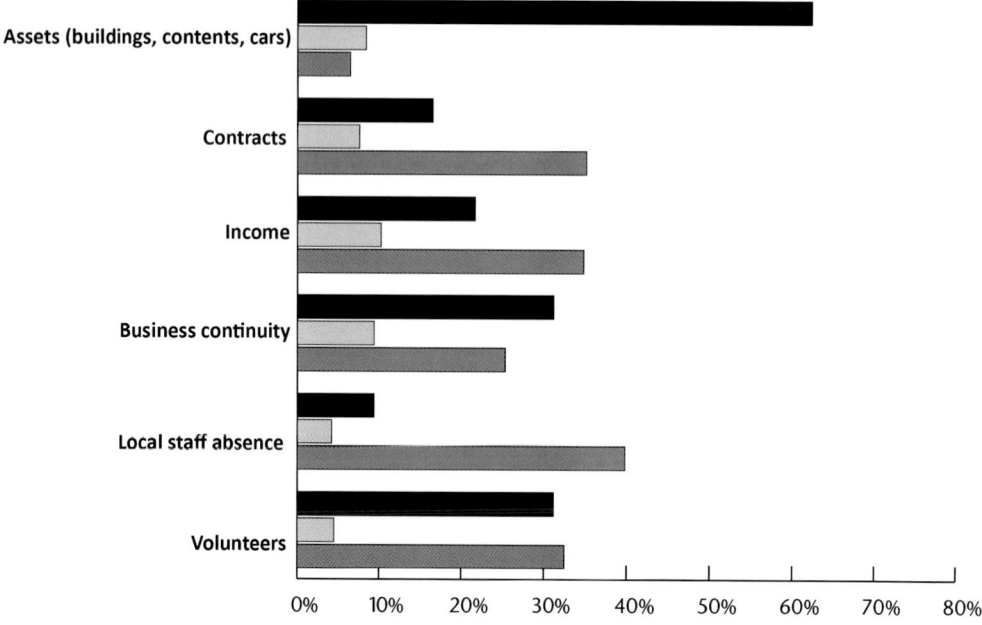

Figure 11.10. Insurance against losses caused by extreme weather events.

Understanding the real impacts

Understanding the impact of climate change on businesses is simple in comparison to the task with CSOs. For business, the key indicator is profitability – the impacts can be measured in dollars and cents. The impacts of disruption to CSOs must be measured in the lived experience of the millions of people who access services each year.

Failure modes and the side effects of collapse

As well as the survey, a series of workshops were conducted with CSO representatives around Australia to carefully work through the consequences of a range of climate change hazards, from carbon price to cyclones. The workshops facilitated the creation of nuanced narratives for the processes of strain and failure on CSOs, their clients and the flowthrough to other parts of society and services.

To create a consistent method of capturing these impacts, the narratives developed were structured into CSOs systems analyses (Fig. 11.11), which articulate the societal and service systems within which CSOs operate. In Figure 11.11, many component organisa-tions are identified – from government departments to the police, from insurers to donors – and more abstract elements like the community and economy are also captured. Different connections between system elements are considered too. In this case the con-nections from clients to the system are compared to the connections stemming from a CSO. This helps us to understand how a shared system is seen from the perspective of different participants.

Using these generalised systems analyses, researchers and workshop participants undertook failure mode analyses, which articulate the processes by which specific services might be strained or experience total failure in response to climate change impacts (Figs 11.12 and 11.13). In Figures 11.12 and 11.13, one class of connecting lines represent mecha-nisms of organisational failure or strain (e.g. local business closure), while another repre-sents the flow-on effects to CSOs (e.g. increased demand of financial assistance) and another set of connecting lines represents the flow-on effects of CSO service failure for other system elements (e.g. increased demand on health services).

These 'failure modes' start to show the complexity of failure and its cascading conse-quences to CSOs, clients, other service providers like the police and hospitals, and the host society. For example, Figures 11.12 and 11.13 demonstrate the potential impact of a cyclone on the provision of emergency relief (material and financial assistance for people

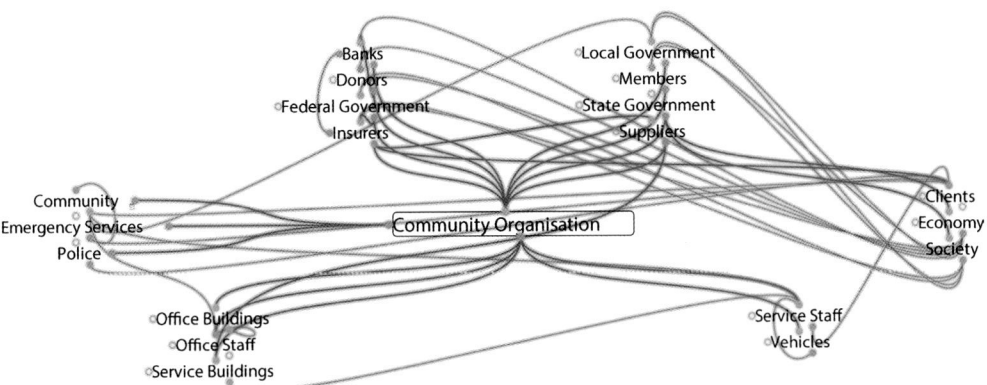

Figure 11.11.   Generalised CSO systems analysis.

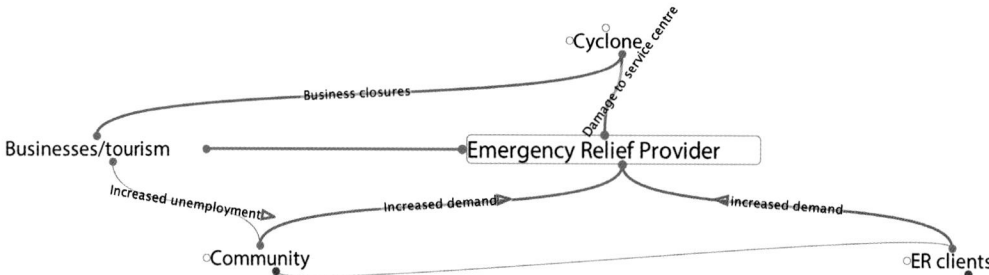

Figure 11.12.   Failure mode: impact of a cyclone on the provision of emergency relief (stage one).

experiencing financial hardship). Figure 11.12 captures a scenario in which an emergency relief provider sustains damage to its service centre during a cyclone, which disrupts the organisation's capacity to distribute material and financial aid. At the same time, demand for emergency relief increases among the organisation's traditional clients due to losses sustained as a result of the cyclone. Demand for emergency relief also increases among the broader community, as more people are forced into financial hardship for the first time due to business closures in the tourist industry, which has been hard hit by a sharp reduction in the number of tourists visiting the region in the cyclone's aftermath.

As the organisation struggles to rebuild its service centre and to meet the increased demand for its services, growing numbers of the community fall into extreme financial hardship, unable to meet mortgage and rent payments and therefore facing the risk of homelessness.

Through the project, CSOs provided detailed feedback about ways in which an extreme event could impact their operations. The following quotes are drawn from qualitative responses provided by survey participants and presented in the final research report (Mallon *et al.* 2013).

Some simply said, 'We would close.' Others explained that they would not only be unable to operate, but would be at risk of failing to meet their contractual commitments with the government:

> *The office would be inoperable without electricity, water and telecommunications. [We] could not operate from these premises and provide the services required according to funding agreement.*

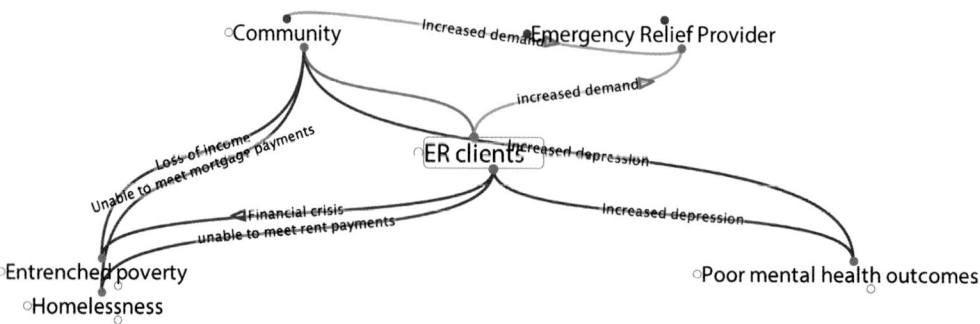

Figure 11.13.   Failure mode: impact of a cyclone on the provision of emergency relief (stage two).

Some pointed out that extreme weather events affected the number of clients able to reach them to start with, as well as the total number of people in need:

When there has been a major disruption in the past we have seen a major drop off of clients walking in and appointments. Then when [the] issue [is] resolved a major influx of clients that were in mild need go to extreme need for assistance.

This comment reflects the quantitative data collected through the survey about the impact of extreme weather events on demand for services, captured in Figure 11.14. Others raised the threat to their staff:

Our clients are in isolated areas in the outback. Disruption to communications and roads means that we cannot contact clients [and the] project could not continue. Some of our staff [are] located in isolated areas. Extreme weather events present real risks to our staff, with some already having close shaves.

Looking at the impacts of disruption on critical infrastructure as a mechanism of failure, a range of unique dependencies emerge:

Telephone crisis support is our core service hence without telecommunications or the power to run them would totally disrupt our service.

All health records are on computer so we would be annihilated if electricity was not functioning.

Without electricity hoists could not be used to provide personal care to support clients and computers would not work which would affect managers and staff ability to do their job effectively.

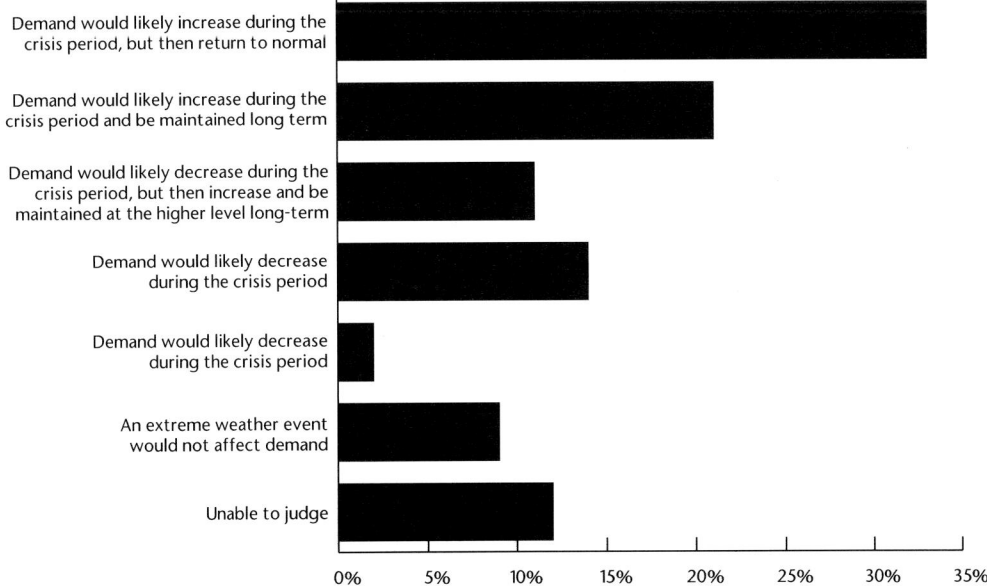

Figure 11.14.   Impact of extreme weather events on demand for services.

No computers would mean no staff pay or invoicing. Freezers full of food would spoil.

We could not accommodate young people if we did not have power and water. It would be unsanitary.

Outage of water supply will make offices unusable from OHS [occupational health and safety] point of view. Showers for the homeless not possible.

[Without roads] workers would not be able to support clients in their homes.

Volunteers and staff could not access clients' homes or the organisation's offices.

Some clients would miss critical treatments (i.e. chemotherapy, blood transfusions). Our area of service delivery is 870 000 km². It would be difficult to get to the affected areas other than by air.

Impacts of service failure on clients

Respondents were also asked to reflect on how clients would be affected if their organisations were unable provide services immediately after a natural disaster. The impacts were clearly dependent on the type of service provided and are catalogued in the full report (Mallon *et al.* 2013). The selection below highlights the range of issues raised:

Homelessness and housing services:

Worst-case scenario is death due to lack of shelter/food for homeless individual [sic]. People would simply remain homeless.

Young people would return to being homeless.

Disability services:

We work with people with a cognitive disability where an understanding of the event may be limited, there may be a heightened sense of fear in general ... My concern would be their lack of knowledge of what is safe, i.e. is it safe to drink the water?

Aged care services:

There is a real possibility of resident death.

They would die.

[N]o eating, no toileting, no showering, no getting out of bed.

Legal services:

We surmise that a client could be imprisoned without advice or representation. Homes could be lost, goods repossessed. Court battles would increase without that advice, representation or mediation.

Women's crisis and domestic violence services (including legal):

Immediate impact might be loss of safety from court orders not being established to loss of children in longer term.

Mental health services:

Those at risk of suicide could be at higher risk. Those with depression could become further depressed.

Any big event often triggers paranoia and an increase in other symptoms experienced by people with mental illness including panic and suicidal ideation.

Health services:

Our health services would not be able to do outreach to remote communities, who would be completely cut off from support.

Emergency relief services:

A lot of our clients who are already vulnerable and have limited resources would have nowhere else to go for emergency relief and would lose lines of communication and advocacy support. Some families would not be able to access basic services for needs including food, shelter, transport.

Aboriginal and Torres Strait Islander services:

I would be most concerned for the elderly, people with disabilities and the homeless people who sleep rough in the river and hills around town. Many of these people have serious and multiple chronic health issues that require regular monitoring and medication.

Barriers: the mind is willing but the budget is weak

Despite the recognition of the impacts from extreme weather and climate change on the ability of the participants to operate, there was a near universal view that the sector lacks the financial, informational and human resources it needs to adapt effectively (Fig. 11.15). This finding is not surprising in a sector in which underfunding and funding uncertainty are routinely identified as critical factors affecting its sustainability and effectiveness (ACOSS 2014).

It is also important to note that large numbers of respondents identified a broad range of barriers to climate change adaptation, including a lack of clear direction from government, a lack of information and awareness about the risks and concern about possible negative client reactions if organisations were to divert resources from service delivery to adaptation. Critically, over 50% of organisations identified the belief that adapting to climate change is beyond the organisation's scope as a key barrier to resilience building. The identification of an almost overwhelming array of barriers to climate change adaptation suggests that supporting the sector to develop engaging awareness-raising and adaptation programs that clearly link climate change impacts with CSOs core business, and helping organisations to implement practical and cost-effective strategies, to plan for and address risks to service delivery, staff and clients, is critical to building intrinsic resilience and adaptive capacity.

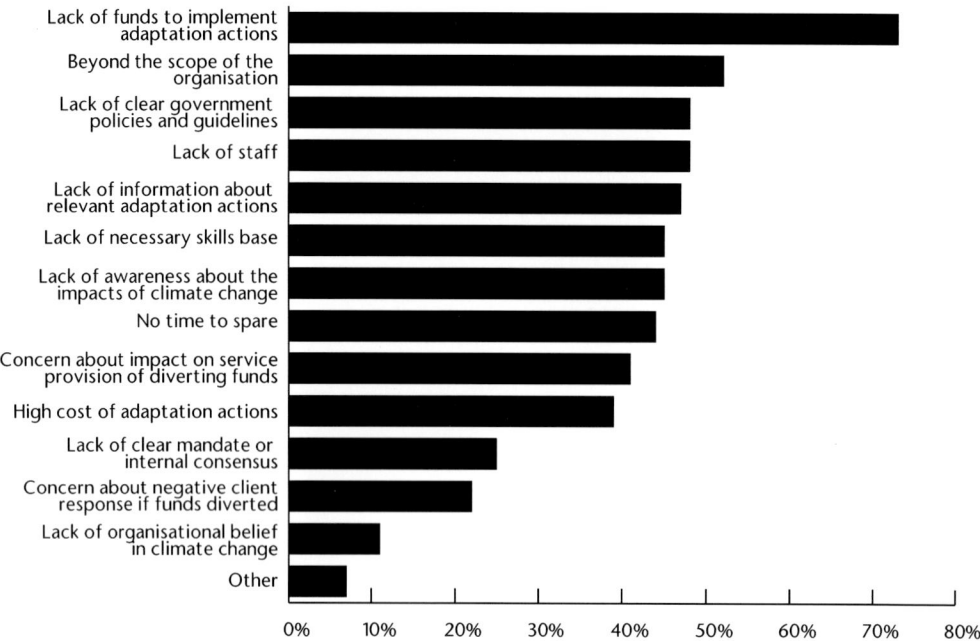

Figure 11.15.   Barriers to adaptation.

Societal spin-offs: beyond internal resilience

In the foregoing discussion the focus has been on the risks to CSOs and the barriers to building resilience. At the outset, the assumed outcome was that increased resilience across the sector would reduce the risks to its particular client groups. However, the research provides strong evidence that a resilient community sector could support significant increases in the resilience in the wider community and in society more broadly.

The drivers of broader resilience co-benefits are:

1. locally available, highly specialised services which are usually under acute demand following disasters caused by extreme weather events
2. increased community resilience through the 'social infrastructure' of connectedness
3. reduced demand on government emergency services and government health services.

Unique specialised services where and when they are needed

Taken as a whole, CSOs are essentially a highly distributed network of specialist people, premises and other assets. The complexity of the sector defies cataloguing at high level, but its depth and complexity are unique: from the equipment to store and transport anti-retro-viral medicine to crisis accommodation; from fleets of disability transport vehicles to networks of volunteers pre-cleared to work with children; from mental health profession-als to logistic professionals. In times of crisis these skills, services, premises and equipment are in exceedingly short supply and are often brought in from around the country.

Yet, if operational as a result of good preparedness, the community sector has this resource base already available on the ground where it is needed and the people to mobilise it within the local community. This has three major effects. First, it increases the resilience of the host community that comes to need these services during extreme events and recovery. Second, it reduces the risk of people slipping into persistent disadvantage following extreme events by limiting the damage of emotional trauma, financial strain and knock on impacts to family and employment stability. Third, it reduces the long-term costs to society through sustained dependence on health services and government financial support.

Building social infrastructure

There is a growing body of knowledge suggesting well-connected communities are resilient communities (Ensor & Berger 2009). They do not need to be connected through disaster preparedness, they can be connected for any reason: through participation in local sports leagues and community events or high rates of volunteering. For example, one CSO that participated in the workshops delivered bushfire awareness to the local community by running a 'firey' burlesque night, which started with a local fireman explaining how to fit and use smoke alarms at home and at work.

When people are socially connected, they know who they can call in a crisis, who in their community might need extra help responding to an evacuation alert, who has the skills and resources to help others in need, and who has a car or generator, or a spare room. They can see when a person might not be coping, or a family is struggling. They have a phone book of people they can call for assistance when the emergency services leave and the recovery process has begun.

Reducing demand on emergency services: being part of the solution

In times of crisis caused by extreme events, it is the emergency services that are usually called upon to help. This can be government agencies and non-government organisations like the Red Cross, and usually these groups are all working in close collaboration.

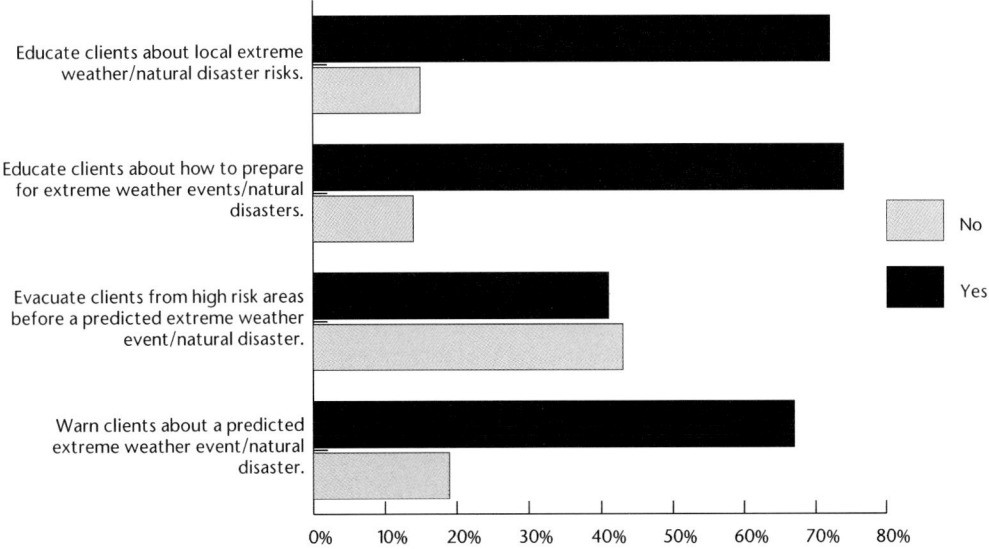

Figure 11.16. Assistance CSOs could provide to clients before an extreme weather event.

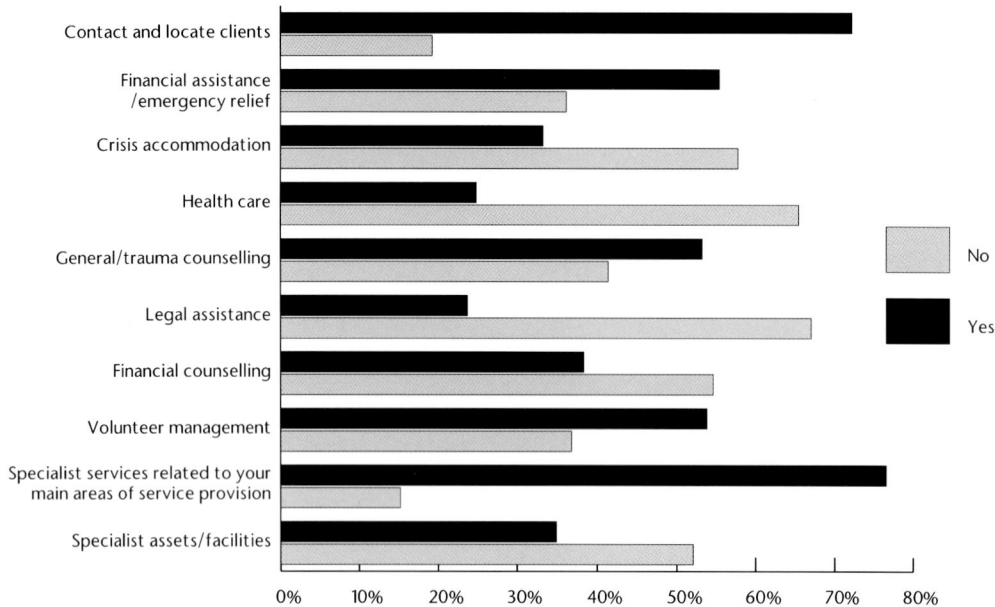

Figure 11.17.   Assistance CSOs could provide to the community after an extreme weather event.

CSOs were invited to consider what assistance they might able to provide ahead of an extreme weather event and in the wake of an extreme weather event (Figs 11.16 and 11.17). The responses show a range of assistance as diverse as the sector itself, a resource that is largely being overlooked in the present institutional arrangements for extreme event management. Harnessing this resources base would both reduce demand on traditional emergency services and provide a new and expanded roll for CSOs in the community.

Conclusion

In our view the research findings presented in this chapter provide an evidence base to support the following conclusions:

1. The community sector plays a critical role in supporting individuals and communities to respond to and recover from extreme events. The CSOs within it are a de facto resilience buffer for people affected by poverty and disadvantage. If CSOs are strained, disrupted or closed by extreme weather events, the coping and recovery capacity of their client groups will be diminished.
2. Based on typical measures of business continuity preparedness, the bulk of CSOs are highly vulnerable to the disruption of their operations through, and in the wake of, extreme weather events. Half can be expected to be unable to operate for up to a week, a quarter could be forced into complete closure. Thus at one of the most critical times for client and community, when many more people are unexpectedly forced into crisis and vulnerability, the community organisations that normally provide support will be absent. There are notable exceptions, among whom are some of the most prepared organisations in the country.

3. The detailed consequences of major disruptions to social service provision for people experiencing poverty and disadvantage are very serious as they impact the basic needs for human survival: access to adequate shelter, sustenance and medical care. In the most extreme cases, there is a real risk of death.

4. Currently, the most reliable indicators of resilience for CSOs are size, knowledge and past experience of extreme events. When looking at how to enhance resilience in a sector with a huge diversity in organisational size and where the cost of gaining experience may be closure, developing and sharing knowledge becomes key.

5. Despite the size of the problem of CSO vulnerability and the severity of its consequences, the literature review clearly shows that to date the community sector has been overlooked in the climate change adaptation policy settings and research agendas of developed economies as evidenced by major gaps in the academic and grey literature.

6. There is a clear recognition in the community sector that extreme weather events and their worsening due to climate change represent a threat to core business, and further, that a failure to build organisational resilience increases the risks to their clients. There is a strong willingness to respond to climate change and extreme weather risks across the sector. However, most organisations have identified a range of barriers to adaptation due to the availability of human, financial and skill resources and an absence of sectoral guidance on how to go about resilience development.

7. The evidence suggests that there are major societal co-benefits to a well-adapted community sector that extend beyond its client base. In particular, the sector is unusual is that it has a concentration of specialist resources, skills and facilities, that are often in high demand and short supply during and after extreme weather events. These services include emergency accommodation, networks of trained volunteers, professionals in mental health, and specialist premises, vehicles and equipment for the disabled. Importantly these are available across the country. Thus a resilient community sector can uniquely contribute to societal resilience and disaster preparedness to assist, and lessen the load on, traditional emergency management services.

8. At present, CSOs perceive an overwhelming range of barriers to action. Key among these is a lack of financial resources and skills and the concern that adaptation is 'beyond the scope' of the sector's core business. The issue of scope is central to establishing if increasingly frequent and intense extreme weather events represent a new 'normal' for CSO operation.

To raise awareness about the risks to service delivery from climate change and to support capacity building within the community services sector, the project produced a series of outputs and resources, which are immediately available for implementation by organisation to assist with the process of identifying, analysing and responding to climate change and extreme weather risks. These are:

- a set of CSO failure mode and adaptation mode exemplars, which identify and codify the mechanisms by which service delivery is disrupted in response to infrastructure failure and, conversely, the processes for and consequences of implementing adaptation for organisations, clients and the broader social system

- community sector risk and adaptation registers, which describe and catalogue over 200 discrete risks and 450 adaptation actions specific to four key areas of CSO activity and operation.

Recommendations

Finally, the research presented a comprehensive set of recommendations about the resources and action required to prepare and adapt CSOs – and the community service sector broadly – to climate change and extreme weather impacts. Their full implementation will ensure that the sector is able to:

- fulfil its service delivery mission to people experiencing poverty and inequality sustainably and over the long-term as the climate changes and the frequency and intensity of extreme weather events worsens
- make a positive contribution to the resilience of the individuals and communities with which they work to climate change and extreme weather impacts
- participate effectively in emergency management planning, relief, response and recovery efforts when disasters occur.

These recommendations fall under three key areas: sector preparedness; building resilience; and sharing risks, and are summarised below (ACOSS 2013).

Sector preparedness

Contracts for service delivery must provide greater flexibility to community service organisations and enable them to participate effectively in disaster response and recovery efforts. Specifically, they should include mechanisms that ensure:

- timely compensation for their contributions to response and recovery efforts
- they are not penalised for failing to meet contractual obligations due to their participation in disaster response and recovery.

The community sector needs to be resourced and supported to:

- raise awareness about the serious risks to its service delivery and to people experiencing poverty and inequality from climate change and worsening extreme weather impacts
- undertake climate change and extreme weather risk assessments and develop and implement disaster management and service continuity plans
- invest in climate change and extreme weather preparedness and response training for staff and volunteers engaged in direct service provision as well as management and administrative roles.

Building resilience

The community sector needs to be resourced and supported to develop:

- a set of easily accessible, practical adaptation and preparedness tools that meet the needs of a broad spectrum of community service organisations and can be implemented and institutionalised within their current operational arrangements
- adaptation and preparedness benchmarks specific to community service provision that enable organisations, their funding agencies and insurers to plot progress towards risk reduction, resilience and adaptive capacity.

Sharing risks

Federal, state and local governments and formal emergency service agencies must:

- recognise the critical role the community services sector plays in emergency management
- resource, facilitate and support its effective participation in planning, response and recovery at all levels.

Acknowledgements

The authors would like to acknowledge the role of Emily Hamilton's employer, the Australian Council of Social Services (ACOSS), who commissioned the research, the National Climate Change Research Facility (NCCARF) who provided research funding and Climate Risk Pty Ltd who provided Dr Mallon as principal researcher.

References

ACOSS (Australian Council of Social Service) (2013) 'Extreme weather, climate change and the community sector – ACOSS submission to the Senate inquiry into recent trends in and preparedness for extreme weather events'. ACOSS Paper 197, January 2013. ACOSS, Strawberry Hills, New South Wales.

ACOSS (Australian Council of Social Service) (2014) *Australia's Vital Community Services Face Funding Uncertainty Crisis: New Report*. ACOSS, Strawberry Hills, New South Wales, <http://www.acoss.org.au/media/release/australias_vital_community_services_face_funding_uncertainty_crisis_new_rep>.

Centre for Risk and Community Safety & Bushfire CRC (2010) 'Review of fatalities in the February 7, 2009, bushfires – final report'. Centre for Risk and Community Safety, RMIT University and Bushfire CRC, Melbourne.

Department of Human Services (2009) 'January 2009 heatwave in Victoria: an assessment of health impacts'. Victorian Government, Melbourne.

Ensor J, Berger R (2009) *Understanding Climate Change Adaptation: Lessons from Community Based Approaches*. Practical Action Publishing, Rugby, UK.

HM Government (2010) 'Shaping our future: the joint ministerial and third sector task force on climate change, the environment and sustainable development'. Task Force Secretariat provided by the third sector through Green Alliance and NCVO. Crown Copyright, Surrey, UK.

Knight P, Gilchrist D (2014) 'Australian Charities 2013: The first report on charities registered with the Australian Charities and not-for-Profit Commission'. Curtin University Not-For-Profit-Initiative, Perth.

Lipton E, Moss M (2012) Housing agencies flaws revealed by storm. *The New York Times*, 9 December 2012, <http://www.nytimes.com/2012/12/10/nyregion/new-york-city-housing-agency-was-overwhelmed-after-storm.html>.

Mallon K, Hamilton E, Black M, Beem B, Abs J (2013) 'Adapting the community sector for climate extremes: Extreme weather, climate change & the community sector – risks and adaptations'. National Climate Change Adaptation Research Facility, Gold Coast.

QCOSS (Queensland Council of Social Service) (2011) 'Submission to the Queensland Floods Commission of Inquiry'. Queensland Floods Commission of Inquiry, Brisbane, <www.floodcommission.qld.gov.au/_data/assets/file/0008/6983/Qld_Council_of_Social_Service_QCOSS.pdf>.

12

Engaging communities in climate change adaptation

Helena Bishop, Aileen Thoms and Wendy Mason

Key points

- The core business of health care is to protect human health, treat illness and disease, and promote health and wellbeing.
- The environment is fundamental to people's health and wellbeing, directly and indirectly.
- Climate change is an environmental determinant of health.
- Taking a health promotion approach to climate change can have positive impacts on the lives of those affected by climate change.
- Community organisations are best placed and have a role to play in supporting consumers and communities through climate change.
- Engaging communities in climate change adaptation is vital if any real and lasting change is to be made.

Introduction

Climate change is affecting human health in multiple ways, both directly, through extreme weather events, food and water insecurity and infectious diseases, and indirectly, through economic instability, migration and as a driver of conflict (Department of Health 2012).

The core business of health care is to protect human health, treat illness and disease, and promote health and wellbeing. This is reflected in international declarations such as the Declaration of the Alma Ata for Primary Health Care (WHO 1978) and the Ottawa Charter for Health Promotion (WHO 1986), which states that the fundamental prerequisites for good health and wellbeing are a stable ecosystem and sustainable resources. As we gain greater knowledge of the health impacts of climate change and environmental degradation, healthcare agencies are gaining awareness of climate justice and the environmental sustainability agenda.

The individual and community impacts of climate change have been explored in previous chapters but in general terms the impacts include decreased opportunities for social connections, social capital and interaction with nature (Kingsley *et al.* 2009). Climate change will also impact an individual's sense of place, identity and attachment, as well as community identity and pride (Kingsley *et al.* 2009). These impacts will be different for different groups, communities and regions; some individuals and communities will be more vulnerable to climate impacts and will require greater support as climate change acts to deepen inequities (Costello *et al.* 2009).

The social model of health identifies individuals within the context of their environment, which is influenced by a range of psycho-socio-cultural determinants such as employment, education, early life influences, social gradient, stress and gender, and the wider structures of political, economic, environmental and cultural determinants. These are interrelated and can enhance or diminish inequities between population groups (Keleher & MacDougall 2009).

The practice of health promotion places a significant emphasis on the social model of health. Despite the fact that climate change has not been traditionally addressed using a health promotion approach, health promotion has the potential to be an effective driver in making significant changes to the lives of those affected by climate change.

Community workers and health promotion practitioners are ideally situated to work with the community to help them to understand potential climate risks and how they can adapt to and mitigate climate impacts, as they possess a unique set of skills and knowledge such as a deeply embedded understanding of their community, effective approaches to community engagement and development and methods that support community capacity building.

Along with the social determinants that are an integral component of the social model of health, environmental sustainability is another factor which needs be taken into consideration when undertaking health promotion in the area of climate change.

Social resilience can be understood as an approach that recognises dynamic systems of interaction between people and their environment (Folke 2006). It is the capacity of a community to cope with disturbances or changes and to maintain adaptive behaviour. A resilient community has a high threshold for stress before it shifts to another state (McGuire & Cartwright 2008). Rather than going into survival mode, a resilient community can respond in positive creative ways that changes the basis of the community, enabling it to grow.

The view of resilience as transformation embraces the dynamic character of communities and human–ecosystem interactions and sees multiple potential pathways within them. A community's vulnerability, resilience and adaptive capacity will be determined in part by their attitudes towards the process of change. The view taken by the community to a problem and how they prepare for change influences how different groups within the community 'cope with' the process of change (Fenton *et al.* 2007 cited in McGuire & Cartwright 2008).

Research into the health impacts of climate change has mostly focused on the impacts of thermal stress, extreme weather events and infectious diseases. However, more research is required on the indirect impacts that result in social, economic and demographic changes. The most vulnerable groups in Australia are likely to include the elderly, the chronically ill, the socio-economically disadvantaged, those with poor access to essential services such as good housing and adequate fresh water, and those whose economic prosperity depends heavily on climatic conditions:

> ... *because climate change acts as an amplifier of existing risks to health, poor and disadvantaged people will experience greater increments in the disease burden than rich, less vulnerable populations.* (Costello et al. *2009, p. 1712)*

Contemporary healthcare practice recognises the importance of social and environmental justice and equity, inter-sectoral action on environmental, political and economic factors, community participation in decision-making, and stable ecosystems as prerequisites for human health (Patrick *et al.* 2012).

An introduction to health promotion

The World Health Organization (WHO 1986) defines health promotion as 'the process of enabling people to increase control over and to improve their health'.

Health promotion focuses not only on the individual's role within their own health, but also places a significant emphasis on the social, environmental, cultural and political structures in which an individual exists. This is known as the social model of health and these factors are known as determinants. Taking a determinants approach recognises that individuals often have very little, if any, control over their environments and determinants and thus, many aspects of their own health. Table 12.1 summarises some of the key determinants affecting an individual's health and highlights the powerless nature of individuals to be able to control many of them. This table has been constructed from a combination of published lists of social determinants and from the authors' combined experiences of working with various community groups in climate change adaptation.

Many individuals and communities experience the effects of multiple determinants and it is rare for them to occur in isolation. For example, a Sudanese woman with low-level English skills who lives on a low income may live in substandard housing that is not insulated against heat in summer and cold in winter. Due to government policies dictating that she must find work or risk losing her pension, she takes on employment in a male-dominated factory where she is on her feet for 12 hours per day, receiving minimum wage. Each of the issues experienced by this woman can have a significant impact on her health.

Health promotion supports the notion that an individual or community cannot experience optimal health if the environment in which they exist does not promote health. Therefore health promotion strategies focus on ways that these conditions can be changed or improved to support positive health outcomes for individuals and communities. This can be referred to as taking an upstream approach to health promotion.

Irving Zola (as cited in McKinlay 1979, p. 9) uses an analogy that articulates upstream health promotion:

> ... sometimes it feels like this. There I was standing by the shore of a swiftly flowing river and I hear the cry of a drowning man. So, I jump into the river, put my arms around him, pull him to shore and apply artificial respiration. Just when he begins to breathe, another cry for help, so back in the river again, reaching, pulling, applying, breathing and then another yell. Again and again, without end, goes the sequence. You know, I am so busy jumping in, pulling them to shore, applying artificial respiration, that I have no time to see who the hell is upstream pushing them all in.

Upstream health promotion looks at the root causes of a problem and places an emphasis on preventing a problem rather than concentrating efforts on fixing problems once they have occurred; it looks at who is pushing the men in the river (preventing the problem) rather than focusing on rescuing them (addressing the problem).

Table 12.1. Determinants of an individual's health

Social	Environmental	Cultural	Political
• Income • Gender • Education • Social support • Housing • Working conditions • Poverty	• Air purity • Sanitation • Water quality • Food security • Infrastructure • Climate • Pollution • Biodiversity • Connections to nature	• Connections to culture • Discrimination • Destruction of culture • Language acquisition	• Participation • Policies • Resident rights and restrictions

Ottawa Charter for Health Promotion

Health promotion can also be understood in the context of the Ottawa Charter for Health Promotion, which is a framework that was developed at the First International Conference on Health Promotion in Ottawa, Canada in November 1986. The Ottawa Charter takes the determinants approach one step further and proposes five key action areas for achieving change, by recognising that individuals do not exist independent from their environments and that there is in fact an interdependent relationship between the two (WHO 1986). The Ottawa Charter states that the fundamental prerequisites for good health and wellbeing are a stable ecosystem and sustainable resources (WHO 1986).

Following are the five key action areas.

1. **Develop personal skills**
   People don't know what they don't know so it is vital to develop their skills. The more knowledgeable and skilful a person is the more likely it is that they will make informed choices about their own health (WHO 1986).
2. **Create supportive environments**
   The environments in which people live, work and play have a significant impact on their health. Supportive environments for health are those which are 'safe, stimulating, satisfying and enjoyable' for people (WHO 1986).
3. **Strengthen community action**
   For effective change to be made, it is important to involve the community in any initiatives undertaken (Ebi & Semenza 2008; Holstein 2010). This is not only a key health promotion strategy but also one of the key principles of community development. By strengthening the community involved, there is an increased likelihood that they will take control and ownership over the factors within their own lives. A collective effort is stronger than one that occurs in isolation (WHO 1986).
4. **Reorient health services**
   The only way to contribute to effective and sustainable change within health is for services to work together towards a common goal. Whether services work specifically in the area of health promotion or not, expanding the scope of work so that everyone is working in a health-promoting way is vital to the health of those who are at risk (WHO 1986).
5. **Build healthy public policy**
   Populations are affected by and governed by policies on a daily basis; both those from the health and non-health-related sectors (Patrick *et al.* 2012). Building healthy

public policy is about putting health on the agenda of all policymakers across all sectors, supporting them to consider the health consequences of their policies on populations (WHO 1986).

These five action areas are underpinned by three common health promotion strategies: advocate, enable and mediate (WHO 1986).

It is understood that for effective change to take place, activities cannot be implemented in isolation; an integrated, multifaceted approach is necessary that covers all five key action areas, with no one component being given more consideration than another (WHO 1986).

Climate change and health promotion

As discussed in previous chapters, climate change affects those populations who are already vulnerable; when this is compounded by a range of other determinants that lead to poor health, disadvantage becomes greater.

Climate change has not traditionally been addressed from a health promotion perspective (Smith & Capon 2011) but given that it is a key determinant impacting on an individual and community's health (Patrick *et al.* 2012) it can be addressed by adopting a health promotion approach. Research suggests that there are many benefits to taking a health promotion approach to climate change (Walker *et al.* 2011; Smith & Capon 2011; Patrick *et al.* 2012).

Taking a health promotion approach to climate change can raise greater awareness of the issue of climate justice, especially when determinants are considered. As discussed earlier, the health of an individual and/or community can be affected by a range of determinants that they have very little to no control over. Unfortunately, those who are already disadvantaged suffer the greatest from the effects of climate change. Taking a health promotion approach recognises this and works towards climate justice for those who have the least impact on their environment yet suffer the greatest from both the direct and indirect impacts of climate change (Department of Health 2012).

Approaching climate change from a health promotion perspective ensures that a proactive approach is taken, where individuals and communities are as well prepared as they can possibly be to deal with changes that will inevitably occur as a result of climate change (Walker *et al.* 2011; Smith & Capon 2011; Ebi & Semenza 2008). This is referred to as social resilience. Building social resilience can help to alleviate some of the anxiety and distress experienced by communities when events such as floods, bushfires and cyclones hit as they are able to respond in a well-prepared way instead of simply reacting to events as they occur (McGuire & Cartwright 2008).

Walker *et al.* (2011, p. S8) describe climate change health promotion as being 'similar to health promotion approaches to chronic disease. Population level behaviour, as well as that of individuals, social and physical environments all need to change'.

As Walker *et al.* (2011, p. S6) describe, 'mitigation and adaptation are both required responses' to climate change and approaches that consider large-scale interventions such as influencing public policy, down to smaller-scale strategies such as teaching individuals how to reduce energy consumption are those that are most effective in the long-term. Given that the topic of climate change stirs up controversial discussions among community and service professionals alike, tackling it from multiple levels ensures that an impact is made regardless of the debate occurring (Walker *et al.* 2011; Smith & Capon 2011).

Given the relevance of the Ottawa Charter to climate change, which highlights the need for stable ecosystems and sustainable resources for the acquisition of good health and

Table 12.2. A framework for action

Key action areas	Suggested activities
Develop personal skills	• Conduct information sessions for members of a particular community that focus on energy efficient practices • Provide general climate change information to members of the community through written and verbal methods • Facilitate risk management workshops for communities that live in bush fire prone areas • Undertake personal planning for extreme weather events • Build skills in resilience • Encourage nature walking
Create supportive environments	• Set up support groups for those who have been affected by natural disaster • Establish community gardens and kitchens that have a focus on food security • Plant trees in local neighbourhoods • Incorporate green spaces within built environments • Improve communication processes during extreme weather events for at risk population groups (e.g. people with a disability, culturally and linguistically diverse communities, elderly people)
Strengthen community action	• Develop community advisory groups who can help to guide climate change initiatives • Conduct focus groups to collect information on community needs in relation to flood risk management • Coordinate a community led sustainability fair
Reorient health services	• Apply an environmental lens to organisation policies and plans • Attend professional networks that have a focus on climate change • Plan for green spaces within and close to health services • Develop professional partnerships that focus on risk reduction in bushfire prone areas • Develop a climate change adaptation community of practice
Build healthy public policy	• Allocate government budget for climate change initiatives • Enact legislation that determines water restrictions • Introduce carbon pricing and renewable energy policies • Develop organisational policies that focus on energy efficiency and resource reduction (e.g. paperless office) • Develop heat policies for employees who work outdoors • Include green infrastructure in design plans

wellbeing, we will use this framework to explore best practice approaches to climate change that can be considered for implementation across and between organisations and sectors that are already working in, or considering working in, the climate change space. Working through the action areas in the Ottawa Charter ensures that those large- and small-scale interventions that act to mitigate and adapt described by Walker *et al.* (2011) are given equal consideration.

When considering these recommendations it must be remembered that these are only suggestions and are not exhaustive. Organisations need to consider what activities are most relevant and appropriate for their individual needs.

Working through the Ottawa Charter framework and implementing key activities such as those described in Table 12.2 contributes to individuals who have well-developed skills and knowledge in the area of climate change, communities that are more prepared and resilient to the risks and effects of climate change, and broader social environments that recognise climate change as a real and serious risk and are supportive of strategies that enable communities to adapt to changes and mitigate against further risks.

The Ottawa Charter emphasises the need to work collectively from the 'top down' and 'bottom up', recognising that the community, service sector and government need to work in collaboration for effective preventative and adaptive measures to be implemented in a meaningful and achievable way (Ebi & Semenza 2008).

Working in the area of climate change

A larger proportion of the community than ever, both professional and non-professional, understand that no one is immune to the effects of climate change. It has the potential to affect all sections of the community; therefore, those who work directly with particular sectors of the community are best placed to work within this area. In this chapter we will explore the role that community service workers and health promotion practitioners play in helping the community to understand climate change and ways that they are able to adapt and mitigate climate impacts, and strategies to engage the community in work around climate change.

Given that there is an increased understanding and recognition of climate change and its effects (Holstein 2010), few services are actively engaged in working with the community directly around the issue. Many organisations have internal policies and practices which act to mitigate against climate change such as reduced energy consumption and 'green' workplaces and some have risk management plans that prepare them for the effects of extreme weather events, yet very few are working with their consumers and communities to help them adapt to and mitigate climate effects.

Climate change is impacting on organisations at many levels. Extreme weather increases the risk to an organisation's infrastructure, and the rising cost of energy is a financial risk. Households and communities face the same risks, raising the issue of how services are adapted to meet these emerging needs.

Increased awareness of climate change as an issue has led to developments within service delivery health organisations and local governments, such as a register of vulnerable community members and increased contact with health workers before and during heatwaves, leading to increased preparedness. Collaboration on common tools and referral systems help organisations to work more cooperatively, increasing capacity for ongoing adaptation and mitigation.

Once an organisation is able to integrate climate change as a priority within their work, they are able to develop strategies to address the concerns of their particular community or consumer group. Understanding your community's exposures for climate change is integral in forming strategies for actions.

For climate change strategies to be effective the community must be engaged (Ebi & Semenza 2008).

Community engagement in climate change follows the same basic principles of all types of community engagement: set goals; know your audience; provide clear, accurate and relevant information; support and encourage open communication; develop mutual trust and respect; encourage participation and collaboration (Holstein 2010; Gardner et al. 2009); support ownership; and seek feedback (Gardner et al. 2009).

Engaging the community in climate change shows people that they can play an effective role in preparing for and addressing the problems associated with climate change (Holstein 2010) but there are always challenges that need to be overcome before this can be effective.

One of the biggest challenges in engaging the community in climate change discussions and strategies is relevance. This is particularly true for those living in metropolitan and regional centres who view climate change as something that only affects those living in rural and remote settings. Possibly even more problematic in the engagement process is resistance caused by people's perceptions about climate change, with many discrepancies existing among the community about the perceived causes of climate change (Riedy *et al.* 2013; Gardner *et al.* 2009; Leviston *et al.* 2014). This can be attributed to a lack of trust in the sources of information (Leviston *et al.* 2014) and fear and uncertainty, with climate change painting a bleak picture of the future (Gardner *et al.* 2009).

The basis of engagement involves getting people on board. Taking into consideration the difficulties outlined earlier, this can be done by presenting accurate and evidence-based information that dispels the myths surrounding climate change and helps people to realise that climate change is something that can affect anyone at any time and in various ways (Gardner *et al.* 2009). However, information alone is not sufficient as a method of engagement. One of the most effective ways is through social pressure, using those who are already engaged to influence others around them (Walker *et al.* 2011).

Presenting real life examples that are relevant to the target community can help to concrete the issue for many. This can be in terms of population group or context (Riedy *et al.* 2013; Ebi & Semenza 2008). As described by Gardner *et al.* (2009, p. 15) 'climate change is highly contextualised ... the adaptation required by a metropolitan coastal community is quite different to the adaptation to climate change required by dairy farmers', therefore relevance is integral in engaging the community on the topic.

One of the key considerations of community engagement is community participation. For any strategy to be successful the community must be active participants throughout all stages. This bottom-up approach, as discussed earlier, ensures that activities are relevant, accepted and sustainable (Gardner *et al.* 2009; Ebi & Semenza 2008; Holstein 2010).

Making the process interesting, interactive and fun can be integral to engaging the community and reducing the fear associated with climate change. This can result in communities who are more receptive and willing to take action (Riedy *et al.* 2013). Some examples of engagement strategies are presented in the following case studies.

Engaging the community in climate change adaptation and mitigation ensures that the community not only owns the problem but is central to the solution.

Case study: considering culture

Women's Health in the South East (WHISE) is the regional women's health service covering the Southern Metropolitan Region of Melbourne. WHISE works in the areas of health information, health promotion and education with the most marginalised women within the community. WHISE assists women to access services that they need and advocates for improved health services so that women can take more effective control of their own health and wellbeing.

WHISE is based in Dandenong, which is the most disadvantaged municipality in Melbourne and the second-most disadvantaged local government area in Victoria (ABS 2013). It is located in Melbourne's south-east, ~24 km from the central business district of

Melbourne. In the City of Greater Dandenong the median weekly gross income is the lowest in Melbourne, crime rates are significantly higher than the average for metropolitan Melbourne, the unemployment rate is ~3% higher than the metropolitan average and it is culturally very diverse (Brown & Smith 2013). The City of Greater Dandenong is the most culturally diverse municipality in Victoria with ~60% of residents born overseas from over 150 different birthplaces (Brown & Smith 2013). Of this 60%, a large majority have recently arrived and include the largest number of humanitarian entrants in Victoria (Brown & Smith 2013).

Given the level of cultural diversity in the City of Greater Dandenong, WHISE chose to focus its climate change initiative on newly-arrived refugee women.

WHISE is in receipt of funding from the Department of Social Services (formerly the Department of Immigration and Citizenship) to provide the Settlement Grants Program. Through the provision of this funding WHISE developed a volunteer home-visiting program that provides basic settlement information and support to newly-arrived refugee women within the setting of their own homes. Referrals are received from a range of ethno-specific, migrant and refugee agencies as well as mainstream services. WHISE staff and volunteers attend the homes of newly-arrived refugee women once a week to assist with tasks that can inhibit effective settlement, and continue to do so until the client reaches a point in which they are able to act independently. Many of the issues that affect these women reflect those discussed in Chapters 7 and 9 of this book.

Being physically located in the central business district of Dandenong, WHISE is also accessed through front-counter services by many women from newly arrived culturally and linguistically diverse backgrounds.

One recurring theme for clients was high energy consumption and as a result, expensive energy bills, which they struggled to pay. Many of the women were unfamiliar with western-style houses and the modern conveniences that come as part of it. WHISE was not only concerned about the inability of clients to pay their bills but also the effect of high energy consumption on the environment.

Given the many complex and competing issues facing newly-arrived refugee women, climate change is not one that is given high priority; however, finances are perhaps the highest priority of all. WHISE saw an opportunity to introduce climate change by focusing primarily on cost reduction strategies.

Given that finances are among the main challenges facing newly-arrived refugee women, staff and volunteers take the opportunity to discuss energy saving strategies with clients as a way of reducing burdensome costs, as many are reliant upon the small income provided by Centrelink (a provider of public financial support) to sustain their everyday living expenses.

Actions that can reduce the cost of utilities include:

- fixing broken items such as dripping taps
- considering energy star ratings when purchasing new appliances
- replacing incandescent light bulbs with energy efficient light bulbs
- employing natural heating and cooling methods such as the use of blockout blinds and open windows.

For those in rented accommodation, which is a large majority of this population group, support can also include advocacy to real estate agents to ensure that repairs are taking place as required by legislation.

Vignette

An older Afghan woman with a long-term illness had been part of the home visiting program for around 18 months. Due to her illness, she suffered considerably from the effects of cold weather and as a result had her heater on day and night, especially through the colder winter months. The heater that she was using was an older style portable heater and at the end of a three-month period she received a bill in excess of $1500.

During the winter, staff attempted to talk with her about the energy efficiency of the heater that she was using and ways to ensure that the heat remained in the room so that she didn't have to leave the heater on at all times. At the time the client was not receptive to the ideas proposed; however, upon receipt of the bill she was a lot more receptive as she was concerned about the cost, and how she was going to pay it, and was open to any strategies to reduce future bills. As a result, the client purchased a new heater with a high energy efficiency rating, placed blockout curtains on her windows and kept them closed to retain the heat, closed off rooms that she was not using and turned the heater off once the desired temperature was reached. She also used warm clothing and blankets rather than relying primarily on the heater to keep her warm.

This example demonstrates that it was not climate change that the client was concerned about but the cost, and only when the cost became a concern was the client receptive to discussions about climate change.

The strategies employed in this example were very simple and can be applied to any client group, in any location, at minimal expense to the organisation and the client. Not only do these strategies have an impact on reducing the effects of climate change, but more importantly they provide an opportunity to introduce discussions with clients about climate change and practical ways that they can mitigate and adapt to changes.

Employing strategies now that will minimise the effects of climate change into the future are essential for those groups who are already vulnerable and will be amongst the most impacted by the effects of climate change (Patrick *et al.* 2012).

Considering climate change within a metropolitan setting

In the previous section we discussed climate change strategies that can be implemented across any client group in any location, at any time, with limited resources. However, it must be remembered that metropolitan areas are not exempt from the effects of adverse weather events, which are usually associated more with rural and remote areas. Therefore discussions and preparations to prevent and reduce impacts are useful regardless of whether clients believe that they will ever be affected.

This was no more apparent than in 2011 when heavy rain and major flooding affected large parts of the metropolitan area of Melbourne. The source of this heavy rainfall was Cyclone Yasi, which occurred ~2500 km away yet resulted in one of the worst floods seen in Melbourne's history (Bureau of Meteorology 2012). This acted as a stark reminder that it is not just those living in rural and remote areas who suffer the effects of adverse weather events. Many people were unprepared for the potential of such an event.

Some of WHISE's clients were impacted by this weather event. Leaking roofs, flooded yards and driveways and parts of homes were flooded, leaving clients not only stranded but also with large damage bills. Despite the fact that the majority of refugee clients live in rented premises and were not responsible for damage caused to the building structure, contents insurance is rarely purchased, resulting in furniture and appliances being ruined.

Living in a metropolitan location away from any main waterway, clients can become complacent about the risk of storms and floods and as a result do not have disaster preparation or management plans for such events.

Every contact with a client presents the opportunity to discuss climate change and their preparation for adverse weather events as well as strategies to reduce their climate footprint.

Case study: how does my garden grow?

'Ku-wirup' is said to mean 'blackfish swimming' from the language of the Boon wurrung Aboriginal people who traditionally used the area as a summer gathering ground. Formerly swampland, it was drained in 1876 to enable agriculture (Roberts 1985). Koo Wee Rup is one of the biggest asparagus producing areas in Australia.

Koo Wee Rup is located in the Cardinia Shire. Cardinia Shire is one of the fastest growing urban communities in Victoria with areas in the north and south of the shire predominantly rural; however, loss of arable and pastoral land, increased use of pesticides and other environmental challenges including climate change have had significant impacts on the community and their way of life.

Connections to local farms for food have been lost and the community is largely reliant on mass produced, packaged food from the supermarkets. Although these issues seem daunting, by acting locally on the things we *can do* helps to empower communities (Shallue 2013).

Stage one

Kooweerup Regional Health Service (KRHS) is a small rural hospital. As a multiple service agency, KRHS provides a range of hospital and community-based services. In 2007, KRHS adopted a proactive approach to health, encompassing primary health care and health promotion. The health service employed a full-time health promotion practitioner and undertook a commitment to integrate health promotion into the organisation, shifting from a medical model to a socio-ecological model of health.

Using a settings-based approach to health promotion, strategies have included:

- preventative health care including environmental justice
- policy development
- community strengthening
- education and skill building
- communication and social marketing.

This process involved a change to core business, using a preventative framework across the organisation, and an increase in community participation.

Strong leadership and clear direction from the Chief Executive Officer and board of the health service enabled staff to work on sustainable initiatives that build social capital and benefit the environment.

A needs analysis included a community consultation, which resulted in the establishment of community networks that included senior hospital management, the health promotion team and local gardeners, community members, aged care staff and residents, local organic farmers, representatives from education settings (such as local schools), youth employment, other health organisations, business, environmental groups and local government. The needs analysis identified community issues such as the lack of access to fresh food, despite being in a rural environment, and a strategy to build a community garden and community kitchen to help educate and support community action.

From these findings, a Community Hub was developed within the hospital grounds. This Hub includes the Men's Shed, a recently retrofitted eco house (Hewitt Eco House) and a community garden, which includes herbs, seasonal vegetables, an orchard, chickens, Indigenous garden, igloo for propagating, pizza oven, smoking tin for seeds and a space for community groups to come together.

Leaders have emerged from the community to support the community garden and Men's Shed, from where much activity evolves.

Stage two

A vibrant active community space was created where people grow, harvest and cook seasonal produce. The garden commenced in 2008, is organic and based on permaculture principles (Mollison 1988). The garden is set up as a communal garden and is managed by the community for the benefit of the community. It commenced with a 'Permablitz'; a permaculture movement where likeminded people come together to create edible gardens, share skills and have fun (Melbourne Permablitz 2006). Some members of the initial group attended a couple of established Permablitzes which gave them the ability to host their own. Over 60 people attended two Permablitzes over two years. This process helped to establish the beginnings of a strong sense of community.

A diverse range of garden beds were built, which include raised beds to ensure access for people of varying abilities, composting areas and a mini orchard with fruit trees, chicken coop and native gardens (Fig. 12.1). The garden also has open areas and a gathering space for community activities to occur.

The gardeners meet twice a week as a community garden group; however, the reality is that the garden is an interactive environment where people and different groups interact each day as weeding and tending of the garden generally happens daily.

The harvest is distributed among local support groups, individuals, or to community members who are in need. Excess produce is available within the hospital waiting room in exchange for a donation. Community cooking groups enjoy a good supply of herbs from several herb gardens. Use of seasonal produce in cooking reduces food miles and keeps the cooks connected with the seasons. Sharing of skills and knowledge across the groups occurs informally, as does discussion about various produce, timing of plantings and harvesting.

The gardens operate on a pick and pay system, with payment either by donation of time working in the garden or a monetary donation. Water is obtained through tanks that collect rainwater from the Men's Shed roof. This encourages water-wise gardening.

Workshops are held on various topics such as permaculture, composting, worm farming, water and sustainable homes. Community events are also held which include markets, social get-togethers such as BBQs, music on the grass, outdoor art shows, and nature play sessions for families, all strengthening social connections.

Figure 12.1. Images of the community garden at Kooweerup Regional Health Service.

(*Continued*)

Figure 12.1.   (*Continued*)

The Men's Shed members have also embraced recycling and re-model and reuse many articles, giving them new lives.

Several manuals are available to help guide communities interested in developing community gardens such as *Sustainable Gardening Australia Community Garden Manual* (Shallue 2013).

Health workers use skills in relationship building, facilitation and conflict management when working with diverse agendas. Applying an environmental sustainability lens to the overall initiatives has led to increased inter-sectoral collaboration and an environmental focus with all community groups.

The use of a socio-ecological model brings people who have a shared interest together, strengthens collective action and encourages greater buy-in from the community. This has enabled the development of mutual support and a sharing of community pride which results in an increase of social capital and community capacity for problem solving, which is most beneficial when the community faces crises, such as during the floods which affected the area in 2011 and 2012 and bushfire support to the region in 2009.

This approach enables the organisation to be concerned with ecological justice (i.e. non-human health and wellbeing). Several diverse groups collaborate to restore and protect habitat for the endangered southern brown-nosed bandicoot and restoration of plants through the creation of a wildlife corridor.

Advocacy to local, state and federal governments in relation to community issues such as active transport, walking paths and public transport has enabled improved access and has resulted in a win-win outcome, not only for people's health and wellbeing but also for the planet. Community information sessions were held to create awareness of coal seam gas exploration and fracking (a technique of gas extraction with serious environmental impacts).

The community garden at Kooweerup Regional Health Service supports many activities, is attractive and is welcoming to the community and diverse groups within it

A range of benefits have been observed from undertaking this process which have been identified through informal and formal interviews, group discussions, participant survey pre- and post-workshops, and organisational audit and quality improvements through hospital accreditation processes, ACHS Equip and the Aged Care Standards. These benefits include:

- improved health and wellbeing
- increased social connectedness
- reduced social isolation
- improved knowledge of coping strategies during disasters and heatwaves
- increased monetary savings
- increased volunteerism
- development of new skills
- positive attitudes to think global but act local (Thoms *et al.* 2010).

Conclusion

Climate change has not traditionally been addressed using a health promotion approach, but as we have explored in this chapter, health promotion can be integral in working with communities on developing effective and sustainable adaptation and mitigation strategies (Walker *et al.* 2011; Smith & Capon 2011; Patrick *et al.* 2012). Climate change disproportionally affects those who are already disadvantaged by a range of determinants beyond their control; therefore, a determinants approach can have one of the greatest impacts (Department of Health 2012).

As the community continues to learn about and comprehend climate change and its impacts, organisations can use their position and relationships within the community to leverage upon this, and engage those who may at one point or another be difficult to engage.

Engagement at the community level is vital as no one person or population group is immune to the affects of climate change, and inevitably engagement will look different for different organisations and different population groups. Within this chapter we have provided some general examples of engagement strategies using a health promotion focus and have shared some of our own processes and learning from engaging communities within climate change.

References

ABS (Australian Bureau of Statistics) (2013) *Media Release: New Data from the 2011 Census Reveals Victoria's Most Advantaged and Disadvantaged areas.* Australian Bureau of Statistics, Canberra, <http://www.abs.gov.au/ausstats/abs@.nsf/Lookup/by%20Subject/2033.0.55. 001~2011~Media%20Release~2011%20Census%20(SEIFA)%20for%20Victoria%20 (Media%20Release)~3>.

Brown H, Smith M (2013) *A Profile of Health and Wellbeing in Greater Dandenong,* City of Greater Dandenong, Victoria.

Bureau of Meteorology (2012) 'Record breaking La Niña events: an analysis of La Niña life cycle and the impacts and significance of the 2010–11 and 2011–12 La Niña events in Australia'. Bureau of Meteorology, Melbourne.

Costello A, Abbas M, Allen A, Ball S, Bell S, Bellamy R, Friel S, Groce N, Johnson A, Kett M, Lee M, Levy C, Maslin M, McCoy D, McGuire B, Montgomery H, Napier D, Pagel C, Patel J, de Oliveira JAP, Redclift N, Rees H, Rogger D, Scott J, Stephenson J, Twigg J, Wolff J, Patterson C (2009) Managing the health effects of climate change. *Lancet* **373**, 1693–1733. doi:10.1016/S0140-6736(09)60935-1

Department of Health (2012) 'Municipal public health and wellbeing planning: having regard to climate change'. Department of Health, Melbourne.

Ebi KL, Semenza JC (2008) Community based adaptation to the health impacts of climate change. *American Journal of Preventive Medicine* **35**(5), 501–507. doi:10.1016/j.amepre.2008. 08.018

Folke C (2006) Resilience: the emergence of a perspective for social–ecological systems analyses. *Global Environmental Change* **16**, 253–267. doi:10.1016/j.gloenvcha.2006.04.002

Gardner J, Dowd A, Mason C, Ashworth P (2009) 'A framework for stakeholder engagement on climate adaptation'. CSIRO Climate Adaptation Flagship Working Paper No.3. <http:// www.csiro.au/resources/CAF-working-papers.html>.

Holstein A (2010) 'GRaBS expert paper 2: participation in climate change adaptation'. Town and Country Planning Association, London.

Keleher H, MacDougall C (2009) Understanding the determinants of health. In *Understanding Health: A Determinants Approach.* 2nd edn. (Eds H Keleher and C McDougall) pp. 41–58. Oxford University Press, South Melbourne, Australia.

Kingsley J, Townsend M, Henderson-Wilson C (2009) Cultivating health and wellbeing: members' perceptions of the health benefits of a Port Melbourne community garden. *Leisure Studies* **24**(4), 525–537.

Leviston Z, Price J, Malkin S, McCrea R (2014) 'Fourth annual survey of Australian attitudes to climate change: interim report'. CSIRO, Perth.

McGuire B, Cartwright S (2008) 'Assessing a community's capacity to manage change: a resilience approach to social assessment'. Bureau of Rural Sciences, Canberra.

McKinlay JB (1979) A case for refocusing upstream: the political economy of illness. In *Patients, Physicians and Illness*. (Ed. EG Jaco) pp. 9–25. The Free Press, New York.

Melbourne Permablitz (2006) *The Lowdown and Dirty on Permablitzing*, Melbourne Permablitz, <http://www.permablitz.net/what-is-a-permablitz>.

Mollison B (1988) *Permaculture – A Designers' Manual*. Tagari Press, Sisters Creek, Tasmania.

Patrick R, Capetola T, Townsend M, Nuttman S (2012) Health promotion and climate change: exploring the key competencies required for action. *Health Promotion International* **27**(4), 475–485. doi:10.1093/heapro/dar055

Riedy C, Herriman J, Ross K, Lederwasch A, Boronyak L (2013) 'Innovative techniques for local community engagement on climate change adaptation'. Global Cities Research Institute, RMIT University, Melbourne.

Roberts D (1985) *From Swampland To Farmland – A History of the Koo-Wee-Rup Flood Protection District*. Rural Water Commission of Victoria. Victoria.

Shallue E (2013) *Sustainable Gardening Australia: Community Gardens Manual*. Sustainable Gardening Australia, Victoria.

Smith JA, Capon A (2011) Addressing climate change through health promotion in Australia. *Health Promotion Journal of Australia* **22**, S3–S4.

Thoms A, Capetola T, Patrick R (2011) Green links: Growing partnerships and knowledge for inter-sectorial collaboration in promoting healthy, sustainable futures. Presentation at *Gippsland Health Promotion Conference – Take the Next Steps*, 11 October 2011, Monash University Gippsland, Victoria.

Walker R, Hassall J, Chaplin C, Congues J, Bajayo R, Mason W (2011) Health promotion interventions to address climate change using a primary health case approach: a literature review. *Health Promotion Journal of Australia* **22**, S6–S12.

WHO (World Health Organization) (1978) 'Declaration of Alma-Ata'. World Health Organization, Geneva.

WHO (World Health Organization) (1986) 'Ottawa Charter for Health Promotion 1986'. World Health Organization, Geneva.

Appendix: climate change adaptation audit tool

Preamble to the audit tool: providing a context

Enliven Victoria is a not-for-profit network of health and social service organisations in the south-east of Melbourne. It works primarily across three local government areas that include established blue collar industrial areas, rapidly growing new residential estates, a large and diverse population of people from culturally and linguistically diverse backgrounds (including many refugees and asylum seekers), and shrinking traditional rural areas. It is very diverse. Every three to four years the participating organisations negotiate up to three priority issues on which they will work. By collaborating they amplify their impact on issues important in their communities.

Since 2007 climate change adaptation has been a priority issue. The climate change adaptation audit tool was developed to support the organisations' efforts to act systematically on climate change adaptation and develop organisational capacity over several years. It was modelled on the quality assurance tools with which all agencies were familiar. The preamble to the tool describes the priority population groups in the Enliven Victoria's locality. These will be different in other localities. Everybody is impacted by climate change, but some more than others. When reviewing the priority population groups in your locality you should find that the chapters in this book identify the most important risks for the most vulnerable members of your population, and offer appropriate actions to address these risks.

If you wish to see the original colour version of the audit tool, and more of Enliven Victoria's work on climate change adaptation, visit the Enliven Victoria website (www.enliven.org.au), where the 'Library' and 'Projects & Plans' pages hold many documents and resources.

Climate change adaptation audit tool

The climate change adaptation audit tool helps health and social service organisations to assess their level of adaptation to climate change, with a particular focus on extreme weather events. The eight criteria are attributes describing a health or social service organisation that has adapted to deal with climate change. It is recognised that organisations differ in size, purpose and resources, suggesting that not all actions under each criterion may be relevant to every organisation.

The issue

The following criteria have been developed to promote a consistent standard of climate change adaptation practice in health and social services, and enable these services to implement adaptation actions to assess and continually improve their performance. By measuring performance against a set of criteria, service providers will be able to judge the quality of their service in implementing climate change adaptation actions. This is a self-audit that can be linked to organisational quality assurance systems, and discussions and review at board (or council), management and team meetings. The attributes are a modified version of those published in *Ten Attributes of a Health Literate Health Care Organisation* (Brach *et al.* 2012) and the examples have, as far as possible, been derived from published literature.

As an effect of climate change, extreme weather events will become more frequent and more intense. Among such events, heatwaves, storms, fires and floods have the most immediate impact on our community. Climate change literature suggests that we need to adjust to climate change (including climate variability and extremes) to moderate potential damage, to take opportunities and to cope with the consequences.

Climate change adaptation is defined by the Intergovernmental Panel on Climate Change as 'adjustments in natural and human systems in response to actual or expected climatic stimuli or their effects, which moderates harm or exploits beneficial opportunities' (IPCC 2007).

Target population

Enliven's climate change adaptation plan 2013/2017 will focus on developing systems and supporting people identified as at 'high risk' in relation to extreme weather events.

High-risk population groups include people with disabilities, serious mental illness and poor English skills, as well as children younger than 15 years, and people over 65 years of age. It also includes people with chronic illnesses (such as heart disease, COPD, severe arthritis or asthma), and morbid obesity, women in advanced stages of pregnancy, and people with transient limitations to function such as an injury.

Approximately half the Australian population (and subsequently the local population) may be at high risk and have special needs prior, during and after adverse weather events.

This broader definition logically leads to inclusive planning that is focused on providing support for functional needs in relation to extreme weather events.

Performance rating scale

Following are commonly-used definitions for each point on the rating scale (Department of Health 2012, p. 5):

Met: clear evidence that performance meets or exceeds the standard.
Partially met: clear evidence that performance meets some, but not all, of the standard.
Not met: clear evidence that performance does not meet the standard.
Not applicable: the item is not applicable.

Criterion 1 Has plans for maintaining service delivery following an extreme weather event (derived from the ACOSS study Mallon *et al.* 2013).	Met	Partially met	Not met	Not applicable
1.1 Service provider has contingency plans if the building became inaccessible.				
1.2 Service provider has contingency plans if the power, telecommunications or water access is affected.				
1.3 The organisation has plans that allow it to cope if demand for services increased due to an extreme weather event.				
1.4 The organisation has plans that allow it to cope if demand for services increased long-term.				
1.5 The organisation has plans that allow it to cope if staff are unable to operate for more than one day.				
1.6 The organisation has plans that allow it to cope if the usual volunteer workforce is unavailable for more than one day.				
1.7 The organisation has adequate insurance cover that would ensure maintenance of services. Potential items for cover include: assets, contracts, income, business continuity, staff absence, and volunteer absence.				
Criterion 2 Has leadership that makes climate change adaptation integral to its mission, structure and operations.	Met	Partially met	Not met	Not applicable
2.1 Has an organisational climate change adaptation policy.				
2.2 Delegates leadership roles on climate change adaptation to appropriate senior managers in the organisation.				
2.3 Climate change adaptation is discussed at Board (or Council) level.				
2.4 Conducts annual audits of progress on climate change adaptation actions.				
2.5 The organisation has documented its role in extreme weather events and its relationship to emergency services.				
Criterion 3 Effectively integrates climate change adaptation action across the organisation.	Met	Partially met	Not met	Not applicable
3.1 Incorporates climate change adaptation into organisational plans. For example: strategic plan, business plan, health promotion plans, program or service plans.				
3.2 Has a structure and/or process in place to coordinate implementation across the organisation.				
3.3 Programs document evidence of climate change adaptation action and that contributes to the organisation's annual audit.				

(Continued)

Criterion 4 Prepares the workforce (including volunteers) to understand climate change adaptation, and monitor adaptation progress.	Met	Partially met	Not met	Not applicable
4.1 Provides training to staff and volunteers to increase their understanding of climate change adaptation in relation to human health and wellbeing.				
4.2 Staff are aware of risks, and their mitigation, in relation to their service provision during heatwaves and extreme weather events.				
Criterion 5 Includes high-risk populations in the design, implementation and evaluation of information and services.	Met	Partially met	Not met	Not applicable
5.1 The organisation has consulted with the local community and clients in the development of climate change adaptation plans.				
5.2 The organisation has considered the heightened needs of high-risk groups in the population. For example: people with disabilities, serious mental illness, poor English skills, and children less than 15 years, and people over 65 years of age, people with chronic illnesses (such as heart disease, COPD, severe arthritis or asthma, for example), morbid obesity, women in advanced stages of pregnancy, and people with transient limitations to function such as an injury.				
5.3 The organisation includes people from high-risk populations as co-facilitators in training sessions.				
Criterion 6 Use health literacy strategies in communication.	Met	Partially met	Not met	Not applicable
6.1 Staff use communication modes appropriate for the high-risk groups e.g. spoken, signing, visual communication and social media.				
6.2 Staff use interpreters with people who do not speak English.				
6.3 Information is available in the common community languages.				
6.4 Staff design and distribute accessible messages on adaptation to extreme weather events, in multiple formats, that are easily understood by the high-risk groups and able to be acted upon.				
Criterion 7 Addresses planning and preparation for extreme weather events in routine service delivery to high-risk groups.	Met	Partially met	Not met	Not applicable
7.1 The organisation advocates for, and supports, preparation of individual plans for high-risk individuals in extreme weather events.				
Criterion 8 Communicates clearly which services and supports can be provided free or if there is a fee for service.	Met	Partially met	Not met	Not applicable
8.1 Information is provided in a way that is consistent with health literacy principles.				

References

Brach C, Keller D, Hernandez LM, Baur C, Parker R, Dreyer B, Schyve P, Lemerise AJ, Schillinger D (2012) *Ten Attributes of Health Literate Health Care Organizations.* Institute of Medicine, National Academy of Sciences, Washington D.C. <http://www.iom.edu/global/perspectives/2012/healthlitattributes.aspx>

Department of Health (2012) 'Continuous improvement framework 2012: a resource of the Victorian service coordination manual'. Primary Care Partnerships, Victoria.

IPCC (Intergovermental Panel on Climate Change) (2007) Appendix 1: Glossary In *Climate Change 2007: Impacts, Adaptation and Vulnerability. Contribution of Working Group II to the Fourth Assessment Report of the Intergovernmental Panel on Climate Change.* (Eds ML Parry, OF Canziani, JP Palutikof, PJ van der Linden and CE Hanson) pp. 869–883. Cambridge University Press, Cambridge, UK.

Mallon K, Hamilton E, Black M, Beem B, Abs J (2013) 'Adapting the community sector for climate extremes: extreme weather, climate change & the community sector – risks and adaptations'. National Climate Change Adaptation Research Facility, Gold Coast.

Index